Larval Settlement on Marine Surfaces: The Role of Physico-Chemical Interactions

Larval Settlement on Marine Surfaces: The Role of Physico-Chemical Interactions

Editor

Francesca Cima

MDPI • Basel • Beijing • Wuhan • Barcelona • Belgrade • Manchester • Tokyo • Cluj • Tianjin

Editor
Francesca Cima
Department of Biology
University of Padova
Padova
Italy

Editorial Office
MDPI
St. Alban-Anlage 66
4052 Basel, Switzerland

This is a reprint of articles from the Special Issue published online in the open access journal *Journal of Marine Science and Engineering* (ISSN 2077-1312) (available at: www.mdpi.com/journal/jmse/special_issues/larval_surfaces_interactions).

For citation purposes, cite each article independently as indicated on the article page online and as indicated below:

LastName, A.A.; LastName, B.B.; LastName, C.C. Article Title. *Journal Name* **Year**, *Volume Number*, Page Range.

ISBN 978-3-0365-7553-7 (Hbk)
ISBN 978-3-0365-7552-0 (PDF)

Cover image courtesy of Francesca Cima

Contents

About the Editor

Francesca Cima

Francesca Cima, Ph.D., is an associate professor of comparative anatomy and cytology at the University of Padova, where she teaches "Cell Biology", "Principles of Animal Biology" and "Evolutionary History of Vertebrates". She oversees the Laboratory of Biology of Ascidians. She is a member of the Ph.D. program in biosciences (cell biology and physiology curriculum) at the University of Padova. Her main scientific interests concern comparative immunology, immunotoxicity, embryotoxicity and bioaccumulation of antifouling compounds on coastal marine invertebrates commonly used as bioindicators (ascidians, bivalve molluscs, sea urchins and sipunculans). She won the "Paolo Gatto's National Prize 1998" of the Accademia Nazionale dei Lincei (Rome) regarding the environmental problems of the Lagoon of Venice due to her pioneer studies on the mechanisms of action of organotin compounds. She was on the Editorial Board of *Applied Organometallic Chemistry* and *European Journal of Histochemistry*, and now is a member of the Editorial Board of the *Journal of Marine Science and Engineering*. She has authored and co-authored more than 80 publications in international scientific, peer-reviewed journals, including 20 book chapters, 1 monograph, 1 book that she edited and 1 book chapter translated from English to Italian.

Preface to "Larval Settlement on Marine Surfaces: The Role of Physico-Chemical Interactions"

This reprint is a collection of 11 scientific articles (8 research papers, 1 communication and 2 reviews) published in the *Journal of Marine Science and Engineering* on the topic of "Larval Settlement on Marine Surfaces: The Role of Physico-Chemical Interactions".

I proposed the above-mentioned topic because advancing our understanding of larvae–surface interactions will certainly be greatly beneficial with respect to addressing the challenges of future innovative eco-engineering designs, yielding the best solutions for industrial biofouling protection and coastal ecosystem preservation. Biofouling occurs when sessile aquatic organisms rapidly settle on artificial hard substrata, which poses a large problem worldwide since their growth often causes severe damage to submerged structures. Accordingly, international efforts are being exerted toward the development of anti-fouling systems because biofouling not only increases static and hydrodynamic loading, but also affects corrosion characteristics and impedes underwater inspection and maintenance.

On the other hand, biofouling leads to the formation of a well-structured community on natural hard substrata characterised by ecological succession and can be considered an important source of biodiversity, whose long-term preservation controls the trophic chain of coastal ecosystems.

In all cases, there is a very close relationship between larval settlement and the type of substratum in question since the ecological succession of the fouling community in coastal marine ecosystems directly depends on interactions between organisms and surfaces. Due to their texture, microstructure, roughness, colour, and chemical composition, the various types of hard substrata affect organisms' settlement by favouring dominant species or preventing the settlement of biofouling organisms. The impact on the ecosystem biodiversity of the resident community caused by the extensive use of anthropogenic substrata requires attention due to the various unpredictable effects that can occur. Under some conditions, artificial substrata can enhance biodiversity and productivity at a local scale in depauperated areas and can positively contribute to regional productivity overall, but they can also lead to the recruitment and selection of invasive species by acting as a collector for larvae that negatively affect the local biodiversity. Therefore, the influence of a substratum's physico-chemical interactions on the settlement of various organisms in the macrofouling community represents an essential factor in choosing an appropriate artificial surface for application in a variety of coastal marine ecosystems. For this reason, I selected studies that focus on various aspects of the evaluation of marine larval settlement on both natural and artificial surfaces, including pro-fouling and anti-fouling systems.

Francesca Cima
Editor

Journal of
Marine Science and Engineering

MDPI

Communication

A Vital Staining Practice That Discerns Ancestry within Groups of Settling Larvae of a Brooding Coral

Dor Shefy [1,2,3,*], Nadav Shashar [2] and Baruch Rinkevich [1]

1 Israel Oceanography and Limnological Research, National Institute of Oceanography, Tel-Shikmona, P.O. Box 8030, Haifa 31080, Israel; buki@ocean.org.il
2 Marine Biology and Biotechnology Program, Department of Life Sciences, Ben-Gurion University of the Negev, Eilat Campus, Beer-Sheva 84105, Israel; nadavsh@bgu.ac.il
3 The Interuniversity Institute for Marine Science, Eilat 88000, Israel
* Correspondence: shefy@post.bgu.ac.il

Abstract: Xenogeneic and allogeneic encounters following aggregated and clustered settlements of coral larvae (planulae) may carry important ecological consequences in shaping coral reefs' communities. However, larval settlement behaviors and settlement location choices in the presence of conspecifics or heterospecifics have not been examined in detail, due to a lack of experimental tools. One potential approach is the employment of vital staining of planulae with dyes that do not impair larval metamorphosis processes, are stable for prolonged periods, and do not diffuse to un-labeled counterpart planulae. For these purposes, we examined the use of neutral red (NR) dye, as an identification marker, on the planulae of *Stylophora pistillata,* a Red Sea branching coral species. To examine possible NR impacts on larval settlement in the presence of conspecific planulae, we followed the settlement ratios of kin, non-kin, and mixed assemblages, as a proxy for metamorphosis success. We found no differences in settlement rates of stained vs. unstained larvae, lack of stain diffusion to other larvae and that NR stain is maintained for more than a week under a still water regimen. Thus, staining with NR may serve as a useful experimental tool, opening new opportunities in studying larval settlement patterns in sessile marine organisms.

Keywords: chimerism; settlement; planulae; neutral red; aggregation; *Stylophora pistillata*; larval behavior

Citation: Shefy, D.; Shashar, N.; Rinkevich, B. A Vital Staining Practice That Discerns Ancestry within Groups of Settling Larvae of a Brooding Coral. *J. Mar. Sci. Eng.* **2021**, *9*, 616. https://doi.org/10.3390/jmse9060616

Academic Editor: Francesca Cima

Received: 28 April 2021
Accepted: 31 May 2021
Published: 3 June 2021

Publisher's Note: MDPI stays neutral with regard to jurisdictional claims in published maps and institutional affiliations.

1. Introduction

A coral's life-cycle includes a pelagic larval stage (planula), larval settlement and metamorphosis to a primary polyp (a colony founder), astogenic growth via budding of polyps and/or branches, and sexual reproduction. When selecting a spatial location, the attachment to the substrate, settlement, and metamorphosis are active responses of the planulae to cues from substrates and the surrounding environments [1–3]. The choice of any specific site, in the context of habitat, substrate, water conditions, turf organisms, and other sessile neighbors, shapes the success of recruitment and the future of the coral reefs [1,4–6]. Although environmental factors such as substrate types, textures, chemical compositions, and physical cues for settlement have been extensively studied, the effect of various traits, such as the possible impacts of conspecifics or heterospecifics on settlement and choice of site, was less investigated.

Allogeneic (with conspecifics) or xenogeneic (with members of other species) encounters following the aggregated settlement of coral larvae have been reported for more than a century in various coral reefs worldwide [7–14]. While xenogeneic encounters of corals always result in interspecific conflicts between the partners, including the development of necrotic zones between interacting young colonies (rejection phenomena; [6,15,16]), allogeneic encounters may not only result in rejection but also tissue fusions between conspecifics, leading to the formation of chimeric colonies [6,15–17]. These outcomes have significant impacts on corals' survivorship, physiology, fitness, and morphology [6,16,18–20]. The wide distribution of coral chimerism in nature is another indication that aggregated settlement of coral larvae is common, beyond what has been thought [21–26] (Figure 1a). However, the lack of appropriate tools limited research activities on aggregated settlements and the impacts of planulae origin.

To follow larval origin within a group of settled planulae, one should obtain the competence to label or tag specific larvae within the assemblage. First, the labeling/tagging method should not interrupt the planktonic phase, the settlement processes, and the metamorphosis into the polyp stage. Second, the identifier mark should be easily detected under regular laboratory conditions without the involvement of sophisticated equipment. A third requirement is the need for prolonged labeling durations targeting successive observations.

Here, we developed and tested a method to label coral larvae using the vital stain neutral red. Neutral red (3-Amino-7-dimethylamino-2-methylphenazine hydrochloride) is a relatively non-toxic dye used for staining of histological sections [27], employed as an intracellular pH indicator [28], as an ecological/biological marker for animal movement and embryogenesis [29,30], and as a textile dye [31]. The use of neutral red began in the 1940s as a means to study the dispersion and movement of larvae of oysters, barnacles, and copepods [32]. The past restricted use of neutral red dye as a color marker in cnidarians was for studying embryo's polarity in the hydrozoan *Clytia gregaria* (previously known as *Phialidium gregarium*) [30]), where again, the development of the embryo to a planula was not affected by the dye.

We labeled the planulae of the branching coral *Stylophora pistillata*, one of the most abundant coral species in the Gulf of Eilat, Red Sea (Figure 1b). *S. pistillata* is a hermaphrodite species characterized by a long reproductive season and with colonies that release to the water column ready-to-settle planulae on a daily basis [33–35]. Most of the planulae of *S. pistillata* settle within the first 48 h upon release, and no substrate fastidiousness has been recorded [36–38]. We tested settlement success and effective metamorphosis by comparing settlement ratios among neutral red-stained and unstained larvae.

Figure 1. (**a**) The allogeneic encounter of two juvenile colonies of *S. pistillata*. These two individuals did not fuse, and they stay in constant conflict interactions [39]. (**b**) An adult colony of *S. pistillata* in Eilat's shallow reef. Both photos were taken in the northern Gulf of Eilat using a Canon G7 camera.

2. Materials and Methods

2.1. Experimental Design

Planulae of *Stylophora pistillata* were collected between February to July 2020 from 15 gravid colonies residing at 3–12 m in front of the Inter-University Institute of Marine Science in Eilat, Israel. Planulae were collected using planulae traps as described in Shefy et al. [35] and transferred to Petri dishes (9.5 cm diameter) in accordance with the experimental design.

Sets of three labeling treatments were established for planulae collected from the same/different maternal colonies, establishing the scenario where each Petri dish in a treatment contains 20 planulae of: (1) Neutral red un-stained controls (Figure 2a); (2) Neutral red labeled planulae (Figure 2b); (3) Neutral red mix treatment: 10 stained planulae originated from one colony and 10 un-stained planulae from a different colony. Each Petri dish was covered with a polyester film ("Maylor paper"; Jolybar, Israel), and preconditioned for one month for the development of microbial films that support coral larvae settlement [12,40]. The dishes were mostly (75%) submerged in the table's water volume, supplied with running seawater to keep ambient temperature, and covered with a lid to minimize evaporation. Water in the Petri dishes was replaced each day with fresh seawater to prevent mucus accumulation and to keep seawater salinity constant, as in natural waters (~40.6 ppt). The dishes were monitored for a week. Settlement rate, as well as the existence of the stain, were recorded daily.

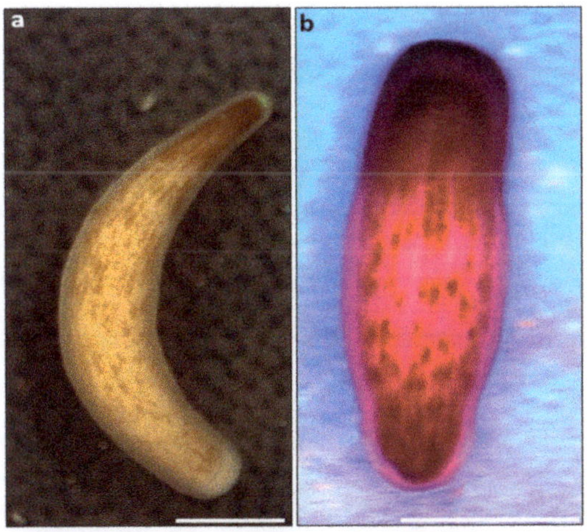

Figure 2. (**a**) An unstained planula. (**b**) A stained planula. Both photos were photographed using a 3.5×–90× trinocular stereo microscope (AmScope) equipped with a 10 MP USB microscope digital camera. Scale bar is 500 μm in both photos.

2.2. Staining Procedures

We first tested methylene blue, cyber green, and neutral red as these dyes were used to stain marine invertebrate larvae [32]. Preliminary results revealed that methylene blue and cyber green did not stain the coral planulae.

Staining was thus executed using the neutral red dye (Aldrich Chemical, Milwaukee, WI, USA). A stock solution of 1% was prepared by mixing 100 mL of filtered seawater with 1 g of neutral red powder. For planulae staining, planulae were immersed for 1–5 min in a Petri dish filled with the stock solution. Before the planulae were transferred to the experimental dishes, they were washed by moving them to a different dish filled with fresh seawater for 5 min, followed by a 10 min immersion in another dish filled with fresh seawater. Planulae were transferred between dishes using a Pasteur pipette to reduce any mechanical damage that may be developed using other washing procedures (such as washing through a mesh; D.S personal observation).

2.3. Statistical Analysis

The statistical analyses of the settlement ratios of *S. pistillata* larvae were performed using R software [41]. After checking for the assumptions (Shapiro-Wilk test $p > 0.05$; Levene's test $p > 0.05$), we used a one-way ANOVA to compare settlement among treatments. Since in the 'mix' treatment, stained and un-stained planulae were in the same dish, comparison between their mean settlement rates was conducted by a paired *t*-test. For comparison of the overall settlement of the stained larvae in the mix and stained treatment vs. un-stained larvae in the mix and un-stained treatment, a two-sample *t*-test was used.

3. Results

We first tested immersion duration in neutral red solution for various time-points of 1,2,3,4, and 5 min. Planulae immersed for 1–4 min resulted in different red shades between planulae or stained with a pale reddish color that dissipated within a few days in the dish. Only 5 min immersion time and prolonged durations resulted in a deep coloration of all stained larvae; thus, the time slot of 5 min immersion duration was inferred as optimal.

Most *S. pistillata* planulae settled within the first 48 h after their release, and observations were followed on larvae and primary polyps in the sealed dished for up to 1 week. The red stain was noticeably retained for the whole observational period, clearly distinguishing the stained from non-stained larvae. Settled stained larvae and metamorphosed spats continued to exhibit the neutral red stain (Figure 3a,b). The stained planktonic larvae were vital, actively changing locations in the dishes, as the controls, until metamorphosed (Figure 2b). Following the one-week observational period, the Petri dishes with the settled primary polyps were transferred to a large container (water-table) with running seawater. At this stage, all primary polyps lost their red color after no longer than three additional days.

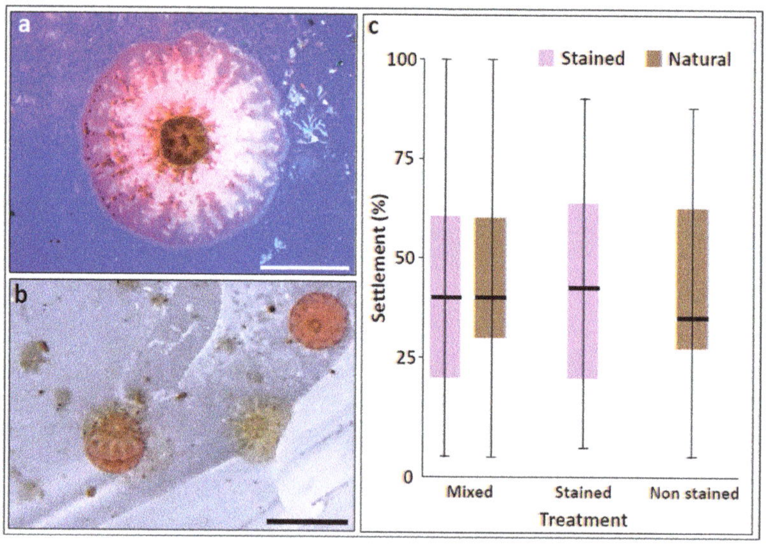

Figure 3. (**a**,**b**) Two days post-settlement. (**a**) A primary polyp developed from a stained planula. Scale bar is 500 μm. (**b**) A clustered settlement of two stained and two un-stained planulae that settled and metamorphosed. One stained and one unstained aggregate. Scale bar is 1000 μm. (**c**) Box plot for settlement percentages. Lower and upper hinges correspond to the 25th and 75th percentiles. The upper/lower whisker extends from the hinge to the largest/smallest value of 1.5 × IQR (interquartile range) from the hinge. The horizontal line corresponds to the median value. Brown boxes refer to unstained planulae and purple boxes to the stained planulae in each treatment.

In total, 42 dishes were assigned to the mixed treatment, 15 to the non-stained treatment, and 19 to the stained treatment. Under these conditions, 43.9 ± 3.6% (mean ± SE) of all planulae (regardless of the treatment) successfully settled and metamorphosed, 42.5 ± 6.3%, 40.7 ± 5.7% and 44.5 ± 3.5% of the non-stained larvae dishes, the stained larvae dishes and the mix dishes, respectively (one way ANOVA: $F_{2,73} = 0.73$, $p = 0.49$;

Figure 3c). In the mix treatment, 46.0 ± 4.7% and 45.1 ± 4.2% of the stained and the non-stained planulae were settled, again, with no significant difference (paired *t*-test: $t_{40} = 0.06$, $p > 0.8$; Figure 3c). Focusing on the stained larvae, the overall mean settled stained planulae in the mix and stained dishes was 44.4 ± 3.7%, as compared to 44.4 ± 3.5% for the non-stained planulae (mix and non-stained; *t*-test: $t_{115} = -0.18$, $p > 0.8$; Figure 3c)

4. Discussion

This study has tested and established the use of neutral red dye as a useful way for labeling coral planulae, amenable for follow-up studies for at least 3 days under constant flow and more than a week under a still water regimen. Using *S. pistillata* as a model case, we demonstrated that staining planulae of a brooding species with neutral red is a safe and reliable labeling procedure that does not harm larval behavior, settlement, and metamorphosis and that the stain is not leaking to unlabelled larvae within the same dish. Stained larvae/primary polyps were clearly distinguished by color from unstained counterparts, using the naked eye and without employing any sophisticated equipment. Using the settlement percentages as a proxy for the potentially harmful effects of the labeling, we found no significant differences in settlement percentages between stained and non-stained planulae, including the mixed dishes where planulae originated from different maternal colonies. An additional advantage to the new labeling practice is its low cost, without necessitating the use of genetic markers or observations under special illumination, such as fluorescent light. Though not tested here, the use of neutral red to label planulae and release them in situ may reveal an additional research pathway in coral biology studies. The same implies to the use of neutral red with planulae from broadcasting coral species, though the literature attests that neutral red stain does not damage the embryogenesis of other cnidarian species [30]. Aggregated and clustered settlements of larvae are continuously reported for various coral species [8–14]. However, despite the ecological importance of larval settlement behavior, the pattern of settlement as shaped by the inter- and intraspecific interactions of larvae in the aggregated settlers remains poorly understood. The literature depicts that newly settled juveniles have higher fusion probabilities between kin [12,42–46]. However, lacking feasible labeling methodologies, previous studies commonly approached the lack of partner identifications by artificially gluing/actively attaching coral spat/colonies next to each other, enforcing the close settlement of planulae without the opportunity to test the natural choices of planulae in a pair.

There has been a resurgence in scientific interest in the importance of the larval settlement phase in the coral population [1,47,48], including studies that considered planulae seeding as a reef restoration strategy approach [13,14,49–54]. Further, corals' xenogeneic and allogeneic interactions in the reef may affect various life-history traits [15,55–57], shaping community structures and population dynamics [55,58]. Using the neutral red labeling method will provide another experimental tool for studying the detailed settlement patterns of kin and non-kin coral larvae for improved yields of coral colonies amenable for the coral transplantation or larval seeding in the reef.

Author Contributions: Conceptualization, D.S., B.R.; methodology, D.S.; software, D.S.; validation, D.S., B.R.; formal analysis, D.S.; investigation, D.S.; resources, B.R.; data curation, D.S.; writing—original draft preparation, D.S. and B.R.; writing—review and editing, D.S., B.R. and N.S.; visualization, D.S.; supervision, B.R., N.S.; project administration, B.R.; funding acquisition, B.R. All authors have read and agreed to the published version of the manuscript.

Funding: This research was funded by the AID-MERC (M33-001), the Israeli-French High Council for Scientific and Technological Research Program (Maïmonide-Israel), the North American Friends of IOLR (NAF/IOLR), and the Barrett Foundation.

Institutional Review Board Statement: Not applicable.

Informed Consent Statement: Not applicable.

Data Availability Statement: The data presented in this study are available in request from corresponding author.

Acknowledgments: We wish to thank O. Ben-Zvi for the creative ideas, to G. Paz for assisting with visualization and Z. Lapidot for assisting with staining.

Conflicts of Interest: The authors declare no conflict of interest.

References

1. Ritson-Williams, R.; Arnold, S.N.; Fogarty, N.D.; Steneck, R.S.; Vermeij, M.J.A.; Paul, V.J. New perspectives on ecological mechanisms affecting coral recruitment on reefs. *Smithson. Contrib. Mar. Sci.* **2009**, 437–457. [CrossRef]
2. Tebben, J.; Motti, C.A.; Siboni, N.; Tapiolas, D.M.; Negri, A.P.; Schupp, P.J.; Kitamura, M.; Hatta, M.; Steinberg, P.D.; Harder, T. Chemical mediation of coral larval settlement by crustose coralline algae. *Sci. Rep.* **2015**, *5*, 10803. [CrossRef] [PubMed]
3. Birrell, C.L.; McCook, L.J.; Willis, B.L. Effects of algal turfs and sediment on coral settlement. *Mar. Pollut. Bull.* **2005**, *51*, 408–414. [CrossRef] [PubMed]
4. Richmond, R.H. Reproduction and recruitment in corals: Critical links in the persistence of reefs. In *Life Death Coral Reefs*; Chapman Hall: New York, NY, USA, 1997; pp. 175–197.
5. Arnold, S.N.; Steneck, R.S. Settling into an Increasingly Hostile World: The Rapidly Closing "Recruitment Window" for Corals. *PLoS ONE* **2011**, *6*, e28681. [CrossRef]
6. Rinkevich, B. Neglected biological features in Cnidarians self—nonself recognition. In *Self-NonSelf Recognition*; López-Larrea, C., Ed.; Springer: New York, NY, USA, 2012; pp. 46–59.
7. Duerden, J.E. Aggregated Colonies in Madreporarian Corals. *Am. Nat.* **1902**, *36*, 461–471. [CrossRef]
8. Goreau, N.I.; Goreau, T.J.; Hayes, R.L. Settling, survivorship and spatial aggregation in planulae and juveniles of the coral *Porites porites* (Pallas). *Bull. Mar. Sci.* **1981**, *31*, 424–435.
9. Neves, E.G.; Silveira, F.L. Release of planula larvae, settlement and development of *Siderastrea stellata* Verrill, 1868 (Anthozoa, Scleractinia). *CEUR Workshop Proc.* **2000**, *1621*, 36–43. [CrossRef]
10. Barki, Y.; Gateño, D.; Graur, D.; Rinkevich, B. Soft-coral natural chimerism: A window in ontogeny allows the creation of entities comprised of incongruous parts. *Mar. Ecol. Prog. Ser.* **2002**, *231*, 91–99. [CrossRef]
11. Puill-Stephan, E.; van Oppen, M.J.H.; Pichavant-Rafini, K.; Willis, B.L. High potential for formation and persistence of chimeras following aggregated larval settlement in the broadcast spawning coral, *Acropora millepora. Proc. R. Soc. Lond. Ser. B Biol. Sci.* **2012**, *279*, 699–708. [CrossRef]
12. Amar, K.O.; Chadwick, N.E.; Rinkevich, B. Coral kin aggregations exhibit mixed allogeneic reactions and enhanced fitness during early ontogeny. *BMC Evol. Biol.* **2008**, *8*, 126. [CrossRef]
13. Doropoulos, C.; Evensen, N.R.; Gómez-Lemos, L.A.; Babcock, R.C. Density-dependent coral recruitment displays divergent responses during distinct early life-history stages. *R. Soc. Open Sci.* **2017**, *4*, 170082. [CrossRef]
14. Cameron, K.A.; Harrison, P.L. Density of coral larvae can influence settlement, post-settlement colony abundance and coral cover in larval restoration. *Sci. Rep.* **2020**, *10*, 5488. [CrossRef]
15. Rinkevich, B.; Shashar, N.; Liberman, T. Nontransitive xenogeneic interactions between four common Red Sea sessile invertebrates. *Proc. Seventh Int. Coral Reef Symp.* **1993**, *2*, 833–839.
16. Rinkevich, B. Allorecognition and xenorecognition in reef corals: A decade of interactions. *Hydrobiologia* **2004**, *530*, 443–450. [CrossRef]
17. Rinkevich, B.; Weissman, I.L. Chimeras in colonial invertebrates: A synergistic symbiosis or somatic-cell and germ-cell parasitism? *Symbiosis* **1987**, *4*, 117–134.
18. Guerrini, G.; Shefy, D.; Shashar, N.; Shafir, S.; Rinkevich, B. Morphometric and allometric rules of polyp's landscape in regular and chimeric coral colonies of the branching species *Stylophora pistillata*. *Dev. Dyn.* **2020**, 1–17. [CrossRef]
19. Shefy, D.; Shashar, N.; Rinkevich, B. Exploring traits of engineered coral entities to be employed in reef restoration. *J. Mar. Sci. Eng.* **2020**, *8*, 1038. [CrossRef]
20. Rinkevich, B. Coral chimerism as an evolutionary rescue mechanism to mitigate global climate change impacts. *Glob. Chang. Biol.* **2019**, *25*, 1198–1206. [CrossRef] [PubMed]
21. Noble, J.P.A.; Lee, D. First report of allogeneic fusion and allorecognition in tabulate corals. *J. Paleontol. Soc.* **1991**, *65*, 69–74. [CrossRef]
22. Puill-Stephan, E.; Willis, B.L.; van Herwerden, L.; van Oppen, M.J.H. Chimerism in wild adult populations of the broadcast spawning coral *Acropora millepora* on the Great Barrier Reef. *PLoS ONE* **2009**, *4*. [CrossRef] [PubMed]
23. Maier, E.; Buckenmaier, A.; Tollrian, R.; Nürnberger, B. Intracolonial genetic variation in the scleractinian coral *Seriatopora hystrix*. *Coral Reefs* **2012**, *31*, 505–517. [CrossRef]
24. Schweinsberg, M.; Weiss, L.C.; Striewski, S.; Tollrian, R.; Lampert, K.P. More than one genotype: How common is intracolonial genetic variability in scleractinian corals? *Mol. Ecol.* **2015**, *24*, 2673–2685. [CrossRef] [PubMed]
25. Rinkevich, B.; Shaish, L.; Douek, J.; Ben-Shlomo, R. Venturing in coral larval chimerism: A compact functional domain with fostered genotypic diversity. *Sci. Rep.* **2016**, *6*, 19493. [CrossRef]

26. Giordano, B.; Bramanti, L. First report of chimerism in Mediterranean red coral (*Corallium rubrum*). *Mediterr. Mar. Sci.* **2021**, *22*, 157–160. [CrossRef]
27. Guérard, M.; Zeller, A.; Singer, T.; Gocke, E. In Vitro genotoxicity of neutral red after photo-activation and metabolic activation in the Ames test, the micronucleus test and the comet assay. *Mutat. Res. Genet. Toxicol. Environ. Mutagen.* **2012**, *746*, 15–20. [CrossRef]
28. LaManna, J.C.; Mccracken, K.A. The use of Neutral Red as an intracellular pH Indicator in rat brain cortex in vivo. *Analyt* **1984**, *142*, 117–125. [CrossRef]
29. New, J.G. Dyes for studying the movements of small mammals. *Am. Soc. Mammal.* **1958**, *39*, 416–429. [CrossRef]
30. Freeman, G. The role of polarity in the development of the hydrozoan planula larva. *Roux's Arch. Dev. Biol.* **1981**, *190*, 168–184. [CrossRef]
31. Zhou, Q. Chemical pollution and transport of organic dyes in water–soil–crop systems of the chinese coast. *Bull. Environ. Contam. Toxicol.* **2001**, *66*, 784–793. [CrossRef] [PubMed]
32. Levin, L.A. A review of methods for labeling and tracking marine invertebrate larvae. *Ophelia* **1990**, *32*, 115–144. [CrossRef]
33. Rinkevich, B.; Loya, Y. The reproduction of the Red Sea coral *Stylophora pistillata*. I. Gonads and planulae. *Mar. Ecol. Prog. Ser.* **1979**, *1*, 145–152. [CrossRef]
34. Rinkevich, B.; Loya, Y. The reproduction of the Red Sea coral *Stylophora pistillata*. II. Synchronization in breeding and seasonality of planulae shedding. *Mar. Ecol. Prog. Ser.* **1979**, *1*, 133–144. [CrossRef]
35. Shefy, D.; Shashar, N.; Rinkevich, B. The reproduction of the Red Sea coral *stylophora pistillata* from Eilat: 4-decade perspective. *Mar. Biol.* **2018**, *165*, 1–10. [CrossRef]
36. Putnam, H.M.; Edmunds, P.J.; Fan, T.-Y. Effect of temperature on the settlement choice and photophysiology of larvae from the reef coral *Stylophora pistillata*. *Biol. Bull.* **2008**, *215*, 135–142. [CrossRef]
37. Baird, A.H.; Morse, A.N.C. Induction of metamorphosis in larvae of the brooding corals *Acropora palifera* and *Stylophora pistillata*. *Mar. Freshw. Res.* **2004**, *55*, 469–472. [CrossRef]
38. Tamir, R.; Eyal, G.; Cohen, I.; Loya, Y. Effects of light pollution on the early life stages of the most abundant northern Red Sea coral. *Microorganisms* **2020**, *8*, 193. [CrossRef]
39. Chadwick-Furman, N.; Rinkevich, B. A complex allorecognition system in a reef-building coral: Delayed responses, reversals and nontransitive hierarchies. *Coral Reefs* **1994**, *13*, 57–63. [CrossRef]
40. Baird, A.H. The Ecology of Coral Larvae: Settlement Patterns, Habitat Selection and the Length of the Larval Phase. Ph.D. Thesis, James Cook University of North Queensland, Queensland, Australia, 2001.
41. R Core Team R: A Language and Environment for Statistical Computing. 2014. Available online: https://www.r-project.org/ (accessed on 29 May 2021).
42. Hidaka, M.; Yurugi, K.; Sunagawa, S.; Kinzie, R.A. Contact reactions between young colonies of the coral *Pocillopora damicornis*. *Coral Reefs* **1997**, *16*, 13–20. [CrossRef]
43. Nozawa, Y.; Loya, Y. Genetic relationship and maturity state of the allorecognition system affect contact reactions in juvenile Seriatopora corals. *Mar. Ecol. Prog. Ser.* **2005**, *286*, 115–123. [CrossRef]
44. Amar, K.O.; Rinkevich, B. Mounting of erratic histoincompatible responses in hermatypic corals: A multi-year interval comparison. *J. Exp. Biol.* **2010**, *213*, 535–540. [CrossRef] [PubMed]
45. Wijayanti, D.P.; Hidaka, M. Is genetic involve in the outcomes of contact reactions between parent and offspring and between siblings of the coral *Pocillopora damicornis*? *ILMU Kelaut. Indones. J. Mar. Sci.* **2018**, *23*, 69. [CrossRef]
46. Frank, U.; Oren, U.; Loya, Y.; Rinkevich, B. Alloimmune maturation in the coral *Stylophora pistillata* is achieved through three distinctive stages, 4 months post-metamorphosis. *Proc. R. Soc. Lond. Ser. B Biol. Sci.* **1997**, *264*, 99–104. [CrossRef]
47. Doropoulos, C.; Ward, S.; Roff, G.; González-Rivero, M.; Mumby, P.J. Linking demographic processes of juvenile corals to benthic recovery trajectories in two common reef habitats. *PLoS ONE* **2015**, *10*, e0128535. [CrossRef] [PubMed]
48. Vermeij, M.J.A.; Sandin, S.A. Density-dependent settlement and mortality structure the earliest life phases of a coral population. *Ecology* **2008**, *89*, 1994–2004. [CrossRef] [PubMed]
49. Rinkevich, B. Ecological engineering approaches in coral reef restoration. *ICES J. Mar. Sci.* **2020**, *2100*. [CrossRef]
50. Rinkevich, B. The active reef restoration toolbox is a vehicle for coral resilience and adaptation in a changing world. *J. Mar. Sci. Eng.* **2019**, *7*, 201. [CrossRef]
51. Linden, B.; Rinkevich, B. Elaborating an eco-engineering approach for stock enhanced sexually derived coral colonies. *J. Exp. Mar. Biol. Ecol.* **2017**, *486*, 314–321. [CrossRef]
52. Cruz, D.W.D.; Harrison, P.L. Enhanced larval supply and recruitment can replenish reef corals on degraded reefs. *Sci. Rep.* **2017**, *7*, 13985. [CrossRef]
53. Linden, B.; Vermeij, M.J.A.; Rinkevich, B. The coral settlement box: A simple device to produce coral stock from brooded coral larvae entirely in situ. *Ecol. Eng.* **2019**, *132*, 115–119. [CrossRef]
54. Amar, K.O.; Rinkevich, B. A floating mid-water coral nursery as larval dispersion hub: Testing an idea. *Mar. Biol.* **2007**, *151*, 713–718. [CrossRef]
55. Tanner, J.E. Interspecific competition reduces fitness in scleractinian corals. *J. Exp. Mar. Biol. Ecol.* **1997**, *214*, 19–34. [CrossRef]
56. Rinkevich, B.; Loya, Y. Intraspecific competitive networks in the Red Sea coral *Stylophora pistillata*. *Coral Reefs* **1983**, *1*, 161–172. [CrossRef]

57. Frank, U.; Brickner, I.; Rinkevich, B.; Loya, Y.; Bak, R.P.M.; Achituv, Y.; Ilan, M. Allogeneic and xenogeneic interactions in reef-building corals may induce tissue growth without calcification. *Mar. Ecol. Prog. Ser.* **1995**, *124*, 181–188. [CrossRef]
58. Buss, L.W.; Jackson, J.B.C. Competitive networks: Nontransitive competitive relationships in cryptic coral reef environments. *Am. Nat.* **1979**, *113*, 223–234. [CrossRef]

Journal of
Marine Science and Engineering

MDPI

Article

Anatomy and Ultrastructure of the Cyprid Temporary Adhesive System in Two Species of Acorn Barnacle

Joshua J. Raine [1,*], Nick Aldred [1,2] and Anthony S. Clare [1]

[1] School of Natural and Environmental Sciences, Newcastle University, Newcastle upon Tyne, Newcastle NE1 7RU, UK; nick.aldred@essex.ac.uk (N.A.); tony.clare@ncl.ac.uk (A.S.C.)
[2] School of Life Sciences, University of Essex, Colchester CO4 3SQ, UK
* Correspondence: j.j.raine@ncl.ac.uk

Received: 30 October 2020; Accepted: 24 November 2020; Published: 27 November 2020

Abstract: Acorn barnacles are sessile as adults and select their settlement site as a cypris larva. Cyprids are well adapted to exploring surfaces in dynamic environments, using a temporary adhesive secreted from the antennules to adhere during this process. The temporary adhesive and the secretory structures are poorly characterized. This study used serial block-face scanning electron microscopy and three-dimensional modelling to elucidate the anatomy related to temporary adhesion. The temporary adhesive glands of two acorn barnacle species, *Balanus amphitrite* and *Megabalanus coccopoma*, were located in the proximal region of the first antennular segment, contrary to previous descriptions that placed them in the more distal second segment. The temporary adhesive systems of these acorn barnacles are therefore similar to that described for the stalked barnacle, *Octolasmis angulata*, although not identical. Knowledge of the true location of the temporary adhesive glands will underpin future studies of the production, composition and secretion of the adhesive.

Keywords: barnacles; larvae; anatomy; adhesion; electron microscopy

---

1. Introduction

Barnacle cypris larvae use a proteinaceous adhesive secreted from their paired antennules to temporarily attach to the substratum, allowing them to effectively explore dynamic environments [1,2]. This adhesive is secreted via pores on the third antennular segment and contains the settlement-inducing protein complex (SIPC), or an analogue [3]. Temporarily adhered cyprids detach via mechanical forces, twisting their antennule and leaving behind a "footprint" of adhesive material [2,4–7]. Neither the temporary adhesive nor the structures related to its synthesis and secretion have been fully characterised in acorn barnacles.

The temporary adhesive system is distinct from the permanent adhesive used by the cyprid at the point of fixation. The morphology of the cyprid permanent adhesive system has been well characterised [8–10], although only basic compositional data have been obtained. The permanent adhesive is produced and stored in a pair of glands situated behind the compound eyes, with each gland serving a single antennule. When secreted, the cement is transported to the attachment disc by a single duct, which radiates at the surface to numerous pores, spreading it over a wider area.

Details regarding the composition and morphology of the temporary adhesive system are equally scant. The first description of the temporary adhesive system and its associated structures was provided by Nott and Foster [11] for *Semibalanus balanoides*; a boreo-arctic, intertidal, acorn barnacle (Cirripedia, Balanomorpha). They noted that the distal region of the second antennular segment contained numerous unicellular glands, arranged around the circumference of the appendage, and postulated that these may be responsible for producing the temporary adhesive [11].

These putative glands extended into ducts that fed into pores on the surface of the third antennular segment (the attachment disc, Figure 1). Supporting rod structures and a sheath cell encapsulated the ducts in places, while the interior of the gland contained the adhesive material packed within vesicles, and numerous microtubules oriented longitudinally (Figure 1). However, these putative glands were only 1–2 μm in diameter and did not appear to contain any of the cellular structures related to protein synthesis or vesicular packaging.

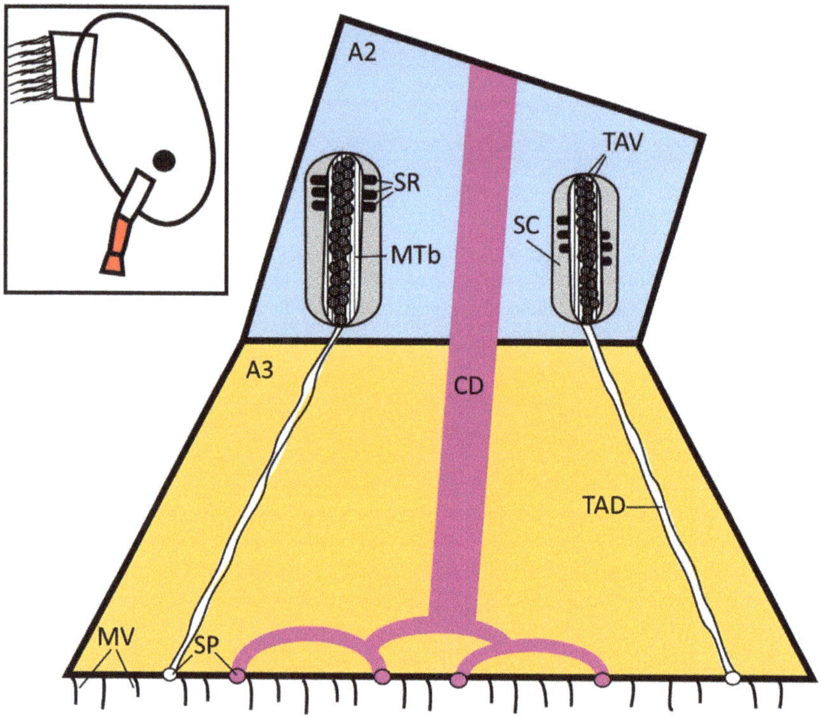

Figure 1. Schematic diagram of the temporary adhesive system located in the distal second antennular segment as proposed by Nott and Foster [11]. Not to scale. The region detailed in the schematic is highlighted in red on the inset. A2 = second antennular segment, A3 = third antennular segment (attachment disc), CD = cement duct, TAV = temporary adhesive vesicle, TAD = temporary adhesive duct, SR = supporting rod, MTb = microtubules, SP = secretory pores, SC = sheath cell, MV = microvilli.

A more recent study on the epibiotic stalked barnacle *Octolasmis angulata* [12] provided evidence of different anatomy. While many aspects of the descriptions of the two species were similar, a major difference was that the temporary adhesive glands of *O. angulata* were located in the anterior mantle of the cyprid behind the compound eyes, rather than in the antennules. The glands (~20 × 25 μm) comprised large, multicellular secretory cells, which contained vesicles. Ducts from the glands extended down the length of the antennules, through which the putative adhesive-containing vesicles were transported to the surface of each attachment disc [12].

As proposed by Yap et al. [12] it is conceivable that the "glands" described by Nott and Foster [11] for *S. balanoides* corresponded to the unicellular gland extensions observed in the distal second and third antennular segments of *O. angulata*. There is, therefore, the possibility that the anatomies of the adhesive systems of the two species are comparable. Certainty over the location of the adhesive glands in acorn barnacles is essential if molecular (e.g., in situ hybridisation) and immunohistochemical techniques are to be successfully applied in studies of adhesive composition.

To address this question, the present study used serial block face-scanning electron microscopy (SBF-SEM) to obtain a comprehensive series of images of the antennule, and to track the temporary adhesive system on its route from the disc surface back to to the glands. In addition, the image series allowed for three-dimensional reconstruction and modelling of the cyprid anatomy in the Microscopy Image Browser (MIB) [13]. Using these techniques, this study explored the temporary adhesive system of two acorn barnacle species, *Balanus amphitrite* (= *Amphibalanus amphitrite*) and *Megabalanus coccopoma*, in order to provide a basis for future investigations into the composition and secretion of the adhesive material.

2. Materials and Methods

2.1. Organism Acquisition

The cypris larvae of two species, *Balanus amphitrite* and *Megabalanus coccopoma*, were prepared for SBF-SEM. Larval production required the culturing of adult populations.

Adult broodstock of *B. amphitrite* was obtained from Duke University Marine Laboratory (DUML), Beaufort, North Carolina (USA). Adults were maintained in ~33 ppt artificial seawater (ASW, Tropic Marin) at 28 °C with aeration. Barnacles were fed daily with freshly hatched *Artemia* sp. (Varicon Aqua Solutions Ltd., Worcester, UK) nauplii supplemented with the diatoms *Tetraselmis suecica* and *Isochrysis galbana* algae at weekends. The water was also changed, and the barnacles cleaned daily.

Hatching and release of *B. amphitrite* nauplii were stimulated by removing the adults from water overnight and reintroducing water the following morning. They were kept in the dark, with a single source of illumination to attract the released nauplii. The nauplii were periodically collected by pipette and pooled in beakers containing *T. suecica*. When their number was sufficient, the nauplii were added to aerated buckets of artificial seawater at a density of no greater than 1000 larvae per litre and cultured at 28 °C on a 16 h:8 h L:D cycle. The nauplii were fed 50 mL per liter of culture of *T. suecica* (~1.3×10^6 cells mL^{-1}) and *I. galbana* (5×10^6 cells mL^{-1}) in a 3:1 ratio initially and after three days. The nauplii developed to the cyprid stage in five to six days under these conditions. Cyprids were collected from cultures by filtration with a 250 μm mesh, then transferred to dishes of ASW for storage at 4 °C until required.

Adult broodstock of *M. coccopoma* was also acquired from DUML and maintained as described for *B. amphitrite*. For the collection of larvae, buckets containing the adult *M. coccopoma* were checked each morning for the presence of nauplii. Nauplii were collected and cultured in the same manner as *B. amphitrite*. The nauplii developed to the cyprid stage in seven to eight days under these conditions. The larvae were filtered from the cultures and returned to clean buckets with fresh ASW and algal feed after four or five days. Cyprids were extracted from cultures by filtration with a 250 μm mesh. Cyprids retained on the mesh were transferred to dishes containing ASW and stored at 4 °C until required.

2.2. Sample Preparation

Cyprids of each species, (*B. amphitrite*, n = 6; *M. coccopoma*, n = 2) were fixed overnight in 2% glutaraldehyde (Agar Scientific) in 0.1 M sodium cacodylate buffer. Once fixed, the samples were processed using a heavy metal staining protocol adapted from Deerinck et al. [14]. Samples were incubated in a series of heavy metal solutions, rinsing three times for five minutes each in distilled water between steps: 3% potassium ferrocyanide in 2% osmium tetroxide for one hour, 10% thiocarbohydrazide for 20 min, 2% osmium tetroxide for 30 min, 1% uranyl acetate overnight at 4 °C, and finally lead aspartate solution (0.02 M lead nitrate in 0.03 M in aspartic acid, pH 5.5) for 30 min at 60 °C. The samples were then dehydrated through a graded series of acetone to 100% and then impregnated with increasing concentrations of Taab 812 hard resin, with three changes of 100% resin. The samples were embedded in 100% resin and left to polymerise at 60 °C for a minimum of 36 h.

The resin blocks were trimmed to approximately 0.75 mm by 0.5 mm and glued onto an aluminium pin. To reduce sample charging within the SEM, the block was painted with silver glue and sputter-coated with a 5 nm layer of gold using a Polaron SEM coating unit.

2.3. Serial Block Face-Scanning Electron Microscopy

The pin was placed into a Zeiss Sigma SEM incorporating the Gatan 3view system with DMGMS3 software, which allows sectioning of the block in situ and the collection of a series of images in the z-direction. Multiple regions of interest were imaged at variable magnification, pixel scan, section thickness and pixel resolution depending on the species (Table 1).

Table 1. Details on the parameters selected for image capture using the Gatan 3view system. "Series" refers to the z-stack of images collected and compiled for three-dimensional model reconstruction, while "Stills" are the images collected for two-dimensional presentation.

Species	Magnification	Pixel Time	Section Thickness	Pixel Resolution
Balanus amphitrite	Series: 1290–1880× Stills: 898–4000× as required	Series: 20 µs Stills: 20 µs	Series: 100 nm Stills: 100 nm	Series: 20 nm, 2000 × 2000–3500 × 3500 Stills: 10 nm 4000 × 4000
Megabalanus coccopoma	Series: 646× Stills: 157–6700× as required	Series: 20 µs Stills: 20 µs	Series: 100 nm Stills: 100 nm	Series: 25 nm, 3000 × 3000 Stills: 5.6 nm–478.5 nm, 2000 × 2000

In the resulting z-stacks, the areas of interest were identified and segmented manually using Microscopy Image Browser (MIB, University of Helsinki) [13]. The segmentations were imported into Amira (FEI) for the construction of the 3-D models. The *B. amphitrite* antennule and *M. coccopoma* third antennular segment overlay models used 3090 and 114 sections, respectively.

3. Results

3.1. Balanus amphitrite

Cyprids of *B. amphitrite* were serially sectioned in the transverse plane. This orientation allowed for the best opportunity to identify the temporary adhesive ducts and track them through the antennule from the gland to the disc. However, the orientation of mounting and curvature of the antennules resulted in some sections being taken at an oblique angle. The temporary adhesive gland (one of a pair) was located in the proximal region of the first antennular segment and can be observed in Figure 2a,b.

The gland was formed of a cluster of 10–15 cells. The gland and component cells were roughly spherocylindrical in shape. The gland was ~40 µm, by 10 µm overall, while individual cells were 8–12 µm by 4–5 µm. Therefore only a few cells sat in parallel. A summary of the dimensions of key structures is presented in Table 2.

The components for protein production and packaging were present in the gland cells (Figure 2a,b). Rough endoplasmic reticulum and Golgi bodies were clearly visible, along with mitochondria. In a number of these cells, newly formed vesicles (50–200 nm) could be observed prior to their transportation down the antennule (Figure 2a,b). While the contents of the vesicles were homogenous in appearance, their size was more variable compared to those observed distally in the first and second antennular segments (Figure 2c,d). Much of the proximal first antennular segment was filled with muscle tissue, in various orientations, and the temporary adhesive gland was surrounded by muscle. The cement duct was also evident in this region and was distinct from the temporary adhesive gland as it proceeded proximally into the cephalon where the cement gland was located (Figure 2a,b). Finally, the bundled axons comprising the antennular nerve were observed, which would eventually connect with the antennular soma cluster and brain if followed proximally (Figure 2b).

Table 2. Summary of the dimensions of the cyprid species imaged, and the key structures of their temporary adhesive systems (± standard deviation). TAV = temporary adhesive vesicle, TAD = temporary adhesive duct, TAG = temporary adhesive gland, A1 = first antennular segment, A2, second antennular segment, A3 = third antennular segment.

Species	TAV Diameter	TAD Diameter	TAG Cells	TAG	Cyprid Length	A1 Length	A2 Length	A3 Diameter
Balanus amphitrite	~250 nm (±89)	~0.5 µm (±0.11)	~4.5 × 4.5 × 10 µm	~10 × 10 × 40 µm	~400 µm	~90 µm	~80 µm	~20 µm
Megabalanus coccopoma	~400 nm (±104)	~0.8 µm (±0.26)	~5 × 5 × 25 µm	~25 × 25 × 50 µm	~500 µm	~150 µm	~140 µm	~35 µm

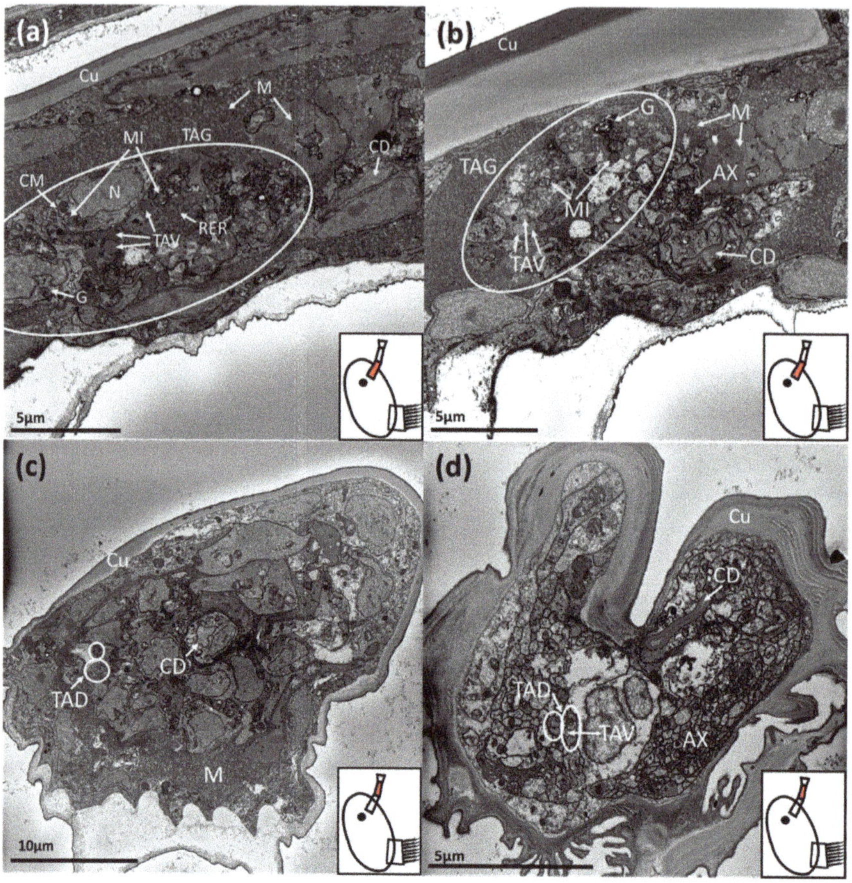

Figure 2. SBF-SEM images of a *Balanus amphitrite* cyprid temporary adhesive gland and ducts. The region where the image was captured is highlighted in red on the schematic inset: (**a**,**b**) = proximal first antennular segment, (**c**) = proximal second antennular segment, (**d**) = distal second antennular segment. TAG = temporary adhesive gland, TAV = temporary adhesive vesicle, TAD = temporary adhesive duct, N = nucleus, MI = mitochondria, G = Golgi body, AX = axons, CM = cell membrane, RER = rough endoplasmic reticulum, M = muscle, CD = cement duct, Cu = cuticle.

Proceeding distally down the first antennular segment, the temporary adhesive gland cells transitioned into narrow ducts. The morphology of the temporary adhesive ducts remained relatively constant through the first and second antennular segments (Figure 2c,d). The temporary adhesive ducts did occasionally vary in width, however, depending on the number of vesicles contained within. For the individual illustrated, the ducts contained few vesicles for much of their length and remained narrow as a result (Figure 2c,d). The cement duct remained similar in size but became flattened as it approached the junction of the second and third antennular segments (Figure 2c,d). The axons of the antennular nerve followed a similar path through these segments, remaining clustered (Figure 2d).

The ducts then entered the third antennular segment, or attachment disc. Here the temporary adhesive ducts expanded noticeably, becoming larger spherocylindrical structures ~1–1.5 μm in diameter with two distinct regions (Figure 3). The inner region seemed to be the transporting duct and contained numerous homogenous vesicles, each ~200 nm in diameter. These vesicles were surrounded by numerous peripheral microtubules (Figure 3b,c). The outer section was identifiable by the presence of supporting rods (Figure 3b,c). The ducts and associated elements did not appear to contain any of the cellular components required to produce and package the temporary adhesive. The number of ducts correlated with the number of cells in the temporary adhesive gland. Towards the disc surface, secretory ducts lost the supporting outer section (Figure 3a) and the secretory pores for both the temporary adhesive and cement duct could be seen terminating among the microvilli that cover the surface of the attachment disc (Figure 3a).

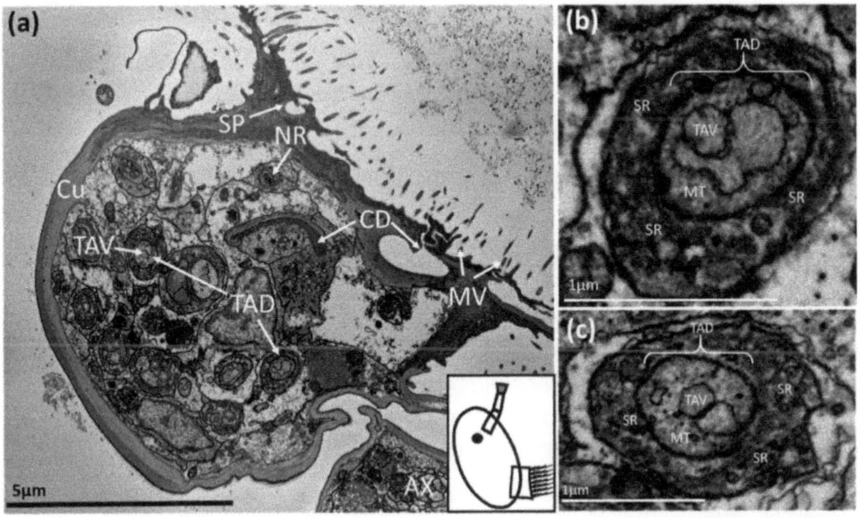

Figure 3. SBF-SEM image of a *Balanus amphitrite* cyprid: (**a**) = third antennular segment, (**b**,**c**) = temporary adhesive ducts within the third antennular segment. The region where the image was captured is highlighted in red on the schematic inset. TAD = temporary adhesive duct, TAV = temporary adhesive vesicle, SP = secretory pore, CD = cement duct, NR = neurone, AX = axon, MT = microtubules, SR = supporting rods, MV = microvilli.

The cement duct remained flattened as it extended into the third antennular segment. As the specimen in Figure 1 had not begun the process of permanent adhesion, the duct was empty and appeared to have narrowed to a slit of ~3 x < 0.5 µm. Presumably, this slit would stretch and open during secretion of the permanent cement (Figure 3a). At the disc surface, this singular permanent cement duct diverged into many smaller channels to dispense the adhesive over a larger area.

Finally, the axons comprising the antennular nerve connected to neurones serving sensilla on the third and fourth antennular segments (Figure 3a).

The two-dimensional images (Figures 2 and 3) allowed for detailed ultrastructural observation of sections of the cyprid adhesive systems. However, the morphology of the complete system is not easy to visualise from two-dimensional sections alone. To this end, a three-dimensional reconstruction of the systems described for *B. amphitrite* was produced (Figure 4). Here the division of the singular cement duct into the smaller channels across the disc surface was visible, as was the position of the cement gland behind the eye. Also modelled were two of the 10–15 temporary adhesive ducts, which led back to the temporary adhesive gland in the proximal first segment, where the widening and elongation of the cells could be seen.

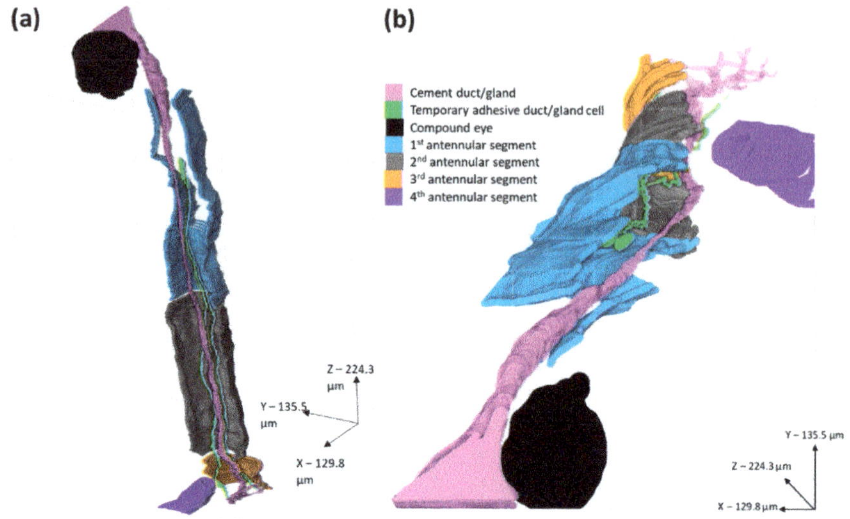

Figure 4. 3-D models of *B. amphitrite* antennule and adhesive systems, reconstructed from an SBF-SEM series. (**a**) = lateral view, (**b**) = dorsal view. The antennular segments were left open to allow viewing of the interior. Lilac = cement duct/gland, green = temporary adhesive duct/gland cell, black = compound eye, blue = 1st antennular segment, grey = 2nd antennular segment, yellow = 3rd antennular segment, purple = 4th antennular segment.

3.2. *Megabalanus coccopoma*

It seemed likely that the adhesive systems of *B. amphitrite* and *M. coccopoma* would be similar, however confirmation of the new observations in a second acorn barnacle species was nevertheless important. Since the temporary adhesive gland of *B. amphitrite* had already been located, the sectioning of *M. coccopoma* cyprids was changed from transverse to longitudinal to observe the structures from an alternate angle.

The initial longitudinal sections revealed the core structures, namely the antennule, the brain and oil cells within the anterior cephalon, and the thoracic appendages within the thorax (Figure 5a). The temporary adhesive gland was not visible in the initial imaging, however, deeper sectioning revealed its location to be consistent with that of *B. amphitrite*, in the proximal region of the first antennular segment (Figure 5a,b).

The temporary adhesive gland was approximately 25 × 50 µm, roughly prolate spheroid in shape and comprised of ~12 elongated cells surrounded by muscle (Figure 5b,c). A summary of the dimensions of key structures is presented in Table 2. A closer view of the gland revealed the cells to be full of homogenous vesicles of 300–400 nm in diameter (Figure 5c,d). These cells were up to 5 µm wide depending on vesicular content and 20–25 µm long. Additionally, the gland cells contained nuclei, rough endoplasmic reticulum and Golgi apparatus; the cellular structures necessary for protein production and vesicular packaging (Figure 5d). The vesicles near the Golgi apparatus were smaller and sparsely distributed, in addition to being situated on the right of the images, furthest from the antennule and transporting ducts (Figure 5c,d).

As was the case in *B. amphitrite*, the gland cells of *M. coccopoma* transitioned into ducts as they proceeded distally, narrowing to ~0.5–1 µm (Figure 6a). The duct diameter appeared to be correlated with the abundance of vesicles, with fewer vesicles resulting in a narrower duct. Other than this, the morphology of the ducts remained constant through the first and second antennular segments (Figure 6a).

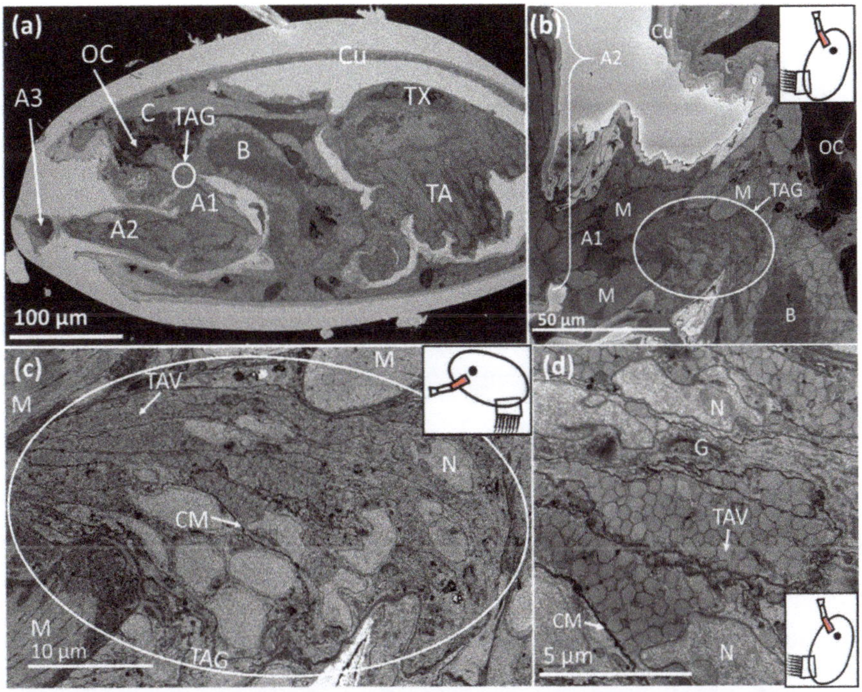

Figure 5. SBF-SEM images of *Megabalanus coccopoma*: (**a**) = cyprid, (**b**) = cyprid anterior mantle, (**c**) = temporary adhesive gland, (**d**) = temporary adhesive gland cells. The region where the image was captured is highlighted in red on the schematic inset. C = cephalon, TAG = temporary adhesive gland, TAV = temporary adhesive vesicle, OC = oil cell, A1 = first antennular segment, A2 = second antennular segment, A3 = third antennular segment (attachment disc), Cu = cuticle, TX = thorax, TA = thoracic appendages, M = muscle, B = brain. CM = cell membrane, N = nucleus, G = Golgi body, RER = rough endoplasmic reticulum.

Once the temporary adhesive ducts reached the third antennular segment, they expanded into larger, spherocylindrical structures on the periphery of the disc close to the surface (Figure 6b). These structures were over 1.5 µm in diameter and formed of two major sections, the vesicle-filled duct at the centre, and the supporting sheath around the outside (Figure 6b). In addition, these structures numbered approximately 12; as for the gland cells (Figure 6b,c). The temporary adhesive ducts were filled with the homogenous vesicles originally observed within the temporary adhesive gland cells. However, no vesicles were present between the spherocylindrical structures and the disc surface, where the ducts narrow and feed into secretory pores (Figure 6b). In contrast to the peripheral, singular, temporary adhesive ducts, the cement duct split into multiple sub-channels and opened out into the centre of the disc (Figure 6b).

Reconstructing the temporary adhesive ducts using three-dimensional modelling reveals that the "collecting structures" occurred in parallel pairs and were peripheral to the cement duct (Figure 6b). Unlike in *B. amphitrite*, the ducts took a circular route immediately following the putative collecting structures (Figure 6c) and surrounding muscle tissue.

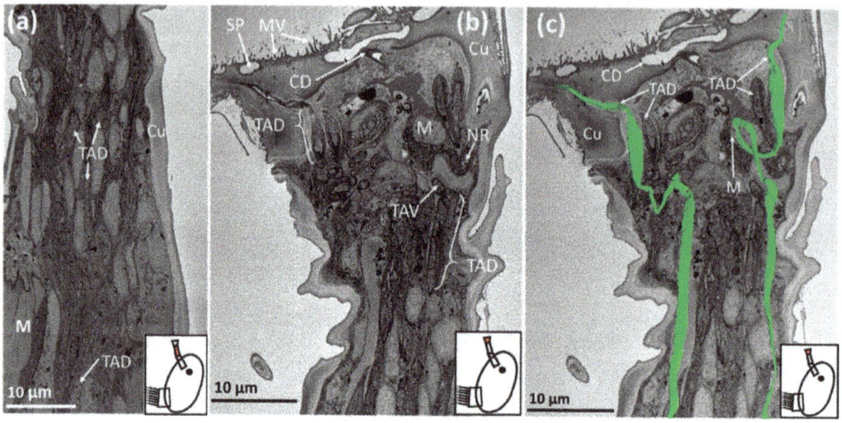

Figure 6. SBF-SEM images of *Megabalanus coccopoma*: (**a**) = second antennular segment, (**b**) = third antennular segment., (**c**) = third antennular segment overlaid with a 3-D model of temporary adhesive ducts (green). The region where the image was captured is highlighted in red on the schematic inset. TAD = temporary adhesive duct, TAV = temporary adhesive vesicle, M = muscle, SP = secretory pore, MV = microvilli, CD = cement duct, Cu = cuticle, NR = neurone.

4. Discussion

This study provides new evidence for the location of the temporary adhesive glands in acorn barnacle species and gives greater insight into the temporary adhesive system. Contrary to the findings of Nott and Foster [11], the temporary adhesive glands of *Balanus amphitrite* and *Megabalanus coccopoma* were not situated in the second antennular segment, but in the first (Figure 2a,b, Figures 4, 5b and 7). Instead, the ultrastructure of the acorn barnacles appeared more consistent with that of the stalked barnacle *Octolasmis angulata* [12], though there were some key differences.

The temporary adhesive glands of *B. amphitrite* and *M. coccopoma* possessed similar morphology and location. The glands were located in the proximal region of the first antennular segment (Figure 2a,b and Figure 5b). Each gland is comprised of ~10–15 elongated cells filled with homogenous vesicles containing the packaged temporary adhesive, and the organelles needed for protein synthesis and secretion (Figure 2a,b and Figure 5c,d). Smaller vesicles were observed in the glands of both species, situated in the vicinity of Golgi apparatus. These vesicles may have been undergoing the process of aggregation, coalescence and condensation described by Yap et al. [12]. In *M. coccopoma*, these smaller and more sparsely distributed vesicles were located at the furthest possible point from the ducts leading to the antennule, indicating an organisation of the cells within the gland to produce and transport the adhesive in a single direction (Figure 5c,d). The glands were roughly spherocylindrical or prolate spheroid in shape, with the overall size correlated to the size of the cyprid of each species (Table 2). However, the glands did not appear to have the morphology of a typical "exocrine gland", as each cell fed a single duct, and these ducts did not combine or diverge as they extended towards the third antennular segment [15–18]. As Yap et al. [12] described in the temporary adhesive system of the stalked barnacle, *Octolasmis angulata*, the ducts that carry the temporary adhesive to the disc surface are simply extensions of the gland cells themselves. In both cases, the glands were surrounded by muscle tissue in various orientations. Antennule locomotion is likely the primary function of these muscles, as described by Lagersson and Høeg [19], however, contraction of these muscles may apply pressure to the gland and aid in the secretion of the adhesive [12].

Following synthesis and vesicular packaging in the temporary adhesive gland, the vesicles containing the proteinaceous adhesive were transported to the third antennular segment via narrow ducts (Figure 2c,d and Figure 6a). Like the gland cells, the diameter of the ducts and the abundance of vesicles were correlated to the size of the cyprid species (Table 2). These vesicles were homogenous in electron density and shape, suggesting that unlike the permanent cement, the temporary adhesive is a single component, or a mixture of components, secreted as a singular phase.

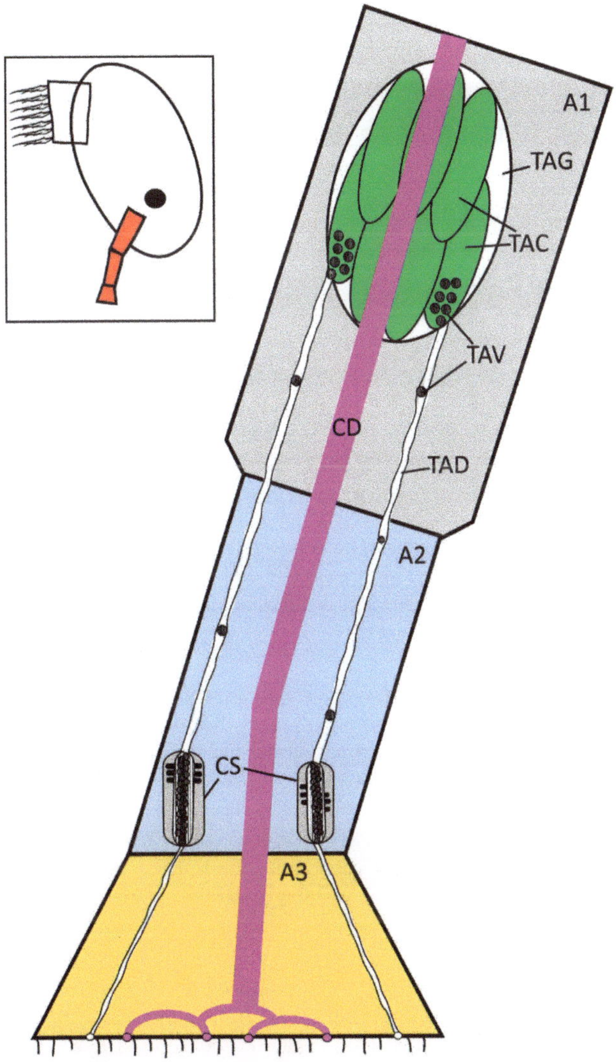

Figure 7. Updated schematic diagram (Figure 1) of the temporary adhesive system, with the gland located in the proximal first antennular segment. Not to scale. The temporary adhesive gland contains numerous cells for the synthesis and packing of the temporary adhesive and each cell feeds one duct. The region detailed in the schematic is highlighted in red on the inset. A1 = first antennular segment, A2 = second antennular segment, A3 = third antennular segment (attachment disc), CD = cement duct, TAG = temporary adhesive gland, TAC = temporary adhesive gland cell, TAV = temporary adhesive vesicle, TAD = temporary adhesive duct, CS = putative collecting structures, previously proposed to be the temporary adhesive gland by Nott and Foster [11].

Close to the surface of the disc, the ducts widened into a spherocylindrical shape, gaining a supportive outer sheath filled with rod-like structures (Figures 3a–c and 6b). These structures were similar to those previously described by Nott and Foster [11], which were interpreted as unicellular glands and the origin of the temporary adhesive. As their ultrastructural descriptions are in close accord with this study, it is proposed that these duct expansions are "collecting structures" for on-demand secretion. The same interpretation was made by Yap et al. [12] for *O. angulata*. In addition, these "collecting structures" appeared to be arranged in parallel pairs, peripheral to the central cement duct. The reason for this is unknown (Figures 3a and 6b). The abundance of these "collecting structures" was consistent with the number of cells observed in the temporary adhesive gland, further indicating that the ducts did not diverge. Finally, between the "collecting structures" and the secretory pores on the disc surface, the ducts lost the outer sheath, contained no vesicles and did not branch (Figures 3a and 6b). Interestingly, 3-D modelling of the ducts revealed that those of *M. coccopoma* encircled a region of muscle tissue in the centre of the third antennular segment, while those of *B. amphitrite* did not, suggesting potential species-specific differences (Figures 4a,b and 6c). Whether this muscle is simply related to articulation of the third and fourth antennular segments, or functions to aid secretion of the temporary adhesive, is unknown.

Although similar in many respects, there are notable differences in the temporary adhesive system in the two acorn barnacle species compared to that of the stalked barnacle *O. angulata* [12]. Primarily, the glands were observed in the proximal region of the 1st antennular segment in the acorn barnacles, as opposed to behind the eye in *O. angulata*. Secondly, no merging or diverging of the temporary adhesive ducts was observed in the acorn barnacles, while in the stalked barnacle some was described (Figure 4a,b). Finally, Yap et al. [12] observed no difference in the content of the glands in relation to cyprid age, implying that production is constitutive throughout the cyprid stage. Cyprids of different ages were not compared here, however, the gland and ducts of the *B. amphitrite* specimens contained considerably fewer vesicles than the *M. coccopoma* specimens, implying either a species-specific difference or that unknown factors can influence temporary adhesive production. It is possible that these differences are attributed to the distance in the evolutionary relationships between stalked and acorn barnacles [20,21].

In conclusion, this study has corrected a decades-old description of the location and structure of the temporary adhesive glands in acorn barnacle cyprids. Furthermore, variations were identified between individuals of the acorn barnacles studied here and the prior observations for *O. angulata*. The importance of these apparent variations will only be determined by further investigation of the ultrastructure of these, and a broader spectrum of barnacle species. Such information would allow for improved future studies of the temporary adhesive systems of barnacle larvae, via both morphology, as some aspects remain unclear still, and molecular biology, as isolation of the gland, will allow for the easier characterisation of the still-unidentified temporary adhesive protein(s).

Author Contributions: Conceptualization, J.J.R., N.A., and A.S.C. Methodology; software; validation.; formal analysis; investigation; resources; data curation; writing—original draft preparation.; visualization; project administration; J.J.R. Writing—review and editing; supervision; funding acquisition, A.S.C. and N.A. All authors have read and agreed to the published version of the manuscript.

Funding: Funded in part by US Office of Naval Research award N00014-16-1-3125 (A.S.C. and N.A.).

Acknowledgments: Thanks to the Newcastle University Electron Microscopy Research Services team for assisting with imaging.

Conflicts of Interest: The authors declare no conflict of interest. The funders had no role in the design of the study; in the collection, analyses, or interpretation of data; in the writing of the manuscript, or in the decision to publish the results.

References

1. Crisp, D.J. Settlement responses in marine organisms. In *Adaptations to Environment: Essays on the Physiology of Marine Animals*; Newell, R.C., Ed.; Butterworths: London, UK, 1976; pp. 83–124.
2. Walker, G.; Yule, A. Temporary adhesion of the barnacle cyprid: The existence of an antennular adhesive secretion. *J. Mar. Biol. Assoc. UK* **1984**, 64, 679–686. [CrossRef]

3. Dreanno, C.; Kirby, R.R.; Clare, A.S. Smelly feet are not always a bad thing: The relationship between cyprid footprint protein and the barnacle settlement pheromone. *Biol. Lett.* **2006**, *2*, 423–425. [CrossRef] [PubMed]
4. Phang, I.Y.; Aldred, N.; Clare, A.S.; Vancso, G.J. Towards a nanomechanical basis for temporary adhesion in barnacle cyprids (*Semibalanus balanoides*). *J. R. Soc. Interface* **2008**, *5*, 397–402. [CrossRef] [PubMed]
5. Phang, I.Y.; Aldred, N.; Ling, X.Y.; Huskens, J.; Clare, A.S.; Vancso, G.J. Atomic force microscopy of the morphology and mechanical behaviour of barnacle cyprid footprint proteins at the nanoscale. *J. R. Soc. Interface* **2010**, *7*, 285–296. [CrossRef] [PubMed]
6. Aldred, N.; Høeg, J.T.; Maruzzo, D.; Clare, A.S. Analysis of the behaviours mediating barnacle cyprid reversible adhesion. *PLoS ONE* **2013**, *8*, e68085. [CrossRef] [PubMed]
7. Guo, S.; Puniredd, S.R.; Jańczewski, D.; Lee, S.S.C.; Teo, S.L.M.; He, T.; Zhu, X.; Vancso, G.J. Barnacle larvae exploring surfaces with variable hydrophilicity: Influence of morphology and adhesion of "footprint" proteins by AFM. *ACS Appl. Mater. Interfaces* **2014**, *6*, 13667–13676. [CrossRef] [PubMed]
8. Walker, G. A study of the cement apparatus of the cypris larva of the barnacle *Balanus balanoides*. *Mar. Biol.* **1971**, *9*, 205–212. [CrossRef]
9. Okano, K.E.; Shimizu, K.A.; Satuito, C.; Fusetani, N.O. Visualization of cement exocytosis in the cypris cement gland of the barnacle *Megabalanus rosa*. *J. Exp. Biol.* **1996**, *199*, 2131–2137. [PubMed]
10. Ödling, K.; Albertsson, C.; Russell, J.T.; Mårtensson, L.G. An in vivo study of exocytosis of cement proteins from barnacle *Balanus improvisus* (D.) cyprid larva. *J. Exp. Biol.* **2006**, *209*, 956–964. [CrossRef] [PubMed]
11. Nott, J.A.; Foster, B.A. On the structure of the antennular attachment organ of the cypris larvae of *Balanus balanoides* (L.). *Philos. Trans. R. Soc. B* **1969**, *256*, 115–134.
12. Yap, F.C.; Wong, W.L.; Maule, A.G.; Brennan, G.P.; Chong, V.C.; Lim, L.H.S. First evidence for temporary and permanent adhesive systems in the stalked barnacle cyprid, *Octolasmis angulata*. *Sci. Rep.* **2017**, *7*, 44980. [CrossRef] [PubMed]
13. Belevich, I.; Joensuu, M.; Kumar, D.; Vihinen, H.; Jokitalo, E. Microscopy image browser: A platform for segmentation and analysis of multidimensional datasets. *PLoS Biol.* **2016**, *14*, e1002340. [CrossRef] [PubMed]
14. Deerinck, T.J.; Bushong, E.A.; Lev-Ram, V.; Shu, X.; Tsien, R.Y.; Ellisman, M.H. Enhancing serial block-face scanning electron microscopy to enable high resolution 3-D nanohistology of cells and tissues. *Microsc. Microanal.* **2010**, *16*, 1138–1139. [CrossRef]
15. Noirot, C.; Quennedey, A. Fine structure of insect epidermal glands. *Annu. Rev. Entomol.* **1974**, *19*, 61–80. [CrossRef]
16. Talbot, P.; Demers, D. Tegumental glands of Crustacea. In *Crustacean Integument. Morphology and Biochemistry Boca Raton*; Horst, M.N., Freeman, J.A., Eds.; CRC Press: Boca Raton, FL, USA, 1993; pp. 151–191.
17. Quennedey, A. Insect epidermal gland cells: Ultrastructure and morphogenesis. In *Microscopic Anatomy of Invertebrates*; Harrison, F., Locke, M., Eds.; Wiley-Liss: Hoboken, NJ, USA, 1998; pp. 177–207.
18. Freeman, S.C.; Malik, A.; Basit, H. *Physiology, Exocrine Gland*; StatPearls Publishing: Treasure Island, FL, USA, 2020; StatPearls [Internet]; (Updated 24 August 2020). Available online: https://www.ncbi.nlm.nih.gov/books/NBK542322/ (accessed on 25 November 2020).
19. Lagersson, N.; Høeg, J. Settlement behavior and antennulary biomechanics in cypris larvae of *Balanus amphitrite* (Crustacea: Thecostraca: Cirripedia). *Mar. Biol.* **2002**, *141*, 513–526.
20. Pérez-Losada, M.; Høeg, J.T.; Crandall, K.A. Unravelling the evolutionary radiation of the thoracican barnacles using molecular and morphological evidence: A comparison of several divergence time estimation approaches. *Syst. Biol.* **2004**, *53*, 278–298. [CrossRef] [PubMed]
21. Ewers-Saucedo, C.; Owen, C.L.; Pérez-Losada, M.; Høeg, J.T.; Glenner, H.; Chan, B.; Crandall, K.A. Towards a barnacle tree of life: Integrating diverse phylogenetic efforts into a comprehensive hypothesis of thecostracan evolution. *Peer J.* **2019**, *7*, e7387. [CrossRef] [PubMed]

Journal of
Marine Science and Engineering

MDPI

Article

Settlement of Bivalve Spat on Artificial Collectors (Net Bags) in Two Commercial Mussel Parks in the North-Western Adriatic Sea

Tihana Marčeta [1,2], Maria Gabriella Marin [1,*], Valentina Francesca Codognotto [1] and Monica Bressan [1]

1 Department of Biology, University of Padova, Via Ugo Bassi 58/B, 35131 Padova, Italy; tihana.marceta@ve.ismar.cnr.it (T.M.); valentina.codognotto@gmail.com (V.F.C.); monica.bressan@unipd.it (M.B.)
2 Institute of Marine Sciences (ISMAR), CNR, Castello 2737/F, 30122 Venezia, Italy
* Correspondence: mgmar@bio.unipd.it

Abstract: Among aquaculture activities, shellfish culture is considered more sustainable and beneficial in terms of food security. Currently, only a few bivalve species are reared and there is a need to explore the possibility to introduce new candidates for shellfish farming. Due to the lack of information on bivalve recruitment in the North-Western Adriatic Sea, in this study, the possibility to collect natural spat of commercial species was investigated. Artificial collectors (net bags) were deployed in two sites, Pellestrina and Caleri (North-Western Adriatic Sea), within two commercial mussel parks, during the spring–summer and summer–autumn periods. At both sites, collectors were placed at a distance of 1 m from each other, from 5 to 14 m depth. The influence of season, site and depth on bivalve recruitment was inspected and the presence of invasive species was also evaluated. In all, 28 bivalve taxa were found, and a higher settlement rate was observed in summer–autumn compared to the spring–summer period. *Mytilus galloprovincialis*, *Flexopecten glaber*, *Mimachlays varia* and *Aequipecten opercularis* were the most abundant species in spring–summer. In the summer–autumn period, in both sites analysed, a very high quantity of *Anadara transversa* and *F. glaber* were found. Indeed, these species were dominant at Pellestrina and Caleri, respectively. Another non-indigenous species, *Arcuatula senhousia*, was also detected. Relevant amounts of Pectinidae spat, *F. glaber* in particular, were collected and the optimal depth range for the scallop spat collection was found to be between 8 and 14 m. Our results highlight the relevant potential of Pectinidae spat collection along the North-Western Adriatic coasts, even though the presence of invasive species needs to be monitored.

Keywords: bivalves; scallops; spat; artificial collectors; settlement; recruitment; mariculture; non-indigenous species; Adriatic Sea

Citation: Marčeta, T.; Marin, M.G.; Codognotto, V.F.; Bressan, M. Settlement of Bivalve Spat on Artificial Collectors (Net Bags) in Two Commercial Mussel Parks in the North-Western Adriatic Sea. *J. Mar. Sci. Eng.* **2022**, *10*, 210. https://doi.org/10.3390/jmse10020210

Academic Editor: Pedro Reis Costa

Received: 31 December 2021
Accepted: 2 February 2022
Published: 5 February 2022

1. Introduction

The natural bivalve populations of the North Adriatic Sea are subject to intense fishing efforts. The main commercial species in the Adriatic Sea (striped venus clam, *Chamelea gallina*, smooth clam, *Callista chione*, Mediterranean scallop, *Pecten jacobaeus* and queen scallop, *A. opercularis*) are fished by hydraulic dredges and rapido trawls [1,2]. These fishing gears have a heavy impact on the seabed, change its morphology, damage benthic organisms and fauna, tend to over-exploit the target species and significantly increase the mortality of non-target species [3,4]. Bivalves are mainly filter and suspension feeders and exploit phytoplankton and organic particles. Additional foods, as well as pharmaceuticals, are not required to sustain bivalve growth in outdoor rearing plants [5]. Within the aquaculture sector, shellfish culture is more beneficial in terms of food security and environment conservation [6]. As for the latter issue, an increase in cultured shellfish products might contribute in reducing the consumption of fished bivalves, with a consequent reduction in fishing pressure and the negative impacts of fishing gears on the sea bottom [1,2].

Between 1961 and 2016, the global supply of fish for human consumption strongly increased, due to the growth of both population and per capita fish food consumption. Increased global fish production was mainly accountable to the steady increase in aquaculture production since the 1980s, rather than to the increase in capture fishery production, which remained relatively static [7]. Italy is the fifth European state for aquaculture production, preceded by Norway, Spain, the United Kingdom and France [8]. As for Spain and France, Italian mariculture is mainly based on bivalve species. In Italy, shellfish culture represents 63% (in weight) of national aquaculture production and is almost entirely limited to the Mediterranean mussel, *M. galloprovincialis* (72%), and Manila clams, *Ruditapes philippinarum* (28%) [9]. Mussels are cultivated mainly in offshore long-line facilities, while clams are grown intensively in the soft-bottoms of confined coastal lagoons, mainly in the Lagoon of Venice [10,11]. The lack of diversification in Italian shellfish culture makes it vulnerable to stochastic events, e.g., new pathology outbreaks, strong sea storms or climatic changes [12]. The availability of seed is one of the basic requirements for shellfish farming [13]. Previous studies showed that wild scallop spat in the Northern Adriatic waters is easy to collect with artificial collectors, where scallop juveniles were the most frequent and abundant among the newly settled bivalves [14,15]. Scallop spat collection is also successful in the Ionian Sea, in particular for *F. glaber* and *M. varia* species [12,16–18]. The collection of wild spat from local bivalve populations reduces production costs and could allow for the avoidance of risking to introduce new, non-indigenous species and invasion of coastal habitats, by spat translocation from different natural sites [19]. Scallops constitute an interesting prospect for rearing, also, in relation to their considerable market value. For the collection of wild seeds intended for scallop culture, artificial collectors, made by synthetic filaments, contained inside a plastic-mesh bag, are mainly used. Competent larvae (pediveliger or larvae with eye spot) enter the bag, attach themselves to the filaments through the byssus and undergo the metamorphosis. Most scallops, after reaching a size of a few millimetres, lose the byssus and detach themselves from the substrate. Therefore, the openings of the plastic bag should not be greater than 3–5 mm, so as to retain the spat inside the collector after the attached phase, while filament type and collector bag dimensions may vary [20].

Several biotic and abiotic factors (life and reproductive traits of the species, type, orientation and heterogeneity of the substrate, water column turbidity and sedimentation rate) influence the settlement and the formation of fouling communities [21–25]. The settlement of bivalve competent larvae occurs some weeks after the species spawning period. Artificial collectors have to be immersed at an appropriate time, in order to allow the formation of the microfouling layer, which facilitates the invertebrate settlement [26]. Besides spat availability, time of collectors' deployment and their suitability, the abundance of collected spat is influenced by the post-settlement mortality, due to competition and/or predation [27,28]. Other factors, such as temperature variations and availability of suitable food, could influence post-settlement mortality [29]. Among abiotic factors, hydrodynamic activity is of crucial importance, for both larval settlement and growth, because it influences larval dispersal and spat distribution [30], as well as the plankton community composition [31] that could be fundamental for successful bivalve larvae metamorphosis and post-settlement survival [13].

Marine aquaculture practices are strongly affected by biofouling, which leads to a wide range of significant impacts on production, involving both the cultured species and the related infrastructures [32,33]. Shell mechanical function of cultured bivalve species could be compromised by fouling colonization, with consequent decreased feeding ability or increased susceptibility to predation. Biofouling could affect shellfish growth and condition, leading to biological competition for food and space and/or reducing the water flow and, consequently, the oxygen level and food availability [32,33]. Furthermore, marine fouling communities of artificial structures are known to represent invasion hotspots for non-indigenous species. The introduction of non-indigenous species could threaten the native communities' stability, through competition for resources, predation, release of toxins, disease transmission and ecosystem engineering [34–36]. The monitoring of the biofouling communities on artificial substrates allows us to detect their dynamics, in relation to time and depth, and represents a valuable method to avoid the peak settlement of non-target species in mariculture infrastructures. Furthermore, early detection of non-indigenous species allows for the implementation of management strategies aimed at reducing non-indigenous species' establishment and spreading [37]. In the Adriatic Sea, recent studies of biofouling communities were focused on natural [38] and artificial substrates [39–42], also in relation to aquaculture infrastructures [43–45]. Along Italian coasts, between 1945 and 2009, the highest number of non-indigenous species were registered in the North Adriatic, with Mollusca as the taxon, having a major number of species [46]. However, the knowledge on bivalve recruitment in the North-Western Adriatic is very limited, even though the bivalve stocks have been heavily exploited for a long time [47].

As previously mentioned, the successful collection of natural seeds depends on many factors. The use of suitable collectors, favourable sites and depths for their immersion and the right recovery time are among the fundamental features to be defined in non-hatchery dependent shellfish culture. In addition, the use of bivalve collectors allows us to fill the gap in data on the presence and abundance of invasive species, and to monitor the presence of invasive species that could alter the structure of local biofouling communities and compromise shellfish farming, by competing for space and resources with reared species. In this study, we analysed the composition and spatial and temporal variation of the bivalve community, present on artificial collectors, located within commercial mussel parks on the north-western coast of the Adriatic Sea. The general hypothesis of this study is that in the North-Western Adriatic Sea, it is possible to collect enough natural spat of commercially relevant bivalve species that could be used to enhance and diversify shellfish rearing. The main purposes of this study were as follows: (a) define the bivalve spat distribution in relation to depth, site and season; (b) inspect the possibility of introducing new local bivalves to be cultured, and (c) highlight the presence of invasive species.

2. Materials and Methods

2.1. Study Area

The artificial collectors used for settlement analyses were deployed in two sites, Pellestrina (P) and Caleri (C), within commercial mussel parks, both located approximately 2 NM off the western coast of the North Adriatic (Figure 1). The geographical coordinates of the mussel farms were 45°15′39″ N, 12°20′13″ E for Pellestrina and 45°05′42″ N, 12°24′12″ E for Caleri. The Northern Adriatic Sea is characterized by shallow waters with maximum depths of 100 m and a mean depth of about 35 m [48]. Its seabed is homogeneous, mostly made of mobile, silty-sandy sediments [49]. The only hard substrates are represented by scattered biogenic outcrops distributed at depths between about 9 and 40 m, at 3 to 13 nautical miles from the North-Western Adriatic coast [50]. The major freshwater input derives from Po river [51] which influences both circulation regime and trophic status of the North-Western Adriatic Sea [52,53]. Pellestrina site is located in front of the Lagoon of Venice and is influenced by outflowing lagoon waters, characterised by higher surface salinity and lower nutrient concentrations than in Caleri site. Indeed, Caleri, located further south, is influenced by large rivers' discharge (Brenta, Adige and Po) and presents generally lower salinity and high concentration of nutrients and chlorophyll [54,55].

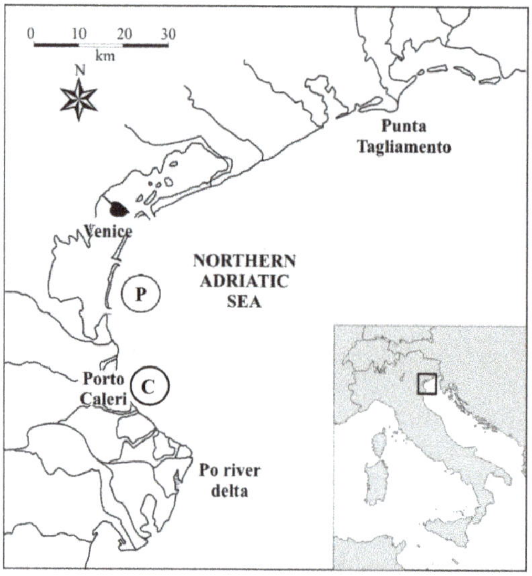

Figure 1. Sites of net bag collectors deployment in North-Western Adriatic Sea: Pellestrina (P) and Caleri (C).

2.2. Spat Collection

The artificial collectors consisted of 28 cm × 53 cm net bags, with a 4 mm mesh opening (Figure 2a). Net bags were filled internally with 5 m long tubular net having 5 cm mesh opening, which is commonly used for mussel rests. This net, called "filler", is folded and rolled up on itself, in order to give volume to the collector. Each set of collectors consisted of a rope on which 10 net bags were placed at a distance of 1 m from each other. The upper end of each rope was tied to a longline located at 4 m depth so that the collectors remained submerged from 5 to 14 m depth. A weight of about 12 kg was hung at the lower end of the rope to keep it stretched towards the seabed. Each set of collectors were 1 m distance from each other. At Pellestrina five sets of collectors were submerged in spring–summer and in summer–autumn. At Caleri the collectors' deployment was granted only in summer–autumn period; due to the accidental loss of a rope, four sets of collectors were recovered. Details are provided in Table 1. During the deployment the collectors were not managed and fouling was not removed until recovery (Figure 2b,c).

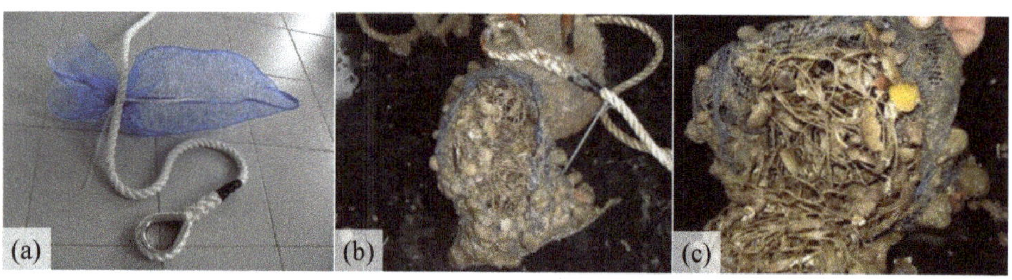

Figure 2. Net bag collectors before (**a**) and after (**b**,**c**) deployment in Pellestrina (PII) commercial mussel park.

Table 1. Sampling design: duration of collectors' immersion periods, and number of collectors used at each site.

	Immersion	Recovery	Total Days	Total Collectors
Pellestrina I	07/May/2009	16/July/2009	70	50
Pellestrina II	27/July/2009	11/December/2009	137	50
Caleri	27/July/2009	11/December/2009	137	40

2.3. Spat Identification

After recovery, collectors were closed in plastic bags, transported to the laboratory and frozen. Before analysis, each collector was thawed and washed with running water, in order to collect all the material contained in the net bag and on the internal filler. The collected material was sieved with a 2 mm mesh net and stored in 70% alcohol, in polyethylene containers. In the case of particularly abundant samples, half of the total material was analysed. Collected samples were examined under the stereomicroscope (Leica S8 AP0) at a maximum magnification of 80×. Several texts were used for the taxonomic identification [56–63]. The filled and closed bags had a total area useful for settlement of approximately 0.092 m^2 (28 cm × 33 cm). The spat abundance was expressed as settlement rate, i.e., number of individuals m^{-2} day^{-1}.

2.4. Statistical Analyses

To properly compare the samples regardless of the duration of collectors' immersion, statistical analyses were performed on the number of individuals per square metre per day. NMDS and PERMANOVA multivariate tests were performed using the software package PRIMER 6 PERMANOVA Plus (PRIMER-e Ltd., Plymouth, UK). Differences in species composition among depths, based on square-root transformed abundance data, were assessed for each sampling survey (PI, PII, C) using multivariate analyses based on a Bray–Curtis similarity, combined with visualization by NMDS plots. To compare the two sites in the same seasonal period (summer–autumn), a PERMANOVA (Permutational Multivariate Analysis of Variance) [64] was applied. A three-factor experimental design was used with 'site' and 'depth' as fixed factors, 'rope' as a random factor.

3. Results

In this study 28 bivalve taxa were identified on the 140 collectors analysed and the total number of taxa was similar between sites (21 in Pellestrina I and II, 22 taxa in Caleri) (Table 2). On average, lower settlement was observed in the spring–summer period (Pellestrina I, 58.2 ind m^{-2} day^{-1}) compared to the summer–autumn period (Pellestrina II and Caleri, 76.8 and 65.1 ind m^{-2} day^{-1}, respectively). Taxa that settled exclusively in the spring–summer period were *P. jacobeus*, *Pectinidae* n.i. and *Tellinidae* n.i., while *Modiolus adriaticus*, *Mytilaster* n.i., *Barbatia barbata*, *Arca noe*, *Limidae* n.i., *Pinna nobilis* and *Varicorbula gibba*, were found only in the summer–autumn samples. *Tellinidae* n.i. were found only in the Pellestrina site in spring–summer samples (Table 2). In all samples, the abundance increased with depth, in particular at PI (Figure 3A–C). Conversely, richness, that is the number of taxa, tended to decrease with increasing depth (Figure 4). In each sample, seven more abundant taxa contributed to about 95% of the total bivalve community abundance (Figure 5). *F. glaber* and *Musculus subpictus* have always been among the most abundant species (Table 2, Figures 5 and 6).

Table 2. List of total bivalve recruits and their relative abundance (m^{-2} day^{-1}), collected by net bags deployed in spring–summer in Pellestrina (PI) and in summer–autumn in Pellestrina (PII) and Caleri (C). n.i. not identified species. For each taxon the most abundant recruitment is in bold.

		Average Abundance m^{-2} Day^{-1}		
Family	**Spieces/Genus**	**P-I**	**P-II**	**C**
Mytilidae				
	Mytilus galloprovincialis Lamarck, 1819	**17.862**	0.180	0.197
	Musculus subpictus (Cantraine, 1835)	**4.421**	2.883	3.388
	Modiolus barbatus Linnaeus, 1758	1.264	**1.426**	0.354
	Musculus costulatus (Risso, 1826)	0.080	**0.493**	0.208
	Arcuatula senhousia (Benson, 1842)	0.003	**2.197**	0.352
	Modiolus adriaticus Lamarck, 1819	-	**0.622**	0.288
	Mytilaster n.i.	-	**0.084**	0.003
Pectinidae				
	Flexopecten glaber (Linnaeus, 1758)	11.111	16.503	**32.196**
	Mimachlamys varia (Linnaeus, 1758)	**8.112**	0.577	1.929
	Aequipecten opercularis (Linnaeus, 1758)	**8.124**	-	0.005
	Pecten jacobaeus (Linnaeus, 1758)	**0.167**	-	-
	Pectinidae n.i.	**0.386**	-	-
Arcidae				
	Anadara transversa (Say, 1822)	0.025	**40.179**	18.943
	Barbatia barbata (Linnaeus, 1758)	-	0.032	**1.372**
	Arca noae Linnaeus, 1758	-	**0.009**	-
	Arcidae n.i.	0.006	**7.724**	-
Anomiidae				
	Anomia ephippium Linnaeus, 1758	**2.683**	0.543	0.874
	Pododesmus pattelliformis (Linnaeus, 1761)	0.037	**0.088**	0.042
Hiatellidae				
	Hiatella arctica (Linnaeus, 1767)	**2.581**	0.605	1.735
Limidae				
	Limidae n.i.	-	**0.719**	0.238
Cardiidae				
	Parvicardium exiguum (Gmelin, 1791)	**0.226**	0.087	0.064
Ostreidae				
	Ostreidae n.i.	1.042	1.768	**2.836**
Veneridae				
	Tapes sp.	**0.009**	-	0.002
	Veneridae n.i.	**0.012**	-	0.003
Mesodesmatidae				
	Atactodea spp	**0.056**	0.005	-
Tellinidae				
	Tellinidae n.i.	**0.037**	-	-
Pinnidae				
	Pinna nobilis Linnaeus, 1758	-	**0.118**	0.045
Corbulidae				
	Varicorbula gibba (Olivi, 1792)	-	-	**0.002**

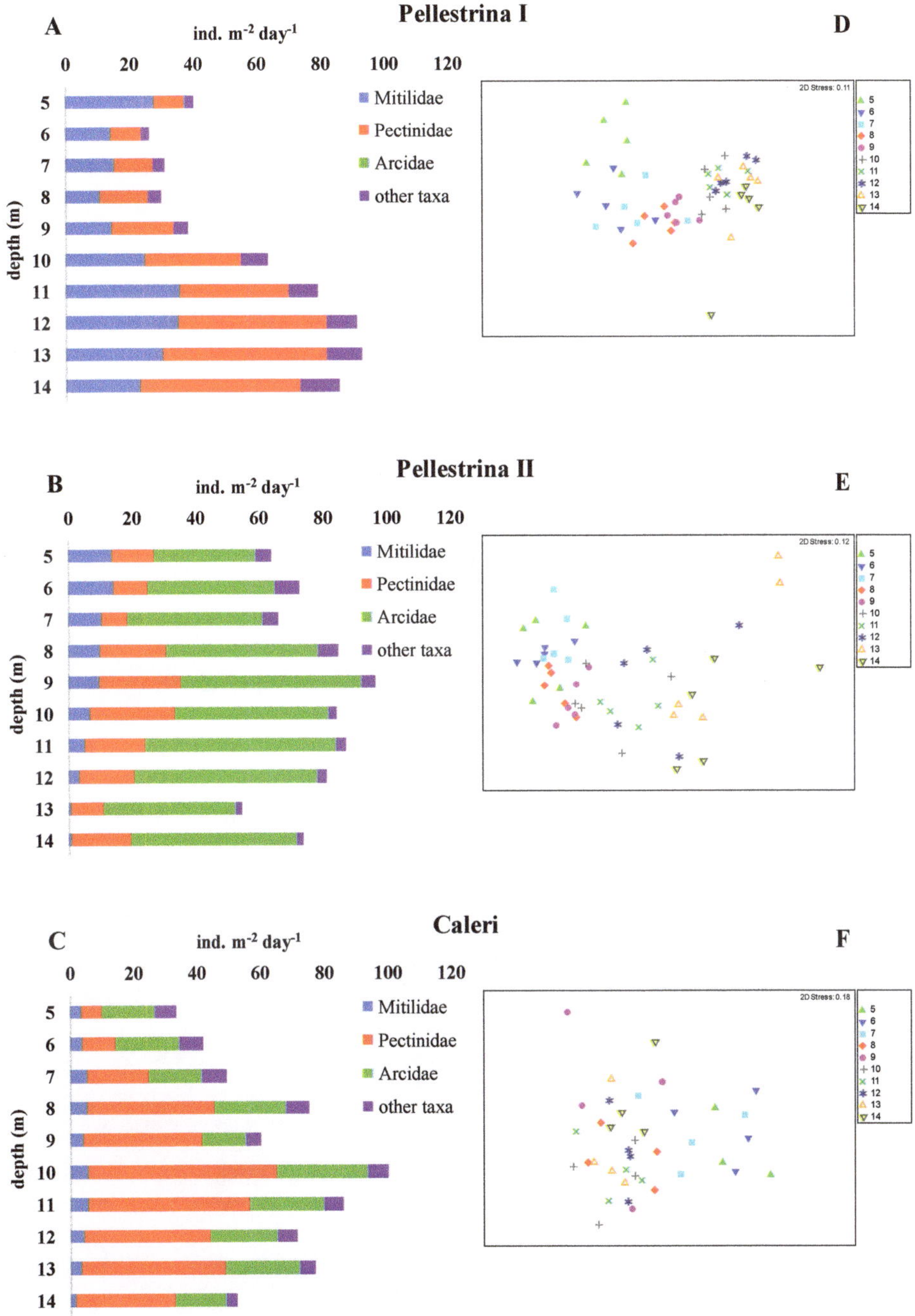

Figure 3. Settlement rate of the main bivalve families at different depths (**A–C**) and 2D-MDS plots of bivalve species composition similarities among depths (**D–F**) over the three sampling surveys.

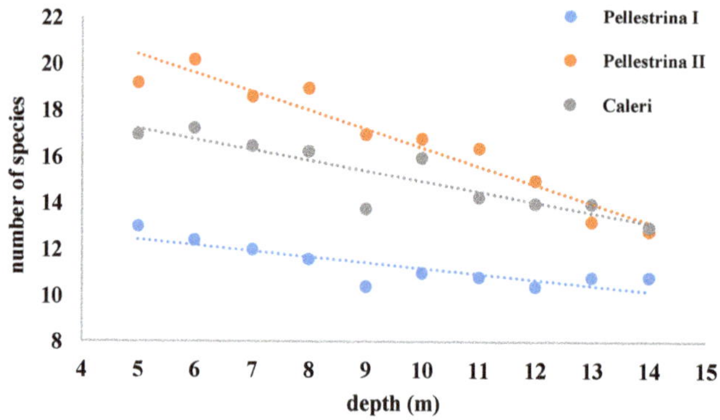

Figure 4. Richness found in Pellestrina I, Pellestrina II and Caleri samples at different depths.

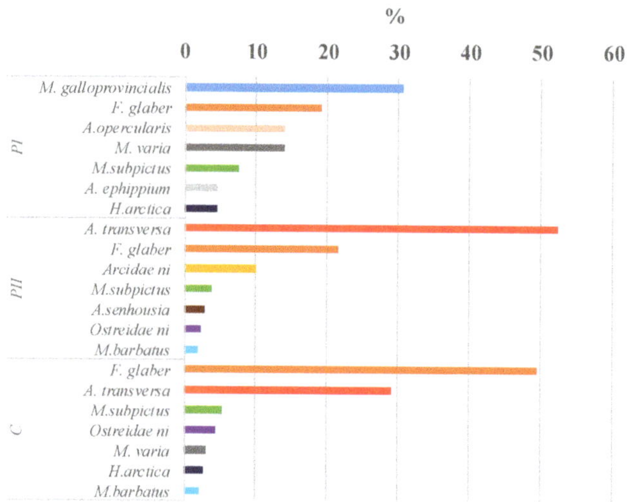

Figure 5. Percentage contribution to total abundance of the first seven more abundant species in each sample: Pellestrina I (PI), Pellestrina II (PII) and Caleri (C).

In Pellestrina I, a clear increase in settlement with depth was observed, in particular, starting from 10 m downwards (Figure 3A). Depth influenced the settlement in Caleri also, where a higher abundance of bivalve spat was found between 8 and 13 m depths (Figure 3C). Conversely, in Pellestrina II, the settlement rate was high and quite constant between depths (Figure 3B).

In Pellestrina I, Pectinidae (47.9%) and Mytilidae (40.6%) were the dominant families, with a slightly greater presence of Pectinidae (Table 2, Figure 3A). Instead, in Pellestrina II and in Caleri, one family was notably dominant, Arcidae (52.4%) and Pectinidae (52.4%), respectively (Table 2, Figure 3B,C). In terms of abundance, the main settlement rate of Mytilidae was found in Pellestrina I (23.6 ind. m^{-2} day^{-1}), while for Pectinidae and Arcidae, the settlement was more abundant in Caleri (34.1 ind. m^{-2} day^{-1}) and Pellestrina II (34.1 ind. m^{-2} day^{-1}).

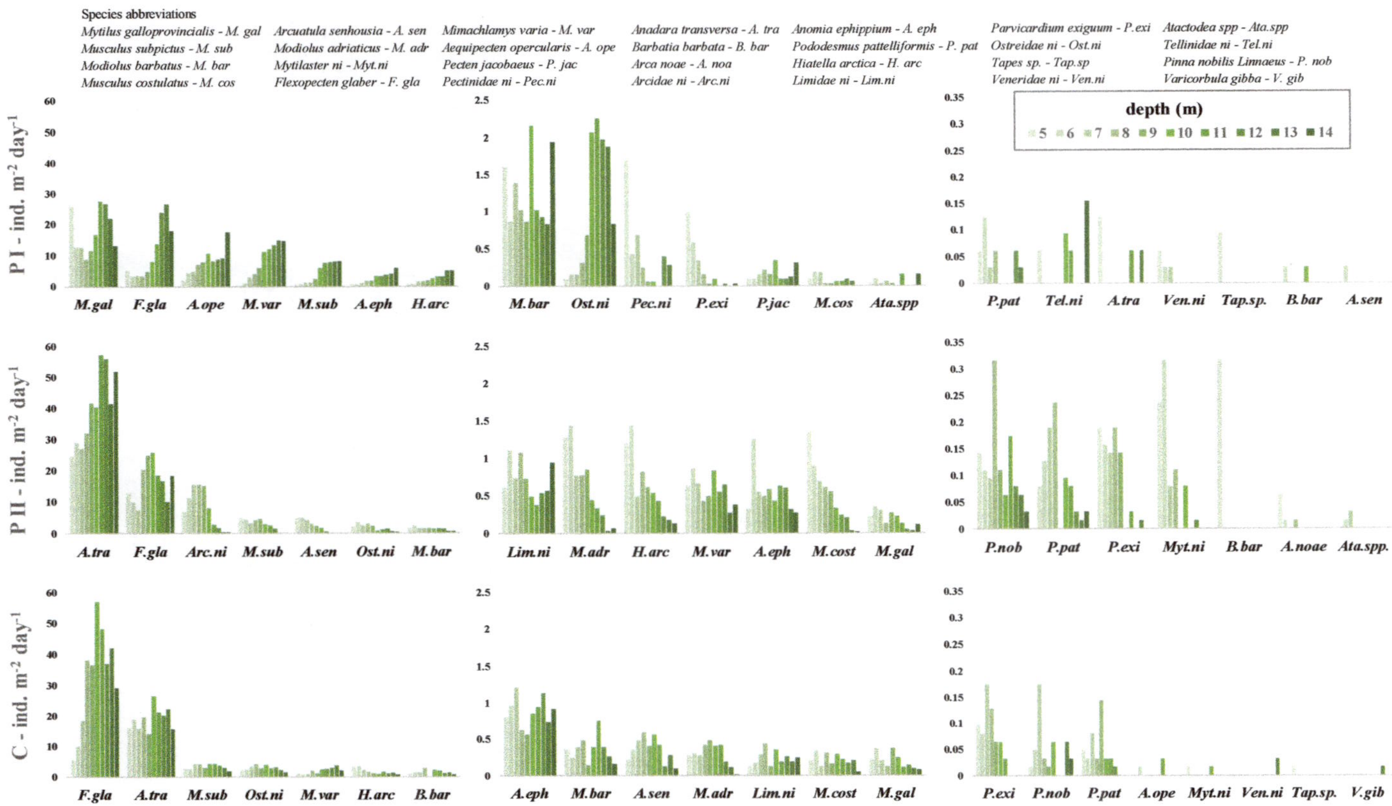

Figure 6. Settlement rates of each taxon at the various depths over the three sampling surveys: Pellestrina I (**PI**), Pellestrina II (**PII**) and Caleri (**C**). From left to right, taxa with decreasing total abundance are shown. Note the different scales on the y-axes.

For each sampling survey, the separation between superficial and deeper samples is evident in MDS plots, due to both lower total settlement in surface layers and different variations in single taxa abundances with depth (Figure 3D–F). The PERMAVOVA, performed on summer–autumn settlement data, from Pellestrina and Caleri, showed significant effects for both 'site' and 'depth' factors (Table 3). The same p-value (0.001) was found for these fixed factors, but the pseudo-F value was much higher for 'site' (37.18) than for 'depth' (4.51), meaning that recruitment differed more between sites than among depths. The interaction between 'site' and 'depth' was also significant, underlining that at the same depths, different taxa with different abundances were present in the two sites. For example, at 9 m depth at PII, the abundances of *F. glaber* and *A. transversa* were 25 and 42 ind. m^{-2} day^{-1}, respectively, while at the same depth at the C site, the abundances of the same species were 36 and 14 ind. m^{-2} day^{-1}. Another species, *B. barbata*, was found only at 5 m, with very small abundance (0.32 ind. m^{-2} day^{-1}) at PII, while it was the seventh most abundant species (up to 2.79 ind. m^{-2} day^{-1}), distributing at all depths at the C site.

Table 3. PERMANOVA test on Bray–Curtis similarities of square root-transformed data according to site (2 levels, fixed: Pellestrina and Caleri), depth (10 levels from 5 to 14 m, fixed) and rope (random).

	gl	Pseudo-F	P	Perms
site	1	37.18	0.001	999
depth	9	4.51	0.001	999
rope	40	1.07	0.329	997
sitexdepth	9	2.5	0.001	994

3.1. *Mytilidae*

In Pellestrina I, the Mytilidae family was very abundant, clearly more abundant than in the Pellestrina II and Caleri samples. In spring–summer samples (PI), *M. galloprovincialis* (17.9 ind. m^{-2} day^{-1}) and *M. subpictus* (4.4 ind. m^{-2} day^{-1}) were the most abundant species, even though in exiguous quantity, other species found in PI were *Modiolus barbatus*, *Musculus costulatus* and *A. senhousia* (Table 2). A higher abundance of *M. galloprovincialis* was observed at 5 and between 11 and 13 m depths, with a mean abundance of 25.69 $\pm$ 2.52 ind. m^{-2} day^{-1}, while *M. subpictus* settled mainly on deeper collectors, starting from 10 m up, to a maximum abundance of 8.35 ind. m^{-2} day^{-1} at 14 m (Figure 6).

In the summer–autumn samples (PII and C), two other taxa were identified, i.e., *M. adriaticus* and *Mytilaster ni* (Table 2). The settlement rates of the single taxa were consistent with those observed in PI, with the exception of *M. galloprovincialis*, which showed very low settlement in summer–autumn (Table 2). In Pellestrina II, the most abundant species were *M. subpictus* (2.88 ind. m^{-2} day^{-1}) and *A. senhousia* (2.20 ind. m^{-2} day^{-1}), while in Caleri, the highest settlement rate was observed for *M. subpictus* (3.39 ind. m^{-2} day^{-1}), followed by *M. barbatus* (0.35 ind. m^{-2} day^{-1}) and *A. senhousia* (0.35 ind. m^{-2} day^{-1}), with a quite significantly lower settlement rate (Table 2, Figure 6). Interestingly, the settlement observed in the two sites appeared more influenced by the depth; in PII, the settlement of the most abundant species was shown to decrease with increasing depth, while in Caleri, it appeared not to be affected by depth (Figure 6).

3.2. *Pectinidae*

In spring–summer (Pellestrina I), *F. glaber* was the most abundant species (11.1 ind. m^{-2} day^{-1}), with a slightly higher settlement rate compared to *M. varia* and *A. opercularis* (8.1 ind. m^{-2} day^{-1} for both species). Although very low in quantity, *P. jacobeus* spat was found only in Pellestrina spring–summer samples (Table 2). Pectinidae species, mainly *F. glaber* and *M. varia*, showed an increased settlement rate with depth, starting from 10 m, specifically (Figure 6).

In summer–autumn samples, *F. glaber* was a dominant and very abundant species, with a higher settlement rate observed at Caleri (32.2 ind. m^{-2} day^{-1}), compared to both

Pellestrina II (16.5 ind. m^{-2} day^{-1}) and Pellestrina I (11.1 ind. m^{-2} day^{-1}) (Table 2). The settlement rate appeared to increase with depth, starting from 8 m at both sites. In particular, higher settlement was observed at 9–10 m depth at Pellestrina II and at 10–11 m depth at Caleri (Figure 6).

3.3. Arcidae

This family was almost totally represented by the invasive species *A. transversa*, which was found in all samples analysed. However, the settlement rate observed in spring–summer (0.02 ind. m^{-2} day^{-1}) was negligible but very high in summer–autumn. *A. transversa* was the dominant species in Pellestrina II (40.2 ind. m^{-2} day^{-1}) and showed the maximum settlement rate observed during this study for a single species (Table 2). In Pellestrina II, a slight increase in settlement with depth was observed, starting from 9 m, while in Caleri the settlement showed similar values among depths, with a peak at 10 m (Figure 6). In addition to *A. transversa*, a second non-indigenous species detected in this study was *A. senhousia*. This species was present in all samples, with higher abundance in summer–autumn and maximum settlement in Pellestrina II (Table 2). Indeed, in Pellestrina II, *A. senhousia* was the fifth most abundant species (Figure 5).

4. Discussion

In the present study, from May to December, 28 bivalve taxa were found on net bag collectors deployed along the north-western coasts of the Adriatic Sea. Most of the species are reported in checklists of sessile bivalves of the North Adriatic [65] and are components of benthic biocenoses, on rocky substrates of biogenic concretions (tegnùe), which are not far from the study area [50,66–68]. A low number of individuals belonging to mobile substrate taxa, such as Cardiidae, Veneridae, Mesodesmatidae and Tellinidae, were found. Indeed, the bivalve community of the North Adriatic Sea, described following a hydraulic dredge survey, showed a higher diversity (54 taxa) [69]. This confirms that net bags are more suitable for sessile bivalve recruitment. Compared to other artificial collectors, such as net panels, "Chinese caps" or tiles, the diversity of the bivalves settled on net bags was higher [14,15,38,39,43,45,70]. The three sampling surveys produced very similar results in total taxa number but differed in total abundance and taxa type, dominance and distribution along depth. Despite the bivalve taxa diversity, in all samples, seven taxa constituted about 95% of total settled individuals.

4.1. Seasonal Settlement Patterns

In spring–summer, the bivalve community was dominated by Mytilidae and Pectinidae, while in summer–autumn, by Arcidae and Pectinidae.

M. galloprovincialis was the most abundant species in spring–summer and its settlement was negligible in summer–autumn. For this species, very limited spawning events are possible, even in July–August; the main spawning occurs in January–February [71,72]. Spring–summer recruitment detected in this study is consistent with the spawning period of *M. galloprovincialis* and with settlement previously observed in the North Adriatic [43,73]. As for the Mytilidae family, *M. subpictus* was very abundant in both periods, with higher settlement in spring–summer.

The second most abundant species in spring–summer was *F. glaber*, followed by *M. varia* and *A. opercularis*. Of the Pectinidae family, *P. jacobeus* was found only in this period, but was among the less abundant species. The recruitment of *F. glaber* was even higher in summer–autumn; indeed, this species was dominant together with *A. transversa*. Our settlement data support the evidence that, for *F. flaber*, there are two spawning events in the Adriatic Sea. A minor one occurs in April and May and the main one between July and September [74]. For this species, the same increased recruitment in summer–autumn, compared with the winter–spring period, was also observed in the Ionian Sea [16]. *M. varia* and *A. opercularis* were recruited in both periods considered, but mainly in spring–summer. For the Mediterranean Sea, the recruitment of both species was observed almost all year round, with a peak in spring–summer. *A. opercularis* recruitment was recorded, also in autumn [16,75] For both species, spawning activity is likely to occur throughout the year, except in winter for *M. varia* [76–79], and our settlement data suggested main spawning events in spring and early summer. In the Adriatic Sea, *P. jacobeus* was observed to have more restricted spawning periods, which occur in May, August and December, with summer as the main spawning season [76]. However, it is to be noted that in the Mediterranean Sea, the settlement was observed in January–March and April–July [16,75]. In our samples, *P. jacobeus* recruitment was very low in spring–summer and was even absent in autumn, while the winter period was not investigated.

The settlement of the arcid clam, *A. transversa*, was very high in summer–autumn, whereas the low recruitment observed in spring–summer suggested the presence of scarce spawning events in spring. These findings are in agreement with previous observations by other authors and support the hypothesis that the main spawning occurs in late summer [43,80,81]. Higher spat abundance of Ostreidae was found in summer–autumn, which is consistent with the spawning period of *Crassostrea gigas* and settlement of *Ostrea edulis* in the Adriatic Sea [43,82]. Based on our observations on low abundant taxa, species detected in all samples, but prevailed mainly in the spring–summer samples, were *Anomia ephippium*, *Hiatella arctica* and *Parvicardium exiguum*, while *M. costulatus* and *A. senhousia* were mainly summer–autumn settlers. Taxa found exclusively in the summer–autumn period were *M. adriaticus*, *Mytilaster* n.i., *B. barbata*, *A. noae*, *Limidae* n.i., *Pinna nobilis* and *V. gibba*, while *Tellinidae* settled exclusively in spring–summer. Data in the literature on *M. barbatus*, *P. nobilis* and *H. arctica* reproduction and settlement are consistent with our findings [83–86].

It is to note that the patterns of seasonal variations in the main hydrological parameters, shown in Figure 7, are consistent with those reported in the literature for the North-Western Adriatic Sea [87,88], and no atypical trend was observed during the study period (data recorded at the CNR Oceanographic Platform "Acqua Alta").

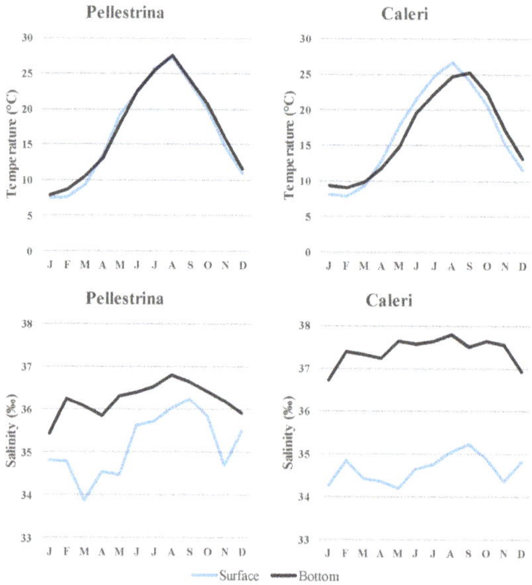

Figure 7. Mean monthly values of temperature and salinity at Pellestrina and Caleri. Data from E.U. Copernicus Marine Service Information (CMEMS) https://marine.copernicus.eu/ accessed on 2 December 2021.

4.2. *Influence of Depth and Site on Settlement Patterns*

In all samples, the total settlement rate increased with depth. Our study was focused on the bivalve community; therefore, other taxa, which usually account for most of the biomass in the Adriatic Sea biofouling communities, such as algae, ascidians and crustaceans [44,45], were not examined. The reduced abundance of bivalve recruits at shallower depths may be due to the lower presence of competent larvae in this part of the water column, but also to the higher competition with other fouling species, including phototrophs, which are more abundant at lower depths [89].

In the spring–summer period, this settlement pattern was evident for dominant taxa, such as *M. galloprovincialis* and Pectinidae (*F. glaber, A. opercularis, M. varia*), and for most of the less abundant species (*M. subpictus, A. ephippium, H. arctica, Ostreidae* n.i., *P. exiguum, P. jacobeus*). *M. galloprovincialis* settlement was observed to occur mainly at shallower depths [90–92]. Our collectors were placed starting from 5 m depth, where high settlement was observed, but we registered an increase in *M. galloprovincialis* spat also at 10–13 m depth. The increased settlement of Pectinidae at higher depths was expected, since high recruitment of scallops is usually ensured by deploying the collectors near the bottom, at a depth ranging from approximately 10 to 30 m [12,16,78,93,94].

Contrarily to what was observed in spring–summer, in summer–autumn samples, the increase in settlement with depth was detected for two dominant species, *A. transversa* and *F. glaber*, while the other less abundant species mainly exhibited the opposite trend. Increased settlement at low depth was particularly noticeable at Caleri, compared to Pellestrina. In this regard, it cannot be excluded that different environmental conditions, mainly concerning stratification of the water column and salinity vertical profile, can influence the settlement at the studied sites (Figure 7). Of the two sites, Caleri was found to be more suitable for scallop spat collection. Indeed, despite the high presence of invasive *A. transversa* at both sites, in the summer–autumn period, *F. glaber* was the most abundant species at Caleri and *M. varia* recruitment was higher than at Pellestrina. In addition, when comparing the settlement rates of the species found in the two sets of summer–autumn samples, *H. arctica* and *B. barbata* were more abundant at Caleri, *A. senhousia* and *M. barbatus* at Pellestrina. Information on the status of natural bivalve stocks in the North-Western Adriatic Sea is scarce and outdated. For this reason, no specific relationship between settlement data in the study areas and broodstock abundance in the proximity can be inferred. In a previous study, performed in 1995, to assess Pectinidae abundance in an area of about 10 km^2, located 40 km south-west of Venice (45°13.5′ N 12°47.1′ E), *A. opercularis* was found to be particularly abundant, with an estimated population density of 2.8 individuals m^{-2}, while *P. jacobeus* was less abundant (0.05–0.08 individuals m^{-2}), and very low amounts of *M. varia* and *F. glaber* were found [95]. Conversely, a dominance (96.52%) of *F. glaber* was observed in the adult pectinid stocks at the Gulf of Manfredonia (South-Western Adriatic Sea) [96]. The relevant abundance of *F. glaber* spat detected in the present study suggests the need for future research addressed to the evaluation of adult Pectinidae abundance in the study areas.

4.3. Occurence of Alien Species

Two alien species, *A. transversa* and *A. senhousia*, were found in this study. *A. transversa* is a Lessepsian Indo-Pacific species, reported for the first time in the Mediterranean, on the Turkish coasts, in the port of Izmir [97], and subsequently found in the North Aegean Sea [98]. In Italy, it was reported for the first time in 2001, at Cesenatico [99], and then in the area of Ancona [100]. The presence of this species in the North Adriatic, and more precisely in the Lagoon of Venice, dates back to 2002 [101], but the significant increase in abundance, confirmed also by our study, indicates the acclimatization of *A. transversa* and spread in the marine environment [43,102,103]. The abundance of this species on the collectors in summer–autumn may represent a serious problem for a fruitful collection of the Pectinidae spat, *F. glaber* in particular, in the same period of the year.

A. senhousia is an allochthonous species, of Asian origin, which caused profound changes in the benthic community of Sacca di Goro, a transitional environment in the river Po delta [104]. As in the case of *A. transversa*, its abundance was greater at Pellestrina than at Caleri. These differences can be related to both variability between sites and competition with other species. Since the abundance of *A. senhousia* detected offshore in this study was lower than in the Sacca di Goro, a milder impact on the other settled species is hopefully expected at sea collection sites.

4.4. Scallop Spat Abundance

In accordance with our observations, the optimal depth range for the scallop spat collection resulted between 8 and 14 m. Unfortunately, in both seasons evaluated, the presence of competing species emerged. It is likely that *M. galloprovincialis*, in spring–summer, and *A. transversa*, in summer–autumn, have reduced the settlement of scallops. Despite the presence of competitors, net bags deployed in May allowed the collection of considerable quantities of scallop spat, belonging to three species (*F. glaber*: 989 $\pm$ 641 sd ind. m^{-2}; *M. varia*: 766 $\pm$ 298 sd ind. m^{-2} and *A. opercularis*: 696 $\pm$ 246 sd ind. m^{-2}), while collectors deployed in July collected the highest quantity of *F. glaber* (PII: 2640 $\pm$ 733 sd ind. m^{-2}, C: 5639 $\pm$ 1245 sd ind. m^{-2}) and minor amounts of *M. varia* spat (PII: 70 $\pm$ 26 sd ind. m^{-2}, C: 325 $\pm$ 113 sd ind. m^{-2}).

In the Ionian Sea, spat of *F. glaber* (average density ranging from 19.3 to 306.1 ind. m^{-2}) were found on net bags immersed for 7 months, near the bottom, together with *M. varia* (3.3–35.6 ind. m^{-2}) and a negligible amount of *A. opercularis* spat [12]. Even if the dominance of the species is consistent with our findings, the spat abundance of *F. glaber* and *M. varia* was clearly higher in the Adriatic Sea. However, in the Western Mediterranean, *A. opercularis* and *M. varia* showed slightly higher recruitment values that were 0.25–99 ind. bag^{-1} and 0–101 ind. bag^{-1}, respectively, corresponding to about 1–412 ind. m^{-2} and 0–421 ind. m^{-2} [75].

5. Conclusions

Our results highlight the relevant potential of Pectinidae spat, *F. glaber* in particular, collection along the North-Western Adriatic coasts, in the perspective of introducing new cultivations of commercial bivalve species. Optimal seasonal and depth ranges have been defined. In this regard, the present work represents an essential premise to analyse potential effects of global change in future research assessing bivalve recruitment in the study area. As recently highlighted [105], global change is among the main threats to aquaculture. Although the presence of invasive species needs to be monitored, scallop spat abundance in the study area was similar to, or higher than, other coastal areas of the Mediterranean. It is important to consider that our collectors were deployed inside commercial mussel farms; therefore, mussel spat abundance and competition is likely to be reduced if areas specifically devoted to the collection of scallop spat can be set up.

Author Contributions: Conceptualization, T.M., M.G.M. and M.B.; methodology, T.M., M.G.M. and M.B.; formal analysis, T.M., M.G.M. and M.B.; investigation, T.M. and M.B.; resources, M.G.M. and M.B.; data curation, T.M. and M.B.; writing—original draft preparation, T.M. and V.F.C.; writing—review and editing, T.M., M.G.M. and M.B.; funding acquisition, M.G.M. and M.B. All authors have read and agreed to the published version of the manuscript.

Funding: This work was part of the project CLODIA, funded by the Veneto Region (Italy) Law 15/2007 (DGR no 4069).

Acknowledgments: The authors wish to acknowledge Francesco M. Falceri for his support in processing data from E.U. Copernicus Marine Service Information.

Conflicts of Interest: The authors declare no conflict of interest. The funders had no role in the design of the study; in the collection, analyses, or interpretation of data; in the writing of the manuscript, or in the decision to publish the results.

References

1. Ardizzone, G.D. An attempt of a global approach for regulating the fishing effort in Italy. *Biol. Mar. Mediterr.* **1994**, *1*, 109–113.
2. Ezgeta-Balić, D.; Peharda, M.; Richardson, C.A.; Kuzmanić, M.; Vrgoč, N.; Isajlović, I. Age, growth, and population structure of the smooth clam Callista chione in the eastern Adriatic Sea. *Helgol. Mar. Res.* **2011**, *65*, 457–465. [CrossRef]

3. Brambati, A.; Fontolan, G. Sediment resuspension induced by clam fishing with hydraulic dredges in the Gulf of Venice (Adriatic Sea). A preliminary experimental approach. *Boll. Ocean. Teor. Appl.* **1990**, *8*, 113–121.
4. Pranovi, F.; Raicevich, S.; Franceschini, G.; Torricelli, P.; Giovanardi, O. Discard analysis and damage to non-target species in the "rapido" trawl fishery. *Mar. Biol.* **2001**, *139*, 863–875.
5. Wijsman, J.W.M.; Troost, K.; Fang, J.; Roncarati, A. Global Production of Marine Bivalves. Trends and Challenges. In *Goods and Services of Marine Bivalves*; Smaal, A., Ferreira, J., Grant, J., Petersen, J., Strand, Ø., Eds.; Springer: Cham, Switzerland, 2019; pp. 7–26.
6. Suplicy, F.M. A review of the multiple benefits of mussel farming. *Rev. Aquacult.* **2020**, *12*, 204–223. [CrossRef]
7. FAO. The State of World Fisheries and Aquaculture 2018. In *Meeting the Sustainable Development Goals*; Food and Agricultural Organization of the United Nations: Rome, Italy, 2018; pp. 2–75.
8. Hough, C. *Regional Review on Status and Trends in Aquaculture Development in Europe—2020*; FAO Fisheries and Aquaculture Circular No. 1232/1; FAO: Rome, Italy, 2022; p. 43.
9. Marino, G.; Crosetti, D.; Petochi, T. *Fisheries and Aquaculture Division [Online]*; FAO: Rome, Italy, 2021. Available online: https://www.fao.org/fishery/en/countrysector/it/en (accessed on 10 December 2021).
10. Parisi, G.; Centoducati, G.; Gasco, L.; Gatta, P.P.; Moretti, V.M.; Piccolo, G.; Roncarati, A.; Terova, G.; Pais, A. Molluscs and echinoderms aquaculture: Biological aspects, current status, technical progress and future perspectives for the most promising species in Italy. *Ital. J. Anim. Sci.* **2012**, *11*, 397–413. [CrossRef]
11. Prioli, G. La molluschicoltura in Italia. In *Estado Actual del Cultivo y Manejo de Moluscos Bivalvos y su Proyección Futura: Factores que Afectan su Sustentabilidad en América Latina. Taller Técnico Regional de la FAO, 20–24 August 2007, Puerto Montt, Chile*; FAO Actas de Pesca y Acuicultura, No. 12; Lovatelli, A., Farías, A., Uriarte, I., Eds.; FAO: Roma, Italy, 2008; pp. 159–176.
12. Prato, E.; Biandolino, F.; Parlapiano, I.; Gianguzza, P.; Fanelli, G. The recruitment of scallops (and beyond) by two different artificial collectors (Gulf of Taranto, Mediterranean Sea). *Aquac. Res.* **2015**, *47*, 3319–3331. [CrossRef]
13. Lagarde, F.; Roque d'orbcastel, E.; Ubertini, M.; Mortreux, S.; Bernard, I.; Fiandrino, A.; Chiantella, C.; Bec, B.; Roques, C.; Bonnet, D.; et al. Recruitment of the Pacific oyster *Crassostrea gigas* in a shellfish-exploited Mediterranean lagoon: Discovery, driving factors and a favorable environmental window. *Mar. Ecol. Prog. Ser.* **2017**, *578*, 1–17. [CrossRef]
14. Chinellato, A.; Bressan, M.; Pellizzato, M. Insediamento, reclutamento ed accrescimento di bivalvi eduli su strutture in sospensione nell'area del campo sperimentale. In *Campo Sperimentale in Mare: Prime Esperienze nel Veneto Relative Ad Elevazioni dal Fondale con Materiale Inerte*; Regione Veneto, ARPAV Osservatorio Alto Adriatico: Venezia, Italy, 2006; pp. 144–164.
15. Chinellato, A.; Pellizzato, M.; Bressan, M. Insediamento di larve di bivalvi su collettori artificiali in un'area a barriere artificiali nel Nord Adriatico. *Biol. Mar. Mediterr.* **2006**, *13*, 1072–1076.
16. Papa, L.; Prato, E.; Biandolino, F.; Parlapiano, I.; Fanelli, G. Strategies for successful scallops spat collection on artificial collectors in the Taranto Gulf (Mediterranean Sea). *Water* **2021**, *13*, 462. [CrossRef]
17. Prato, E.; Biandolino, F.; Parlapiano, I.; Papa, L.; Denti, G.; Fanelli, G. Estimation of growth parameters of the black scallop Mimachlamys varia in the gulf of taranto (Ionian Sea, Southern Italy). *Water* **2020**, *12*, 3342. [CrossRef]
18. Tsotsios, D.; Tzovenis, I.; Katselis, G.; Geiger, S.P.; Theodorou, J.A. Spat settlement of the smooth scallop *Flexopecten glaber* (Linnaeus, 1758) and variegated scallop *Chlamys varia* (Linnaeus, 1758) in Amvrakikos Gulf, Ionian Sea (Northwestern Greece). *J. Shellfish Res.* **2016**, *35*, 467–474. [CrossRef]
19. Mckindsey, C.W.; Landry, T.; O'Beirn, F.X.; Davies, I.M. Bivalve aquaculture and exotic species: A review of ecological considerations and management issues. *J. Shellfish Res.* **2007**, *26*, 281–294. [CrossRef]
20. Lodeiros, C.; Davidson, L.-A.; Dadswell, M.; Rupp, G.S.; Mazón-Suástegui, J.M. Scallop aquaculture. In *Molluscan Shellfish Aquaculture: A Practical Guide*; Shumway, S., Ed.; 5m Publishing: Great Easton, UK, 2021.
21. Khalaman, V.V. Life strategies of marine sessile organisms as an approach for exploration of structure and succession of fouling communities. In *Biofouling: Types, Impact and Anti-Fouling*; Chan, J., Wong, S., Eds.; Nova Science Publishers, Inc.: New York, NY, USA, 2010; pp. 1–33.
22. Glasby, T.M.; Connell, S.D. Orientation and position of substrata have large effects on epibiotic assemblages. *Mar. Ecol. Prog. Ser.* **2001**, *214*, 127–135. [CrossRef]
23. Ushiama, S.; Smith, J.A.; Suthers, I.; Lowry, M.; Johnston, E.L. The effects of substratum material and surface orientation on the developing epibenthic community on a designed artificial reef. *Biofouling* **2016**, *32*, 1049–1060. [CrossRef] [PubMed]
24. Lapointe, L.; Bourget, E. Influence of substratum heterogeneity scales and complexity on a temperate epibenthic marine community. *Mar. Ecol. Prog. Ser.* **1999**, *189*, 159–170. [CrossRef]
25. Maughan, B.C. The effects of sedimentation and light on recruitment and development of a temperate, subtidal, epifaunal community. *J. Exp. Mar. Biol. Ecol.* **2001**, *256*, 59–71. [CrossRef]
26. Wieczorek, S.K.; Todd, C.D. Inhibition and facilitation of settlement of epifaunal marine invertebrate larvae by microbial biofilm cues. *Biofouling* **1998**, *12*, 81–118. [CrossRef]
27. Hunt, H.L.; Scheibling, R.E. Role of early post-settlement mortality in recruitment of benthic marine invertebrates. *Mar. Ecol. Prog. Ser.* **1997**, *155*, 269–301. [CrossRef]
28. Nydam, M.; Stachowicz, J.J. Predator effects on fouling community development. *Mar. Ecol. Prog. Ser.* **2007**, *337*, 93–101. [CrossRef]
29. Avendaño, M.; Cantillánez, M.; Thouzeau, G. Effects of water depth on survival and growth of *Argopecten purpuratus* (Lamarck, 1819) spat in northern Chile. *Aquacult. Int.* **2008**, *16*, 377–391. [CrossRef]

30. Bandelj, V.; Solidoro, C.; Laurent, C.; Querin, S.; Kaleb, S.; Gianni, F.; Falace, A. Cross-scale connectivity of macrobenthic communities in a patchy network of habitats: The Mesophotic Biogenic Habitats of the Northern Adriatic Sea. *Estuar. Coast. Shelf Sci.* **2020**, *245*, 106978. [CrossRef]
31. Millet, B.; Cecchi, P. Wind-induced hydrodynamic control of the phytoplankton biomass in a lagoon ecosystem. *Limnol. Oceanogr.* **1992**, *37*, 140–146. [CrossRef]
32. Adams, C.M.; Shumway, S.E.; Whitlatch, R.B.; Getchis, T. Biofouling in marine molluscan shellfish aquaculture: A survey assessing the business and economic implications of mitigation. *J. World Aquac. Soc.* **2011**, *42*, 242–252. [CrossRef]
33. Fitridge, I.; Dempster, T.; Guenther, J.; de Nys, R. The impact and control of biofouling in marine aquaculture: A review. *Biofouling* **2012**, *28*, 649–669. [CrossRef]
34. Stachowicz, J.J.; Witlatch, R.B.; Osman, R.W. Species diversity and invasion resistance in a marine ecosystem. *Science* **1999**, *286*, 1577–1579. [CrossRef]
35. Galil, B.S.; Zenetos, A. A sea change. Exotics in the Eastern Mediterranean. In *Invasive Aquatic Species of Europe: Distribution, Impacts and Management*; Leppäkoski, E., Gollasch, S., Olenin, S., Eds.; Kluwer Academic Publishers: Dordrecht, The Netherlands, 2002; pp. 325–326.
36. Zenetos, A.; Koutsoumbas, D.; Vardala-Theodorou, E. Origin and vectors of introduction of exotic molluscs in Greek waters. *Belg. J. Zool.* **2005**, *135*, 279–286.
37. Gartner, H.N.; Clarke Murray, C.; Frey, M.A.; Nelson, J.C.; Larson, K.J.; Ruiz, G.M.; Therriault, T.W. Non-indigenous invertebrate species in the marine fouling communities of British Columbia, Canada. *BioInvasions Rec.* **2016**, *5*, 205–212. [CrossRef]
38. Ponti, M.; Fava, F.; Abbiati, M. Spatial-temporal variability of epibenthic assemblages on subtidal biogenic reefs in the northern Adriatic Sea. *Mar. Biol.* **2011**, *158*, 1447–1459. [CrossRef]
39. Fortič, A.; Mavrič, B.; Pitacco, V.; Lipej, L. Temporal changes of a fouling community: Colonization patterns of the benthic epifauna in the shallow northern Adriatic Sea. *Reg. Stud. Mar. Sci.* **2021**, *45*, 101818. [CrossRef]
40. Jelić-Mrčelić, G.; Slišković, M.; Antolić, B. Biofouling communities on test panels coated with TBT and TBT-free copper based antifouling paints. *Biofouling* **2006**, *22*, 293–302. [CrossRef] [PubMed]
41. Jelić-Mrčelić, G.; Slišković, M.; Antolić, B. Macroalgae fouling community as quality element for the evaluation of the ecological status in Vela Luka Bay, Croatia. *Acta Soc. Bot. Pol.* **2012**, *81*, 159–165. [CrossRef]
42. Spagnolo, A.; Auriemma, R.; Bacci, T.; Balković, I.; Bertasi, F.; Bolognini, L.; Cabrini, M.; Cilenti, L.; Cuicchi, C.; Cvitković, I.; et al. Non-indigenous macrozoobenthic species on hard substrata of selected harbours in the Adriatic Sea. *Mar. Pollut. Bull.* **2019**, *147*, 150–158. [CrossRef] [PubMed]
43. Nerlović, V.; Perić, L.; Slišković, M.; Jelić-Mrčelić, G. The invasive *Anadara transversa* (Say, 1822) (Mollusca: Bivalvia) in the biofouling community of northern Adriatic mariculture areas. *Manag. Biol. Invasions* **2018**, *9*, 239–251. [CrossRef]
44. Slišković, M.; Jelić-Mrčelić, G.; Antolić, B.; Aničić, I. The fouling of fish farm cage nets as bioindicator of aquaculture pollution in the Adriatic Sea (Croatia). *Environ. Monit. Assess.* **2011**, *173*, 519–532. [CrossRef]
45. Pica, D.; Bloecher, N.; Dell'Anno, A.; Bellucci, A.; Pinto, T.; Pola, L.; Puce, S. Dynamics of a biofouling community in finfish aquaculture: A case study from the South Adriatic Sea. *Biofouling* **2019**, *35*, 696–709. [CrossRef]
46. Occhipinti-Ambrogi, A.; Marchini, A.; Cantone, G.; Castelli, A.; Chimenz, C.; Cormaci, M.; Froglia, C.; Furnari, G.; Gambi, M.C.; Giaccone, G.; et al. Alien species along the Italian coasts: An overview. *Biol. Invasions* **2011**, *13*, 215–237. [CrossRef]
47. Kennedy, V.S.; Bolognini, L.; Dulčić, J.; Woodland, R.J.; Wilberg, M.J.; Harris, L.A. Status of Fish and Shellfish Stocks. In *Coastal Ecosystems in Transition*; Malone, T.C., Malej, A., Faganeli, J., Eds.; American Geophysical Union: Washington, DC, USA, 2020.
48. Artegiani, A.; Bregant, D.; Paschini, E.; Pinardi, N.; Raicich, F.; Russo, A. The Adriatic Sea general circulation. Part I: Air-sea interactions and water mass structure. *J. Phys. Oceanogr.* **1997**, *27*, 1492–1514. [CrossRef]
49. Russo, A.; Artegiani, A. Adriatic Sea hydrography. *Sci. Mar.* **1996**, *60*, 33–43.
50. Casellato, S.; Stefanon, A. Coralligenous habitat in the northern Adriatic Sea: An overview. *Mar. Ecol.* **2008**, *29*, 321–341. [CrossRef]
51. Marini, M.; Jones, B.H.; Campanelli, A.; Grilli, F.; Lee, C.M. Seasonal variability and Po River plume influence on biochemical properties along western Adriatic coast. *J. Geophys. Res. Oceans* **2008**, *113*, C05S90. [CrossRef]
52. Artegiani, A.; Bregant, D.; Paschini, E.; Pinardi, N.; Raicich, F.; Russo, A. The Adriatic Sea general circulation. Part II: Baroclinic circulation structure. *J. Phys. Oceanogr.* **1997**, *27*, 1515–1532. [CrossRef]
53. Degobbis, D.; Precali, R.; Ivancic, I.; Smodlaka, N.; Fuks, D.; Kveder, S. Long-term changes in the northern Adriatic ecosystem related to anthropogenic eutrophication. *Int. J. Environ. Pollut.* **2000**, *13*, 495–533. [CrossRef]
54. Bernardi Aubry, F.; Berton, A.; Bastianini, M.; Socal, G.; Acri, F. Phytoplankton succession in a coastal area of the NW Adriatic, over a 10-year sampling period (1990–1999). *Cont. Shelf Res.* **2004**, *24*, 97–115. [CrossRef]
55. Solidoro, C.; Bandelj, V.; Barbieri, P.; Cossarini, G.; Fonda Umani, S. Understanding dynamic of biogeochemical properties in the northern Adriatic Sea by using self-organizing maps and k-means clustering. *J. Geophys. Res.* **2007**, *112*, C07S90. [CrossRef]
56. Cossignani, T.; Di Nisio, A.; Passamonti, M. *Atlante Delle Conchiglie del Medio Adriatico*; L'Informatore Piceno Editore: Ancona, Italy, 1992; pp. 1–120.
57. Cesari, P. *I Molluschi Della Laguna di Venezia*; Arsenale Editrice: Venice, Italy, 1994; pp. 1–189.
58. Doneddu, M.; Trainito, E. *Conchiglie del Mediterraneo*; Il Castello Editore: Milano, Italy, 2005; pp. 1–256.
59. Schiaparelli, S. Checklist della flora e della fauna dei mari italiani (Parte I). Bivalvia. *Biol. Mar. Mediterr.* **2008**, *15*, 296–314.
60. Sullivan, C.M. Bivalve larvae of Malpeque Bay. *J. Fish. Res. Board Can.* **1948**, *77*, 1–36.

61. Loosanoff, V.L.; Davis, H.C.; Chanley, P.E. Dimensions and shapes of larvae of some marine bivalve mollusks. *Malacologia* **1966**, *4*, 351–435.
62. Rose, R.A.; Dix, T.G. Larval and juvenile development of the doughboy scallop *Chlamys asperrimus* (Lamark) (Mollusca: Pectinidae). *Aust. J. Mar. Freshw. Res.* **1984**, *35*, 315–323. [CrossRef]
63. Hodgson, C.A.; Burke, R.D. Development and larval morphology of the spiny scallop, *Chlamys hastate*. *Biol. Bull.* **1988**, *174*, 303–318. [CrossRef]
64. Anderson, M.J.; Gorley, R.N.; Clarke, K.R. *PERMANOVA+ for PRIMER: Guide to Software and Statistical Methods*; PRIMER-E: Plymouth, UK, 2008; pp. 15–85.
65. Hrs-Brenko, M.; Legac, M. Inter- and intra-species relationships of sessile bivalves on the eastern coast of the Adriatic Sea. *Nat. Croat.* **2006**, *15*, 203–230.
66. Casellato, S.; Sichirollo, E.; Cristofoli, A.; Masiero, L.; Soresi, S. Biodiversità delle "tegnúe" di Chioggia, zona di tutela biologica del Nord Adriatico. *Biol. Mar. Mediterr.* **2005**, *12*, 69–77.
67. Casellato, S.; Masiero, L.; Sichirollo, E.; Soresi, S. Hidden secrets of the Northern Adriatic: "Tegnúe", peculiar reefs. *Cent. Eur. J. Biol.* **2007**, *2*, 122–136. [CrossRef]
68. Gabriele, M.; Bellot, A.; Gallotti, D.; Brunetti, R. Sublittoral hard substrate communities of the northern Adriatic Sea. *Cah. Biol. Mar.* **1999**, *40*, 65–76.
69. Peharda, M.; Ezgeta-Balić, D.; Vrgoč, N.; Isajlović, I.; Bogner, D. Description of bivalve community structure in the Croatian part of the Adriatic Sea—Hydraulic dredge survey. *Acta Adriat.* **2010**, *51*, 141–158.
70. Fava, F.; Ponti, M.; Abbiati, M. Role of recruitment processes in structuring coralligenous benthic assemblages in the northern adriatic continental shelf. *PLoS ONE* **2016**, *11*, e0163494. [CrossRef]
71. Da Ros, L.; Bressan, M.; Marin, M.G. Reproductive cycle of the mussel (*Mytilus galloprovincialis* Lmk) in Venice Lagoon (North Adriatic). *Boll. Zool.* **1985**, *52*, 223–229. [CrossRef]
72. Orban, E.; Di Lena, G.; Nevigato, T.; Casini, I.; Marzetti, A.; Caproni, R. Seasonal changes in meat content, condition index and chemical composition of mussels (*Mytilus galloprovincialis*) cultured in two different Italian sites. *Food Chem.* **2002**, *77*, 57–65. [CrossRef]
73. Ceccherelli, V.U.; Rossi, R. Settlement, growth and production of the *Mytilus galloprovincialis*. *Mar. Ecol. Prog. Ser.* **1984**, *16*, 173–184. [CrossRef]
74. Marčeta, T.; Da Ros, L.; Marin, M.G.; Codognotto, V.F.; Bressan, M. Overview of the biology of *Flexopecten glaber* in the North Western Adriatic Sea (Italy): A good candidate for future shellfish farming aims? *Aquaculture* **2016**, *462*, 80–91. [CrossRef]
75. Peña, J.B.; Canales, J.; Adsuara, J.M.; Sos, M.A. Study of seasonal settlements of five scallop species in the western Mediterranean. *Aquac. Int.* **1996**, *4*, 253–261. [CrossRef]
76. Castagnolo, L. La pesca e la riproduzione di *Pecten jacobaeus* (L). e di *Aequipecten opercularis* (L.) nell'alto Adriatico. *Boll. Malacol.* **1991**, *27*, 39–48.
77. Reddiah, K. The sexuality and spawning of Manx pectinids. *J. Mar. Biol. Assoc. U. K.* **1962**, *42*, 683–703. [CrossRef]
78. Román, G.; Campos, M.J.; Acosta, C.P. Relationships among environment, spawning and settlement of queen scallop in the Ria de Arosa (Galicia, NW Spain). *Aquacult. Int.* **1996**, *4*, 225–236. [CrossRef]
79. Shafee, M.S. Seasonal changes in the biochemical composition and calorific content of the black scallop *Chlamys varia* (L) from Lanveoc, Bay of Brest. *Oceanol. Acta* **1981**, *4*, 331–341.
80. Morello, E.B.; Solustri, C.; Froglia, C. The alien bivalve *Anadara demiri* (Arcidae): A new invader of the Adriatic Sea, Italy. *J. Mar. Biol. Ass. U. K.* **2004**, *84*, 1057–1064. [CrossRef]
81. Solustri, C.; Morello, E.; Froglia, C. Osservazioni su *Anadara demiri* (Piani, 1981) (Bivalvia: Arcidae) epibionte di alcune specie di molluschi. *Biol. Mar. Mediterr.* **2003**, *10*, 622–625.
82. Ezgeta Balić, D.; Radonić, I.; Bojanić Varezić, D.; Zorica, B.; Arapov, J.; Stagličić, N.; Jozić, S.; Peharda, M.; Briski, E.; Lin, Y.; et al. Reproductive cycle of a non-native oyster, *Crassostrea gigas*, in the Adriatic Sea. *Mediterr. Mar. Sci.* **2020**, *21*, 146–156. [CrossRef]
83. Mladineo, I.; Peharda, M.; Orhanović, S.; Bolotin, J.; Pavela-Vrančić, M.; Treursić, B. The reproductive cycle, condition index and biochemical composition of the horse-bearded mussel *Modiolus barbatus*. *Helgol. Mar. Res.* **2007**, *61*, 183. [CrossRef]
84. Cabanellas-Reboredo, M.; Deudero, S.; Alós, J.; Valencia, J.; March, D.; Hendriks, I.; Álvarez, E. Recruitment of *Pinna nobilis* (Mollusca: Bivalvia) on artificial structures. *Mar. Biodivers. Rec.* **2009**, *2*, E126. [CrossRef]
85. Richardson, C.A.; Peharda, M.; Kennedy, H.; Kennedy, P.; Onofri, V. Age, growth rate and season of recruitment of *Pinna nobilis* (L.) in the Croatian Adriatic determined from Mg: Ca and Sr: Ca shell profiles. *J. Exp. Mar. Biol. Ecol.* **2004**, *299*, 1–16. [CrossRef]
86. Guijarro Garcia, E.; Thorarinsdóttir, G.G.; Ragnarsson, S.A. Settlement of bivalve spat on artificial collectors in Eyjafjordur, North Iceland. *Hydrobiologia* **2003**, *503*, 131–141. [CrossRef]
87. Solidoro, C.; Bastianini, M.; Bandelj, V.; Codermatz, R.; Cossarini, G.; Canu, D.M.; Ravagnan, E.; Salon, S.; Trevisani, S. Current state, scales of variability, and trends of biogeochemical properties in the northern Adriatic Sea. *J. Geophys. Res.* **2009**, *114*, C07S91. [CrossRef]
88. Grilli, F.; Accoroni, S.; Acri, F.; Bernardi Aubry, F.; Bergami, C.; Cabrini, M.; Campanelli, A.; Giani, M.; Guicciardi, S.; Marini, M.; et al. Seasonal and Interannual Trends of Oceanographic Parameters over 40 Years in the Northern Adriatic Sea in Relation to Nutrient Loadings Using the EMODnet Chemistry Data Portal. *Water* **2020**, *12*, 2280. [CrossRef]

89. Cronin, E.R.; Cheshire, A.C.; Clarke, S.M.; Melville, A.J. An investigation into the composition, biomass and oxygen budget of the fouling community on a tuna aquaculture farm. *Biofouling* **1999**, *13*, 279–299. [CrossRef]
90. Fuentes, J.; Molares, J. Settlement of the mussel *Mytilus galloprovincialis* on collectors suspended from rafts in the Ría de Arousa (NW of Spain): Annual pattern and spatial variability. *Aquaculture* **1994**, *122*, 55–62. [CrossRef]
91. Halla, M.I.; Kassila, J.; Chattou, E.M.A.; Ouaggajou, Y.; El Aamri, F.; Benbani, A.; Nhhala, H. Depth and seasonal effects on the settlement density of two mussel species (*Perna perna* and *Mytilus galloprovincialis*) in offshore, Agadir (Morocco). *Eur. Sci. J.* **2018**, *14*, 229–240. [CrossRef]
92. Yildiz, H.; Berber, S. Depth and seasonal effects on the settlement density of *Mytilus galloprovincialis* L. 1819 in the Dardanelles. *J. Anim. Vet. Adv.* **2010**, *9*, 756–759. [CrossRef]
93. Pearce, C.M.; Gallager, S.M.; Manuel, J.L.; Manning, D.A.; O'Dor, R.K.; Bourget, E. Settlement of larvae of the giant scallop, *Placopecten magellanicus*, in 9-m deep mesocosms as a function of temperature stratification, depth, food, and substratum. *Mar. Biol.* **1996**, *124*, 693–706. [CrossRef]
94. Thouzeau, G. Experimental collection of postlarvae of *Pecten maximus* (L.) and other benthic macrofaunal species in the Bay of Saint-Brieuc, France. II. Reproduction patterns and postlarval growth of five mollusc species. *J. Exp. Mar. Biol. Ecol.* **1991**, *148*, 181–200. [CrossRef]
95. Hall-Spencer, J.M.; Froglia, C.; Atkinson, R.J.A.; Moore, P.G. The impact of Rapido trawling for scallops, *Pecten jacobaeus* (L.), on the benthos of the Gulf of Venice. *ICES J. Mar. Sci.* **1999**, *56*, 111–124. [CrossRef]
96. Vaccarella, R.; Paparella, P.; Bello, G.; Marano, G. The smooth scallop, *Chlamys glabra*, fishery in the Gulf of Manfredonia (south-western Adriatic Sea). *Rapp. Comm. Int. Mer. Médit.* **1998**, *35*, 500–501.
97. Demir, M. On the presence of *Arca (Scapharca) amygdalum* Philippi, 1847 (Mollusca: Bivalvia) in the harbour of Izmir, Turkey. *J. Fac. Sci. Istanb. Univ.* **1977**, *42*, 197–202.
98. Zenetos, A. *Scapharca demiri* (Piani, 1981): Primo ritrovamento nel nord Egeo. *La Conchiglia* **1994**, *271*, 37–38.
99. Rinaldi, E. Segnalazione faunistica n. 41. *Quad. Studi Nat. Romagna* **2001**, *14*, 127–128.
100. Morello, E.; Solustri, C. First record of *Anadara demiri* (Piani 1981) (Bivalvia: Arcidae) in Italian waters. *Boll. Malacol.* **2001**, *37*, 231–234.
101. Mizzan, L. Biodiversita' della Laguna di Venezia. Segnalazioni (1–143) –18. *Anadara demiri. Boll. Mus. Civ. St. Nat. Venezia* **2002**, *53*, 265–266.
102. Mizzan, L.; Vianello, C. Biodiversità della Laguna di Venezia e della costa nord adriatica veneta. Segnalazioni 189–201. *Boll. Mus. Civ. St. Nat. Venezia* **2007**, *58*, 319.
103. Nerlović, V.; Doğan, A.; Perić, L. First record of *Anadara transversa* (Mollusca: Bivalvia: Arcidae) in Croatian waters (Adriatic Sea). *Acta Adriat.* **2012**, *53*, 139–144.
104. Mistri, M. The non-indigenous mussel *Musculista senhousia* in an Adriatic lagoon: Effects on benthic community over a ten year period. *J. Mar. Biol. Assoc. U. K.* **2003**, *83*, 1277–1278. [CrossRef]
105. Halley, E.F.; Koehn, J.Z.; Holsman, K.K.; Halpern, B.S. Emerging trends in science and news of climate change threats to and adaptation of aquaculture. *Aquaculture* **2022**, *549*, 737812.

Journal of
Marine Science and Engineering

MDPI

Article

Morphological Study and 3D Reconstruction of the Larva of the Ascidian *Halocynthia roretzi*

Lucia Manni [1,*], Federico Caicci [1], Chiara Anselmi [2,3], Virginia Vanni [1], Silvia Mercurio [4] and Roberta Pennati [4,*]

1 Department of Biology, University of Padova, 35131 Padova, Italy; federico.caicci@unipd.it (F.C.); virginia.vanni@phd.unipd.it (V.V.)
2 Department of Biology, Hopkins Marine Station, Stanford University, Pacific Grove, CA 93950, USA; chiara90@stanford.edu
3 Institute for Stem Cell Biology and Regenerative Medicine, Stanford University School of Medicine, Stanford, CA 94305, USA
4 Department of Environmental Science and Policy, University of Milano, 20133 Milano, Italy; silvia.mercurio@unimi.it
* Correspondence: lucia.manni@unipd.it (L.M.); roberta.pennati@unimi.it (R.P.)

Abstract: The swimming larva represents the dispersal phase of ascidians, marine invertebrates belonging to tunicates. Due to its adhesive papillae, the larva searches the substrate, adheres to it, and undergoes metamorphosis, thereby becoming a sessile filter feeding animal. The larva anatomy has been described in detail in a few species, revealing a different degree of adult structure differentiation, called adultation. In the solitary ascidian *Halocynthia roretzi*, a species reared for commercial purposes, embryogenesis has been described in detail, but information on the larval anatomy is still lacking. Here, we describe it using a comparative approach, utilizing 3D reconstruction, as well as histological/TEM observations, with attention to its papillae. The larva is comparable to those of other solitary ascidians, such as *Ciona intestinalis*. However, it displays a higher level of adultation for the presence of the atrium, opened outside by means of the atrial siphon, and the peribranchial chambers. It does not reach the level of complexity of the larva of *Botryllus schlosseri*, a phylogenetically close colonial ascidian. Our study reveals that the papillae of *H. roretzi*, previously described as simple and conform, exhibit dynamic changes during settlement. This opens up new considerations on papillae morphology and evolution and deserves to be further investigated.

Keywords: adhesive papillae; adultation; *Botryllus schlosseri*; *Ciona intestinalis*; tunicates

Citation: Manni, L.; Caicci, F.; Anselmi, C.; Vanni, V.; Mercurio, S.; Pennati, R. Morphological Study and 3D Reconstruction of the Larva of the Ascidian *Halocynthia roretzi*. *J. Mar. Sci. Eng.* **2022**, *10*, 11. https://doi.org/10.3390/jmse10010011

Academic Editor: Nick Aldred

Received: 30 November 2021
Accepted: 20 December 2021
Published: 24 December 2021

1. Introduction

Ascidians are sessile marine invertebrates belonging to the tunicates, a group that is evolutionarily close to vertebrates [1]. Some species are gregarious and form complex clusters of several organisms living in close proximity [2]. They can colonize different natural and artificial substrates and can modify the primary substrate providing habitats for other organisms. Thus, ascidians are considered essential members of the benthic communities contributing to increased ecosystem complexity and biodiversity [3] and have been proposed as valuable model organisms to test coastal water pollution [4–6]. Ascidians also have economic importance since some species, such as *Halocynthia roretzi*, are commercially reared for human consumption [7,8]. For this reason, *H. roretzi* is one of the most studied ascidians: its genome is compact with about 16,000 protein-coding genes [9] and its development is known in detail [10]. Other species (such as *Ciona intestinalis*) are potential food and biofuel sources [4]. Some ascidians (such as *Botryllus schlosseri* and *Botrylloides violaceus*) are considered invasive organisms in marinas and also fouling organisms in marine aquaculture, and their presence results in economic loss [11].

Ascidians show an extraordinary range of life history traits, with some species being solitary and others showing colonial organisation; the latter evolved several times during the diversification of the group [12,13]. Tunicate taxonomy has long been questioned; the group comprises approximately 3000 species that have traditionally been divided into three classes: Ascidiacea (sea squirts), Thaliacea (pelagic salps, doliolids and pyrosomes), and Appendicularia (larvaceans) [14]. Molecular based phylogenies revealed that Ascidiacea was paraphyletic [15,16] and the following three clades were proposed: (1) Stolidobranchia, (2) Appendicularia, and (3) Phlebobranchia plus Aplousobranchia plus Thaliacea [17].

Sessile adults develop from swimming lecithotrophic larvae that represent the dispersal stage of their life cycle. Larvae of different species show the same general body organisation corresponding to the basal body plan of chordates. The larva is formed by a trunk and a muscular tail used for locomotion. The trunk contains the anterior nervous system equipped with two pigmented sensory organs (the ocellus and otolith), mesenchyme cells and the endodermal pharynx, whereas the tail houses the notochord, the posterior neural tube dorsal to the notochord and lateral bands of muscle cells.

The larva is sheltered by the epidermis, which secretes the larval tunic, the characteristic external fibrous matrix of the phylum. The larval tunic exhibits two layers (compartments). The outer layer covers the trunk and the tail, forming, on the latter, continuous fins for swimming; this layer is lost at larval metamorphoses. The inner layer, called the definitive tunic, covers the trunk and is maintained during metamorphosis to constitute the tunic covering the adult body. Two cuticular layers thicken the layer surfaces [18–21].

Despite the gross anatomical similarities, ascidian larvae have been remodelled in different ways during evolution [22]. Colonial species, which produce yolked eggs, are ovoviviparous (or even viviparous) [23,24] and exhibit prolonged embryogenesis. Their larvae often undergo adultation, a mode of development in which adult structures differentiate precociously in the tadpole trunk. Adultation can manifest to different degrees, from the early appearance of siphons, a partial digestive tract, a few gill slits, a rudimentary heart, or one or more buds prior to asexual reproduction, up to involving more extensive development of adult structures [22,25]. For example, the larva of the colonial *Botryllus schlosseri* exhibits both open siphons, five pairs of perforated stigmata, a hollow heart, the rudiment of the adult nervous system close to the larval one, and two buds [26]. Caudalisation is another alternative mode of development present in ascidians consisting of the addition of muscle cells to the tadpole tail without changing the number of other larval cells. This enhances the swimming ability of the larva and its dispersal [22]. In comparison to colonial ascidians, solitary species, such as the model ascidian *Ciona intestinalis*, usually develop smaller eggs, are oviparous, and exhibit rapid embryogenesis producing relatively simpler larvae [21]. These reproductive and larval anatomical differences are surprising when exhibited by closely related species, such as the solitary *H. roretzi* and the colonial *B. schlosseri*, both of which are stolidobranch ascidians.

The larva plays a crucial role in ascidian life since it selects the substrate to adhere to permanently. Indeed, after hatching, the tadpole larva swims for up to a few days, then attaches to a substrate and undergoes metamorphosis: the tail is retracted into the trunk, the larval tissues are resorbed, and adult structures differentiate from rudiments in the trunk [27]. The critical step of substrate selection and adhesion is mediated by three mucus-secreting organs, the adhesive papillae (or palps) that most ascidian larvae bear at their anterior tip. Generally, the papillae are composed of elongated mucus-secreting cells, called collocytes (CCs), primary sensory neurons (PSNs) and supporting axial columnar cells (ACCs) [27]. It was proposed that papillary sensory cells are mechanosensory neurons playing a central role in substrate selection [28]. Actually, the morphology of adhesive papillae is quite variable among species, and they can be classified into 10 types according to their histological characteristics [27,29]. A major distinction can be traced between simple conic papillae that are present in most solitary ascidians, and the eversible papillae characteristic of some colonial species, which are capable of rapid eversion to expose sticky mucus. Due to their complexity and diversity, it is not surprising that larvae are of key importance for the taxonomic recognition of ascidian species. Indeed, ascidians are the only chordates identified also by larval characteristics [30,31].

In the model species *C. intestinalis*, the fine anatomy of the adhesive papillae has been extensively analysed. Each one of these organs are formed by exactly 20 cells: four ACCs surrounded by 12 CCs and flanked by two pairs of PSNs. The central ACCs, which have been suggested to have a chemosensory function, extend their apical endings into a hyaline cap. In *C. intestinalis*, they can be identified by the expression of the *βγ-Crystallin* gene [32]. CCs secrete the adhesive mucus responsible for performing the temporary adhesion to the substrate. PSNs are glutamatergic neurons that sense the substrate [33].

Although *H. roretzi* is a well-established model species in developmental biology, the overall morphology of the larva and the structure of its papillae have been poorly described. Therefore, in the present article, we present the first 3D reconstruction of the swimming larva of *H. roretzi* to gain insight into its morphology. We compare it with the 3D reconstruction of the close colonial species *B. schlosseri* and of the solitary species *C. intestinalis*, discussing the degree of adultation in the three models. Moreover, we focus on the description of the adhesive papillae of *H. roretzi* for their crucial role during larval settlement. Our data show that, despite the simple larval anatomy of *H. roretzi*, similar to that of other solitary ascidians, the papillae are dynamic structures that undergo retraction and remodelling during adhesion.

2. Materials and Methods

2.1. Animals

Adults of *Halocynthia roretzi* were collected near the Marine Biological Station "Asamushi" of the Tohoku University (Aomori, Japan). Embryos were obtained by in vitro fertilisation and were reared at 13 °C for 35 h until the larva hatched. Images of 47 larvae were taken using a Leica MZ16 Stereomicroscope and analysed by Photoshop software to determine the following morphometric parameters: total length of the larva, trunk length, tail length. The pool of larvae consisted of typical swimming larvae with protruded anterior papillae and larvae at the beginning of metamorphosis, with shorter papillae.

2.2. Histology

Larvae were fixed for 9 h in 1.5% glutaraldehyde in 0.2 M sodium cacodylate buffer at pH 7.2. at 4 °C. They were rinsed several times in 0.2 M sodium cacodylate and 1.7% NaCl buffer and then post-fixed for 1 1/2 h in 1% OsO4 in 0.2 M cacodylate buffer at 4 °C. Samples were dehydrated and then soaked in Epon and propylene solution at 37 °C, 45 °C, and 60 °C. They were then embedded in epoxy resin (Sigma-Aldrich), oriented and sectioned using a Leica Ultramicrotome UC7. Five individuals at larval stage and three individuals at adhering larval stage were cut for histological analysis. Sections, 1 μm thick, were stained with toluidine blue in borax. Ultrathin sections (80 nm thick) were stained with uranyl acetate and lead citrate to provide contrast. Photomicrographs were taken with a FEI Tecnai G^2 electron microscope operating at 100 kV. Images were captured with a Veleta (Olympus Soft Imaging System) digital camera.

2.3. Three-Dimensional Reconstructions

A selected swimming larva of *H. roretzi* was embedded in resin as previously described and serially transversely cut using a Histo Jumbo Knife (Diatome). Sections, 1 μm thick, were arranged in chains of about 20 sections each and stained with toluidine blue. All the sections were then photographed with a Leica DMR optical microscope. Images were aligned using Adobe Photoshop CS on a Windows 10 computer. Based on the resulting stack of images, the 3D model of the anatomy of all organs but tunic was created in Amira software (Thermofisher scientific, Waltham, MA, USA). For the comparison with the 3D reconstructions of larvae of *Ciona intestinalis* and *Botryllus schlosseri*, we used previously prepared datasets of serial sections [26,32].

3. Results

3.1. The 3D Reconstruction of Halocynthia roretzi Larva

The *H. roretzi* larvae, 10 h post hatching at 12 °C, were found to be 1744 ± 18 μm long; the tail length is 1252 ± 15 μm and the trunk is 490 ± 6 μm (Figure 1, Table S1). At the anterior tip of the trunk, three adhesive papillae are arranged according to the vertices of an equilateral triangle (two dorsal, one ventral) (Figure 1A,B′,G).

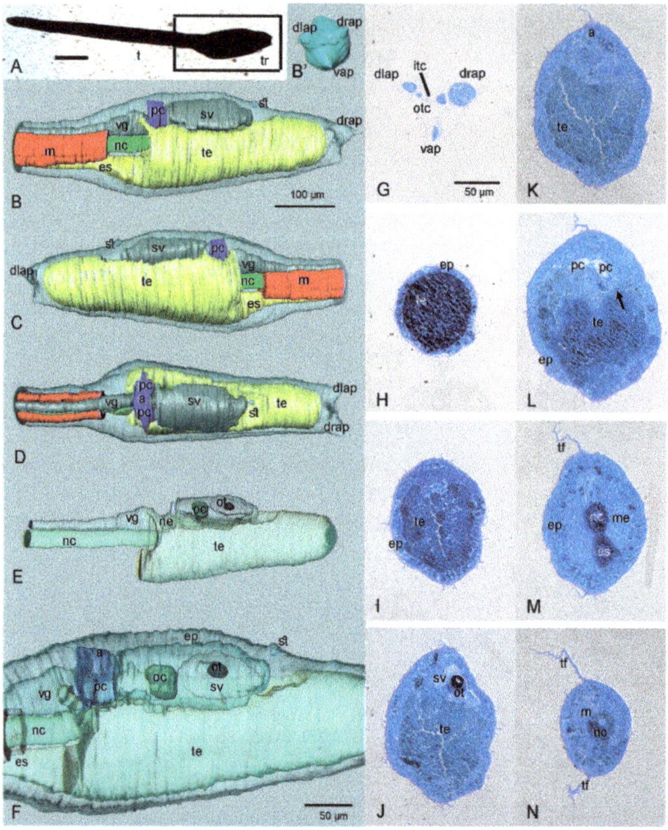

Figure 1. Anatomy of the larva of *H. roretzi* (see also the 3D-PDF of Figure S1; requires Acrobat Reader DC or higher). (**A**) Whole mount larva (in block of resin, contrasted with OsO_4). Right side view. t: tail; tr: trunk. The black square indicates the trunk region illustrated in (**B–E**). (**B–F**) Three-dimensional reconstructions. Colour code: blue, atrial chamber; dark green, ocellus; grey, nervous system; light green, notochord; red, muscles; yellow, endoderm (i.e., prospective pharynx). In B–D, the epidermis is transparent; in (**E**,**F**), the organs are transparent and their internal structures can be seen. (**B**) Right side view. (**B′**) Frontal view of the trunk showing the three papillae. (**C**) Left side view. (**D**) Dorsal view. (**E**) View of internal organs: trunk endoderm, nervous system and notochord. (**F**) Close view of the trunk region. (**G–N**) Some cross histological sections, from anterior to posterior, utilized for the 3D reconstruction. Toluidine blue. Arrow in L indicates the area where the protostigmata will open during metamorphosis. a: atrium; drap, dlap, vap: dorsal right-, dorsal left-, and ventral-adhesive papilla, respectively; ep: epidermis; es: endodermal strand; itc: inner tunic compartment; m: muscle; me: mesenchyme; nc: notochord; ne: neck; otc: outer tunic compartment; pc: peribranchial chambers; oc: ocellus; ot: otolith; st: stomodeum (i.e., oral siphon primordium); sv: sensory vesicle; te: trunk endoderm; tf: tail fin; vg, visceral ganglion. The enlargement is the same in (**B–E**) and in (**G–N**).

With respect to the histological sections, the 3D reconstruction allows a better understanding of the reciprocal positions of internal structures, the relative development of different organs and their anatomical details (Figure 1A–F; Figure S1; Video S1). Larval organs and tissues can be visualized in the MorphoNet browser at https://morphonet.org/9vILe4SK, accessed on 23 December 2021 [34]. The oral siphon primordium overlays the prospective pharynx that is differentiating from the trunk endoderm. The latter presents a narrow lumen (Figure 1I–K) and occupies most of the trunk, just under the nervous system (Figure 1B–D). The central nervous system is dorsal to the endoderm; in the trunk, it swells to form the sensory vesicle and, posterior to this, the visceral ganglion (Figure 1E–F). The two bulges are separated by a narrow neck as in other species, such as *Ciona intestinalis* (Figure 1E) [35]. The sensory vesicle is asymmetric and more enlarged on the right. It exhibits two pigmented sensory organs, the ocellus (occupying its posterior wall) and the otolith (emerging from its floor) (Figure 1F,J). The otolith is formed by a single large, pigmented cell (Figure 1J). A small, anteriorly blind, neurohypophyseal duct (or neurohypophysis) extends anteriorly from the sensory vesicle wall (Figure S1). In ascidians, it contributes to the formation of the adult brain and the associated neural gland [36]. A large mass of mesenchyme cells is present in the posterior ventral part of the trunk, anteriorly to the tail insertion (Figure 1M).

In the posterior dorsal part of the trunk, the endoderm of the pharynx is in contact with the developing ectodermal peribranchial chambers; here, the protostigmata will open during metamorphosis (Figure 2). As is clearly visible in serial histological sections, the two peribranchial chambers are joined dorsally to the nervous system to form a small chamber representing the prospective atrial chamber of the adult (Figure 2D–I). The atrial chamber opens outside through the atrial siphon in the form of a shallow dent in the dorsal trunk Figures 1L and 2A). However, at this stage, both the siphons are still covered with a thin layer of tunic, preventing circulation of seawater within the larva, which is lecithotrophic and does not filter (Figure 2D). During metamorphosis, the two peribranchial chambers and the atrial chamber will elongate as the pharynx grows, and the stigmata number will increase.

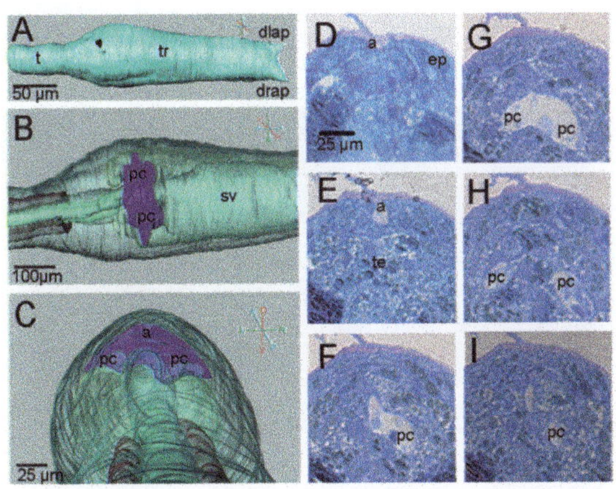

Figure 2. Atrial chamber in *H. roretzi.* See Figure 1 for colour code and abbreviations. (**A–C**) Three-dimensional reconstruction. In B,C the organs are transparent. (**A**) Dorsal external view showing the dint in correspondence of the atrial siphon (arrowhead). (**B**) Dorsal view. (**C**) Frontal-posterior view enlightening the position of the atrial chamber relative to the notochord. (**D–I**) Serial cross histological sections from anterior to posterior at the level of the atrial and peribranchial chambers. Toluidine blue. The enlargement is the same in (**D–I**).

The tail, only partially represented in Figure 1, houses the notochord, which is flanked by locomotory musculature on both sides. The notochord extends anteriorly in the trunk up to the posterior side of the atrial primordium (Figure 1E,F). In the tail, dorsal to the notochord, the neural tube runs up to the posterior end on the larva.

In Figure 3, the 3D reconstructions of three ascidian larvae are compared: *H. roretzi* (solitary, stolidobranch; Figure 3A,B), *Botryllus schlosseri* (colonial, stolidobranch; Figure 3C,D) and *Ciona intestinalis* (solitary, phlebobranch; Figure 3E,F).

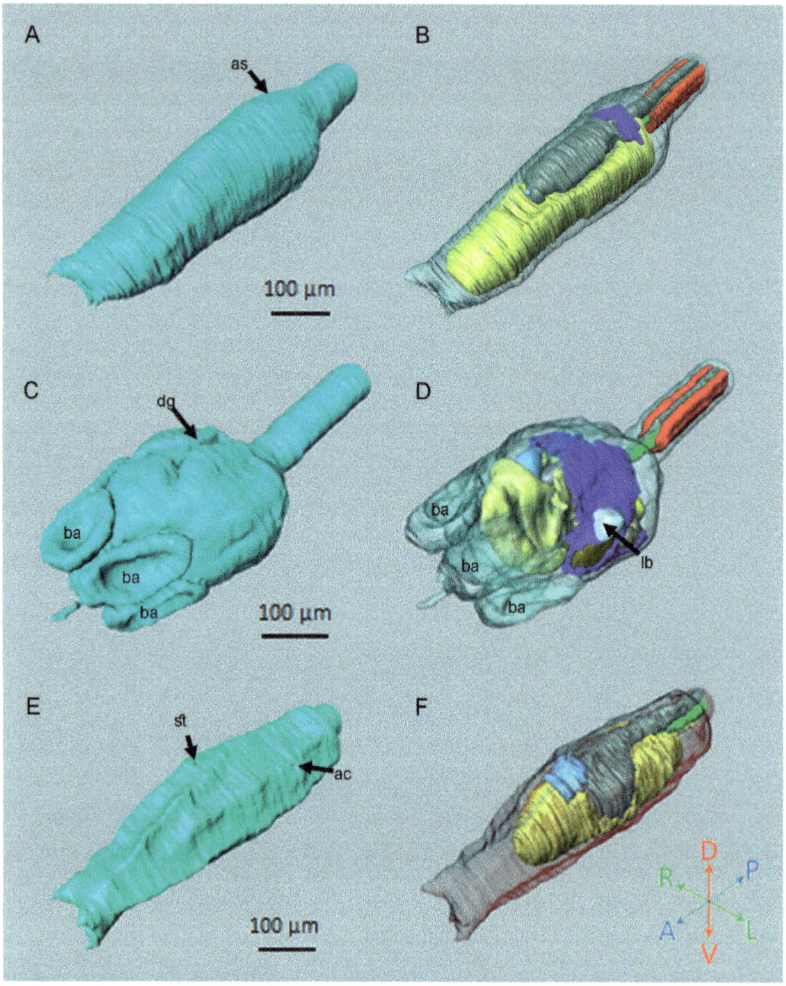

Figure 3. Comparison of the 3D reconstructions of three larvae. (**A**,**B**) *H. roretzi*. (**C**,**D**) *B. schlosseri*. (**E**,**F**) *C. intestinalis*. (**A**,**C**,**E**) External views. (**B**,**D**,**F**) Internal views (organs are transparent). Colour code as in Figure 1. Note the different levels of adultation. ac, left atrial chamber rudiment; as, atrial siphon rudiment; ba, blood ampulla; dg, dorsal groove; lb, left bud; st, stomodeum.

Externally, the trunk of *H. roretzi* (Figure 3A) is more similar to that of the *C. intestinalis* (Figure 3E) as it is elongated, fusiform, and ends at its anterior-most section with three conic papillae. The two stolidobranch species, *H. roretzi* and *B. schlosseri*, share signs of adultation that are absent in *C. intestinalis* (Figure 3B,D,F; Table 1). In fact, in both stolidobranch species, the atrial chamber forms a well-developed uneven structure; in *C. intestinalis*, the atrial rudiments are in the form of two dorsal-lateral shallow depressions of the epidermis. However, in comparison to *H. roretzi*, *B. schlosseri* shows the highest level of adultation, exemplified by the large and perforated pharynx and the differentiated gut (Figure 3D; Table 1). Moreover, the overall morphology of *H. roretzi* is more similar to that of *C. intestinalis*.

Table 1. List of anatomical structures represented in the 3D reconstructions of the three selected species. v: present; -: absent.

Organ	*H. roretzi*	*B. schlosseri*	*C. intestinalis*
Atrial chamber	v	v	v
Atrial siphon rudiment	v	v	v
Cerebral ganglion rudiment	-	v	-
Endodermal strand	v	v	v
Epidermis	v	v	v
Esophagus	-	v	-
Heart	-	v	-
Intestine	-	v	-
Left bud rudiment	-	v	-
Neurohypophysial duct	v	v	v
Notochord	v	v	v
Ocellus	v	-	v
Otolith	v	-	v
Oral siphon rudiment	v	v	v
Peribranchial chamber	v	v	-
Pharynx	v	v	v
Photolite	-	v	-
Pyloric caecum and gland	-	v	-
Sensory vesicle	v	v	v
Stomach	-	v	-
Right bud rudiment	-	v	-
Tail musculature	v	v	v

3.2. Adhesive Papillae

The swimming larva of *H. roretzi* bears three elongated conic papillae at the anterior tip of the trunk. The papillae lay over the basal lamina and are formed by a monolayer of differentiated ectodermal cells (Figure 4). Below the basal lamina, a conspicuous mesenchymal cell corresponding to each papilla can be consistently recognised (Figure 4B,D,I). In swimming larvae, the papillae are covered by the two layers of tunic (Figure 4A–C).

To determine the relative position and ultrastructural characteristics of the different cell types, we analysed the *H. roretzi* papillae by Transmission Electron Microscopy (TEM) in longitudinal sections. TEM analysis revealed that they are formed by three different cell types that can be identified by their positions and ultrastructural characteristics. The ACCs form the core centre of the papillae; they are elongated cells and bear finger-like protrusions of 2 μm in diameter at their distal ends (Figure 4B). Their nuclei are located close to the basal plasmalemma, and the cytoplasm is rich in vesicles with different electron densities (Figure 4D). Several CCs encircle the ACCs; they have basal nuclei and are easily recognisable by the numerous electron-dense vesicles of different sizes (Figure 4C). Their distal endings bear numerous apical microvilli (Figure 4F). PSNs occupy the lateral part of the papillary body (Figure 4C). They are characterized by a spindle shape, the nuclei being in a central position, and the presence of few small vesicles (Figure 4C,E). Unlike those of the ACCs and CCs, their distal endings do not reach the tip of the papilla, but rather, they stop in a more backward position (Figure 5A).

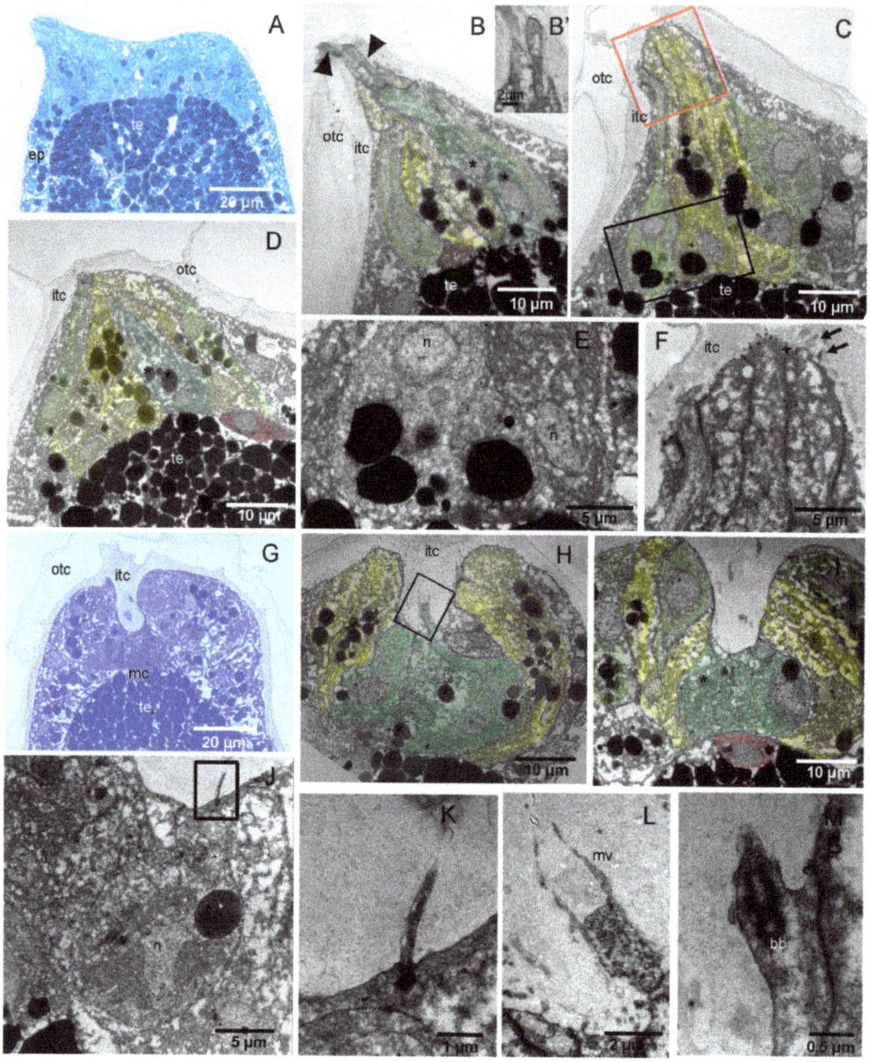

Figure 4. Histological and ultrastructural analysis of adhesive papillae in *H. roretzi*. (**A–F**) Swimming larva: the papillae are conic. (**G–M**) Adhering larva: the papillae have a central groove. (**A**,**G**) Longitudinal sections of the anterior trunk. Toluidine blue. (**B–F**,**H–M**) TEM images at different magnifications and different levels of papillae. Colour code: dark green, ACCs; light green, PSNs; red, mesenchyme cells; yellow, CCs. (**B–D**): Three longitudinal sections of the same papilla. Note that, in (**B**), the finger-like protrusions (arrowheads) of ACCs are also visible in (**B′**). The black square area in (**C**) is enlarged in (**E**) to show a detail of a PNS basal cytoplasm. The red square area in (**C**) is enlarged in (**F**) to show a detail of CCS distal endings with microvilli. Asterisks: vesicles in ACC. (**H–I**): Two longitudinal sections of the same papilla. The square area in (**H**) is enlarged in (**L**) to show ACC apical endings with microvilli, whereas the square area in (**J**) is enlarged in (**K**) to show a cilium of PSN. (**M**): Cilium basal body belonging to a PSN. bb: basal body of a cilium; mc: mesenchymal cell; mv: microvillus; n: nucleus. See also Figure 1 for abbreviations.

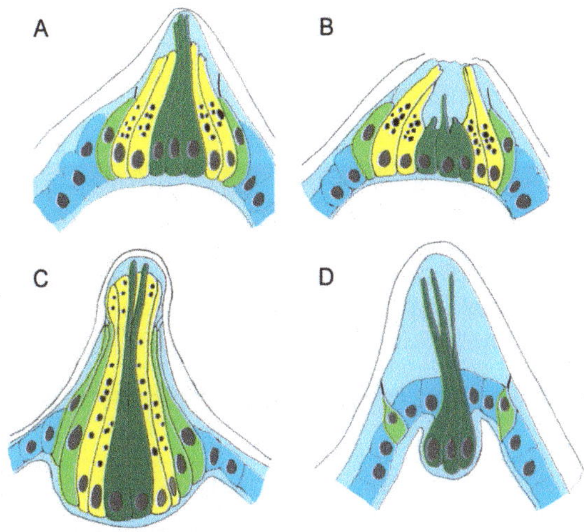

Figure 5. Schematic illustrations of adhesive papillae. (**A**) Swimming larva before metamorphosis of *H. roretzi*. (**B**) Adhering larva at the onset of metamorphosis of *H. roretzi*. (**C**) *C. intestinalis* swimming larva (redrawn from Zeng et al., 2019). (**D**) *B. schlosseri* swimming larva (redrawn from Caicci et al., 2010). Colour code: dark blue, ectodermal cells; dark green, ACCs; light blue, inner compartment of the tunic; light green, PSNs; yellow, CCs.

In adhering larvae, the papillae change shape and form a central groove (Figure 4G). The central ACCs, still recognisable by their finger-like protrusions, are shorter than in swimming larvae and have nuclei with an irregular shape (Figure 4H,I). Their apical endings are located at the base of the groove (Figure 4L). Vesicle-rich CCs form the walls of the groove, and PSNs, exhibiting cilia with atypical morphology, reach the anterior margin of the papilla (Figures 4I,K,M and 5B).

4. Discussion

4.1. *H. roretzi Exhibits Minimal Adultation*

In this paper, we present the first 3D reconstruction of the larva of *H. roretzi*, an ascidian that has long been established as a model organism in developmental biology studies with more than 400 scientific articles published in the last 10 years (indexed in Scopus). *H. roretzi* is also an important commercial species in Korea and Japan. Since 1995, mass mortality of cultured *H. roretzi*, due to soft tunic syndrome caused by the flagellate *Azumiobodo hoyamushi*, has resulted in significant losses to the ascidian farming industry in Korea [37]. Thus, the detailed knowledge of its biology is an important issue from many points of view. In fact, despite the commercial and scientific importance of this species, the morphology of its larva has been poorly described. In particular, it has been long considered very similar to the well-described larvae of the genera *Ciona* and *Phallusia*, which are phlebobranchi model species [38]. These larvae are simple in their anatomy with no sign of adultation; the pharynx primordium occupies most of the trunk where the mesenchyme is organized in two lateral-ventral pockets [39]. The tail contains 40 notochord cells organized in a central row, flanked by 36 muscle cells on both sides. Dorsally to the notochord, the posterior neural tube is formed by a hollow of ependymal cells [27].

Indeed, it was already known that the larva of *H. roretzi* presents minimal caudalisation, since the tail has 42 rather than the conventional 36 or 38 tail muscle cells [40]. In this species, muscle cells are added to the posterior tip of the tail through the induction of more secondary muscle cells [41]. In this study, we show that the larva also presents minimal adultation as the atrial primordia are recognisable and the atrial siphon already developed, even if a thin layer of tunic occludes its opening. This peculiar condition has never been described before in the larva of *H. roretzi* but was documented in the larva of *B. schlosseri* [42]. During the embryogenesis of *B. schlosseri*, the atrial chamber originates as a dorsal invagination of the ectoderm, which sinks into the body and bifurcates over the nervous system to form the primordia of the peribranchial chambers. The latter descend ventrally, on both sides, adjacent to the pharyngeal walls. In the contact area, the gill slits perforate and extend to form parallel fissures, the protostigmata. In the phlebobranch *C. intestinalis*, the atrial primordia are paired; they develop as two dorsal, symmetrical invaginations of the ectoderm, one on the left and one on the right, and contact the posterior-most wall of the pharyngeal rudiment. In the mature larva, the two atrial invaginations are opened outward, representing two small chambers, at the bottom of which the protostigmata will perforate [25]. Each atrial chamber forms a rudiment of the atrial siphon that, at the end of metamorphosis, will eventually fuse together and open dorsally to form the unique atrial siphon of the adult. In the juvenile, the two atrial chambers will also converge dorso-medially and fuse with each other, thus creating the unpaired atrial cavity. This important difference in atria formation, together with the gonad location in the lateral body wall, differentiates stolidobranch species from other ascidians.

4.2. The Adhesive Papillae of H. roretzi Undergo Dynamic Changes during Adhesion

The papillae of *H. roretzi* were previously described as simple conic ones, similar to those of *C. intestinalis* and *P. mammillata*, even though a detailed description was never reported [27]. Indeed, our TEM analysis revealed that in swimming larvae, the papillae of *H. roretzi* are formed by the same cell types of those of *C. intestinalis*. In particular, the ACCs bear finger-like protrusions and vesicle-rich CCs encircle the ACCs. The PNSs do not reach the papilla apex but are in a more proximal position. This is in accordance with what has been observed in other ascidians species such as *Clavelina lepadiformis* and *Botrylloides leachii* [43,44]. It was proposed that the neuron protrusions reach the substrate and are, therefore, stimulated only once the papillae are attached to the substrate and retracted [44]. Differently from *C. intestinalis*, the hyaline cap, which is formed by the inner tunic compartment on the tip of the papillae, is not recognisable in *H. roretzi*.

Interestingly, our morphological analysis reveals that during settlement, the papillae of *H. roretzi* undergo dynamic changes in shape, assuming a cup-like appearance characterized by a central groove. The ACCs become shorter than in swimming larvae and have nuclei with irregular shapes, suggesting that they undergo a deep rearrangement. Overall, the papillae are less extended and the PNSs now reach the anterior endings in a position where they can contact the substrate. We hypothesize that some cell components are also degenerating. These observations, which recall those reported for species with "scyphate papillae with axial protrusions" [27], suggest that the adhesive organs of *H. roretzi* are more complex than previously supposed and deserve to be further analysed also with a molecular approach.

Supplementary Materials: The following are available online at https://www.mdpi.com/article/10.3390/jmse10010011/s1, Figure S1: Interactive PDF file of *H. roretzi* 3D reconstruction. Table S1: measurements of trunk length and width of *H. roretzi* larvae; Video S1: video of *H. roretzi* 3D reconstruction.

Author Contributions: Conceptualization, L.M. and R.P.; 3D reconstruction, F.C.; histology and TEM, L.M. and F.C.; morphometric analysis, S.M.; result interpretation and discussion, L.M., F.C., C.A., V.V., S.M. and R.P.; writing, R.P. and L.M. All authors have read and agreed to the published version of the manuscript.

Funding: This research received no external funding.

Institutional Review Board Statement: Not applicable for studies on tunicates.

Informed Consent Statement: Not applicable.

Acknowledgments: The authors thank Paolo Burighel for his helpful discussions, Hiroki Nishida for specimen collection assistance, Karla Palmieri for English revision and Emmanuel Faure for hosting the 3D model on MorphoNet.org.

Conflicts of Interest: The authors declare no conflict of interest.

Abbreviations

List of abbreviations present in the text.

a	atrium
ac	left atrial chamber rudiment
ACC	axial columnar cell
as	atrial siphon rudiment
ba	blood ampulla
bb	basal body of a cilium
CC	collocyte
dlap	dorsal left adhesive papilla
drap	dorsal right adhesive papilla
dg	dorsal groove
ep	epidermis
es	endodermal strand
itc	inner tunic compartment
lb	left bud
m	muscle
mc	mesenchymal cell
me	mesenchyme
mv	microvillus
n	nucleus
nc	notochord
ne	neck
oc	ocellus
ot	otolith
otc	outer tunic compartment
pc	peribranchial chambers
PSN	primary sensory neuron
te	trunk endoderm
tf	tail fin
st	stomodeum
sv	sensory vesicle
vap	ventral adhesive papilla
vg	visceral ganglion

References

1. Delsuc, F.; Brinkmann, H.; Chourrout, D.; Philippe, H. Tunicates and not cephalochordates are the closest living relatives of vertebrates. *Nature* **2006**, *439*, 965–968. [CrossRef]
2. Sepúlveda, R.; Rozbaczylo, N.; Ibáñez, C.; Flores, M.; Cancino, J. Ascidian-associated polychaetes: Ecological implications of aggregation size and tube-building chaetopterids on assemblage structure in the Southeastern Pacific Ocean. *Mar. Biodivers* **2014**, *45*, 733–741. [CrossRef]
3. Voultsiadou, E.; Kyrodimou, M.; Antoniadou, C.; Vafidis, D. Sponge epibionts on ecosystem-engineering ascidians: The case of Microcosmus sabatieri. *Estuar. Coast. Shelf Sci.* **2010**, *86*, 598–606. [CrossRef]
4. Mercurio, S.; Messinetti, S.; Manenti, R.; Ficetola, G.F.; Pennati, R. Embryotoxicity characterization of the flame retardant tris(1-chloro-2-propyl)phosphate (TCPP) in the invertebrate chordate *Ciona intestinalis*. *J. Exp. Zool. Part A Ecol. Integr. Physiol.* **2021**, *335*, 339–347. [CrossRef] [PubMed]
5. Messinetti, S.; Mercurio, S.; Pennati, R. Effects of bisphenol A on the development of pigmented organs in the ascidian *Phallusia mammillata*. *Invertebr. Biol.* **2018**, *137*, 329–338. [CrossRef]
6. Messinetti, S.; Mercurio, S.; Pennati, R. Bisphenol A affects neural development of the ascidian *Ciona robusta*. *J. Exp. Zool. Part A Ecol. Integr. Physiol.* **2019**, *331*, 5–16. [CrossRef]
7. Lambert, G.; Karney, R.C.; Rhee, W.Y.; Carman, M. Wild and cultured edible tunicates: A review. *Manag. Biol. Invasions* **2016**, *7*, 59–66. [CrossRef]
8. Coleman, F.C.; Williams, S.L. Overexploiting marine ecosystem engineers: Potential consequences for biodiversity. *Trends Ecol. Evol.* **2002**, *17*, 40–44. [CrossRef]

9. Wang, K.; Dantec, C.; Lemaire, P.; Onuma, T.A.; Nishida, H. Genome-wide survey of miRNAs and their evolutionary history in the ascidian, *Halocynthia roretzi*. *BMC Genom.* **2017**, *18*, 314. [CrossRef]
10. Brozovic, M.; Martin, C.; Dantec, C.; Dauga, D.; Mendez, M.; Simion, P.; Percher, M.; Laporte, B.; Scornavacca, C.; Di Gregorio, A.; et al. ANISEED 2015: A digital framework for the comparative developmental biology of ascidians. *Nucleic Acids Res.* **2016**, *44*, D808–D818. [CrossRef]
11. Lambert, G. Invasive sea squirts: A growing global problem. *J. Exp. Mar. Biol. Ecol.* **2007**, *342*, 3–4. [CrossRef]
12. Brown, F.D.; Swalla, B.J. Evolution and development of budding by stem cells: Ascidian coloniality as a case study. *Dev. Biol.* **2012**, *369*, 151–162. [CrossRef] [PubMed]
13. Alié, A.; Hiebert, L.S.; Scelzo, M.; Tiozzo, S. The eventful history of nonembryonic development in tunicates. *J. Exp. Zool. B Mol. Dev. Evol.* **2021**, *336*, 250–266. [CrossRef] [PubMed]
14. Berrill, N.J.; Watson, D.M.S.; II-Studies in Tunicate development. Part V-The evolution and classification of Ascidians. *Philos. Trans. R. Soc. Lond. B Biol. Sci.* **1936**, *226*, 43–70. [CrossRef]
15. Rubinstein, N.D.; Feldstein, T.; Shenkar, N.; Botero-Castro, F.; Griggio, F.; Mastrototaro, F.; Delsuc, F.; Douzery, E.J.P.; Gissi, C.; Huchon, D. Deep Sequencing of Mixed Total DNA without Barcodes Allows Efficient Assembly of Highly Plastic Ascidian Mitochondrial Genomes. *Genome Biol. Evol.* **2013**, *5*, 1185–1199. [CrossRef]
16. Tsagkogeorga, G.; Turon, X.; Hopcroft, R.R.; Tilak, M.-K.; Feldstein, T.; Shenkar, N.; Loya, Y.; Huchon, D.; Douzery, E.J.P.; Delsuc, F. An updated 18S rRNA phylogeny of tunicates based on mixture and secondary structure models. *BMC Evol. Biol.* **2009**, *9*, 187. [CrossRef]
17. DeBiasse, M.B.; Colgan, W.N.; Harris, L.; Davidson, B.; Ryan, J.F. Inferring Tunicate Relationships and the Evolution of the Tunicate Hox Cluster with the Genome of *Corella inflata*. *Genome Biol. Evol.* **2020**, *12*, 948–964. [CrossRef]
18. Lübbering, B.; Goffinet, G. Ultrastructural survey of tunic morphogenesis in the larval and young adult ascidian *Ascidiella aspersa* (Tunicata, Ascidiacea). *Belg. J. Zool.* **1991**, *121*, 39–53.
19. Cloney, R.A.; Cavey, M.J. Ascidian larval tunic: Extraembryonic structures influence morphogenesis. *Cell Tissue Res.* **1982**, *222*, 547–562. [CrossRef]
20. Cavey, M.J.; Cloney, R.A. Development of the larval tunic in a compound ascidian: Morphogenetic events in the embryos of *Distaplia occidentalis*. *Can. J. Zool.* **1984**, *62*, 2392–2400. [CrossRef]
21. Hotta, K.; Dauga, D.; Manni, L. The ontology of the anatomy and development of the solitary ascidian *Ciona*: The swimming larva and its metamorphosis. *Sci. Rep.* **2020**, *10*, 17916. [CrossRef]
22. Jeffery, W.R.; Swalla, B.J. Evolution of alternate modes of development in ascidians. *BioEssays* **1992**, *14*, 219–226. [CrossRef] [PubMed]
23. Zaniolo, G.; Manni, L.; Burighel, P. Ovulation and embryo-parent relationships in Botrylloides leachi (Ascidiacea, Tunicata). *Invertebr. Reprod. Dev.* **1994**, *25*, 215–225. [CrossRef]
24. Zaniolo, G.; Manni, L.; Brunetti, R.; Burighel, P. Brood pouch differentiation in *Botrylloides violaceus*, a viviparous ascidian (Tunicata). *Invertebr. Reprod. Dev.* **1998**, *33*, 11–23. [CrossRef]
25. Manni, L.; Lane, N.J.; Joly, J.-S.; Gasparini, F.; Tiozzo, S.; Caicci, F.; Zaniolo, G.; Burighel, P. Neurogenic and non-neurogenic placodes in ascidians. *J. Exp. Zool. B. Mol. Dev. Evol.* **2004**, *302*, 483–504. [CrossRef]
26. Kowarsky, M.; Anselmi, C.; Hotta, K.; Burighel, P.; Zaniolo, G.; Caicci, F.; Rosental, B.; Neff, N.F.; Ishizuka, K.J.; Palmeri, K.J.; et al. Sexual and asexual development: Two distinct programs producing the same tunicate. *Cell Rep.* **2021**, *34*, 108681. [CrossRef]
27. Burighel, P.; Cloney, R.A. Urochordata: Ascidiacea. In *Microscopic Anatomy of Invertebrates, Vol 15. Hemichordata, Chaetognatha, and the Invertebrate Chordates*; Harrison, F.W., Rupert, E.E., Eds.; Wiley-Liss, Inc.: New York, NY, USA, 1997; Volume 15.
28. Dolcemascolo, G.; Pennati, R.; De Bernardi, F.; Damiani, F.; Gianguzza, M. Ultrastructural comparative analysis on the adhesive papillae of the swimming larvae of three ascidian species. *Invertebr. Surviv. J.* **2009**, *6*, S77–S86.
29. Caicci, F.; Zaniolo, G.; Burighel, P.; Degasperi, V.; Gasparini, F.; Manni, L. Differentiation of papillae and rostral sensory neurons in the larva of the ascidian *Botryllus schlosseri* (Tunicata). *J. Comp. Neurol.* **2010**, *518*, 547–566. [CrossRef]
30. Pennati, R.; Ficetola, G.F.; Brunetti, R.; Caicci, F.; Gasparini, F.; Griggio, F.; Sato, A.; Stach, T.; Kaul-Strehlow, S.; Gissi, C.; et al. Morphological Differences between Larvae of the *Ciona intestinalis* Species Complex: Hints for a Valid Taxonomic Definition of Distinct Species. *PLoS ONE* **2015**, *10*, e0122879. [CrossRef]
31. Brunetti, R.; Gissi, C.; Pennati, R.; Caicci, F.; Gasparini, F.; Manni, L. Morphological evidence that the molecularly determined *Ciona intestinalis* type A and type B are different species: *Ciona robusta* and *Ciona intestinalis*. *J. Zool. Syst. Evol. Res.* **2015**, *53*, 186–193. [CrossRef]
32. Zeng, F.; Wunderer, J.; Salvenmoser, W.; Ederth, T.; Rothbächer, U. Identifying adhesive components in a model tunicate. *Philos. Trans. R. Soc. B Biol. Sci.* **2019**, *374*, 20190197. [CrossRef]
33. Horie, T.; Kusakabe, T.; Tsuda, M. Glutamatergic networks in the *Ciona intestinalis* larva. *J. Comp. Neurol.* **2008**, *508*, 249–263. [CrossRef] [PubMed]
34. Leggio, B.; Laussu, J.; Carlier, A.; Godin, C.; Lemaire, P.; Faure, E. MorphoNet: An interactive online morphological browser to explore complex multi-scale data. *Nat. Commun.* **2019**, *10*, 2812. [CrossRef] [PubMed]
35. Manni, L.; Pennati, R. Tunicata. In *Structure and Evolution of Invertebrate Nervous Systems*; Schmidt-Rhaesa, G.P.A., Harzsch, S., Eds.; Oxford University Press: Oxford, UK, 2016. [CrossRef]

36. Manni, L.; Agnoletto, A.; Zaniolo, G.; Burighel, P. Stomodeal and neurohypophysial placodes in *Ciona Intestinalis: Insights* into the origin of the pituitary gland. *J. Exp. Zool. Part B Mol. Dev. Evol.* **2005**, *304*, 324–339. [CrossRef]
37. Kumagai, A.; Ito, H.; Sasaki, R. Detection of the kinetoplastid *Azumiobodo hoyamushi,* the causative agent of soft tunic syndrome, in wild ascidians *Halocynthia roretzi*. *Dis. Aquat. Organ.* **2013**, *106*, 267–271. [CrossRef] [PubMed]
38. Hudson, C.; Yasuo, H. Similarity and diversity in mechanisms of muscle fate induction between ascidian species. *Biol. Cell.* **2008**, *100*, 265–277. [CrossRef] [PubMed]
39. Katz, M.J. Comparative anatomy of the tunicate tadpole, *Ciona intestinalis*. *Biol. Bull.* **1983**, *164*, 1–27. [CrossRef]
40. Passamaneck, Y.J.; Di Gregorio, A. *Ciona intestinalis*: Chordate development made simple. *Dev. Dyn.* **2005**, *233*, 1–19. [CrossRef] [PubMed]
41. Nishida, H. Determinative mechanisms in secondary muscle lineages of ascidian embryos: Development of muscle-specific features in isolated muscle progenitor cells. *Development* **1990**, *108*, 559–568. [CrossRef] [PubMed]
42. Manni, L.; Lane, N.J.; Zaniolo, G.; Burighel, P. Cell reorganisation during epithelial fusion and perforation: The case of ascidian branchial fissures. *Dev. Dyn.* **2002**, *224*, 303–313. [CrossRef]
43. Pennati, R.; Zega, G.; Groppelli, S.; De Bernardi, F. Imunohistochemical analysis of the adhesive papillae of *Botrylloides leachi* (Chordata, Tunicata, Ascidiacea): Implications for their sensory function. *Ital. J. Zool.* **2007**, *74*, 325–329. [CrossRef]
44. Pennati, R.; Groppelli, S.; De Bernardi, F.; Mastrototaro, F.; Zega, G. Immunohistochemical analysis of adhesive papillae of *Clavelina lepadiformis* (Müller, 1776) and *Clavelina phlegraea* (Salfi, 1929) (Tunicata, Ascidiacea). *Eur. J. Histochem.* **2009**, *53*, e4. [CrossRef] [PubMed]

Journal of
Marine Science and Engineering

MDPI

Article

Searching for the Origin and the Differentiation of Haemocytes before and after Larval Settlement of the Colonial Ascidian *Botryllus schlosseri*: An Ultrastructural Viewpoint

Francesca Cima

Laboratory of Ascidian Biology, Department of Biology (DiBio), University of Padova, Via U. Bassi 58/B, 35131 Padova, Italy; francesca.cima@unipd.it; Tel.: +39-49-827-6198

Abstract: The colonial ascidian *Botryllus schlosseri* possesses an innate immunity, which plays fundamental roles in its survival, adaptability, worldwide spread and ecological success. Three lines of differentiation pathways of circulating haemocytes are known to be present in the haemolymph, starting from undifferentiated haemoblasts: (i) the phagocytic line (hyaline amoebocytes and macrophage-like cells), (ii) the cytotoxic line (granular amoebocytes and morula cells) and (iii) the storage cell line (pigment cells and nephrocytes). Many questions remain about their origin, and thus, observations during various stages of development were undertaken in this study. Haemocytes were detected beginning from the early tailbud embryo stage. Haemoblasts were always present and morula cells were the first differentiated haemocytes detected. In both the next stage, just before hatching, and the swimming tadpole larva stage, hyaline amoebocytes and pigment cells were also recognisable. Some morula cells containing active phenoloxidase migrated from the haemolymph into the tunic after having crossed the epidermis, and this behaviour could be related to the preparation of a defensive function for spatial competition. During larval metamorphosis, macrophage-like cells appeared with their phagosomes positive to acid phosphatase activity and containing apoptotic cells from tail tissue degeneration. After metamorphosis, in the filter-feeding oozoid stage, nephrocytes involved in nitrogen catabolism finally appeared. In both the subendostylar sinus and the peripheral blind-sac vessels (ampullae), clusters of haemoblasts were recognisable, some of which showed incipient specialisations, considering the hypothesis of the presence of putative niches of haemolymph stem cells.

Keywords: ascidians; *Botryllus schlosseri*; development; enzyme histochemistry; haematopoiesis; haemocytes; metamorphosis; tunicates; ultrastructure

Citation: Cima, F. Searching for the Origin and the Differentiation of Haemocytes before and after Larval Settlement of the Colonial Ascidian *Botryllus schlosseri*: An Ultrastructural Viewpoint. *J. Mar. Sci. Eng.* **2022**, *10*, 987. https://doi.org/10.3390/jmse10070987

Academic Editor: Francesco Tiralongo

Received: 30 May 2022
Accepted: 17 July 2022
Published: 19 July 2022
Corrected: 29 August 2022

1. Introduction

The haemocytes of many ascidian species that are protagonists of soft fouling—such as the cosmopolitan colonial species *Botryllus schlosseri* [1,2]—have long been studied and characterised, from a functional point of view, as being responsible for the defence responses of innate immunity, from which the survival of individuals and the ability to compete for the substratum after settlement derive [3,4]. In ascidians, which are the closest relatives to vertebrates [5,6], haemocytes are responsible for all vital activities because they are involved in nutrition [7], excretion [8], asexual reproduction [9–13] and part of tunic formation [14]. In defence responses [15], haemocytes are able to phagocytise foreign materials [16–20]; release opsonins [21], cytokines [22–24], agglutinins [25,26], complement factors [27,28] and antimicrobial peptides [29]; actively participate with cytotoxic substances in inflammatory reactions [30]; and are involved in allorecognition between contacting colonies that are genetically incompatible [31–35].

The haemocytes of ascidians, due to their high polymorphism, have led to a very complex and uncertain classification by various authors. However, in the case of *B. schlosseri*, the numerous cell types have finally been grouped into differentiation pathways on the basis of their ultrastructural features [36,37] and cytochemical and morpho-functional properties [38]. On this basis, three haemocyte lines have been defined, the origin of which has been hypothesised from undifferentiated cells (Figure 1). The first two are involved in defence responses ('immunocytes'), and the latter is involved in catabolite storage and excretion: (i) phagocytic cell lines, represented by hyaline amoebocytes (previously known as 'microgranular amoebocytes' [36]) and macrophage-like cells; (ii) cytotoxic cell lines, represented by granular amoebocytes (previously known as 'macrogranular amoebocytes' [36]) and morula cells; and (iii) storage cell lines, represented by pigment cells and nephrocytes.

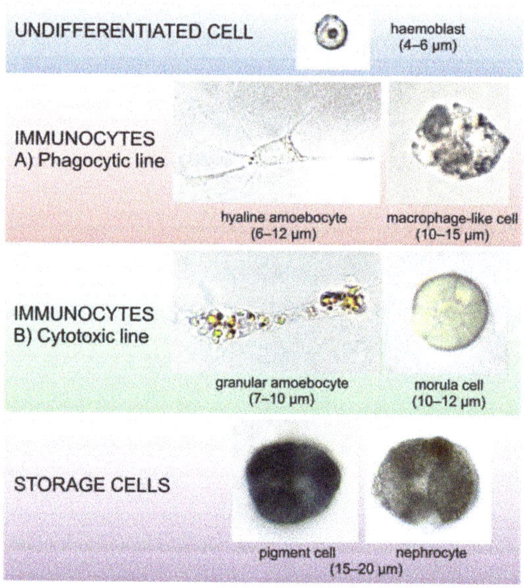

Figure 1. Classification of circulating morphotypes in the haemolymph of adult individuals (i.e., zooids) of the colonial ascidian *Botryllus schlosseri*. Representative living haemocytes shown with their size range.

The hypotheses concerning the origin of the various differentiation pathways in ascidians, *B. schlosseri* included, are currently still controversial [8,39–42], especially regarding both the moment of appearance of the various cell types during embryo and larva development and the search for stem cells. The most likely candidate of the haemocyte stem cell seems to be represented by the haemoblast (Figure 1), a circulating undifferentiated cell recognised in *B. schlosseri* by the anti-CD34 antibody, which has been employed as a marker because CD34 is a transmembrane phosphoglycoprotein of the haematopoietic stem and progenitor cells of vertebrates [38]. During ascidian embryogenesis, haemocytes originate from trunk mesenchymal cells derived from the A7.6 blastomere [43], and the differentiation of haemocytes begins from the developmental stage of the early tailbud embryo. For example, in *Clavelina picta* [44] and *Ecteinascidia turbinate* [45] haemocytes have been observed migrating and distributing around the coelom in contact with the endoderm. Starting from the late tailbud embryo stage, before the hatching of the tadpole larva, the real differentiation of haemocytes begins, which is completed during the period of free life and metamorphosis. The activation of many innate immunity genes during metamorphosis, and trans-epidermal haemocyte migrations into the tunic during both the swimming larval period and the settlement, have been previously reported in *Boltenia villosa* [46,47]. These events were probably related to (i) the programmed maturation of the adult immune system, (ii) the high phagocytosis activity during tissue remodelling at metamorphosis, and (iii) the potential inflammatory response to environmental settlement cues. The latter are mainly due to the influence of the surface characteristics, e.g., the microbial film or biofilm, as observed in ascidians and other fouling invertebrates during settlement on the substratum [48–51].

After metamorphosis, a true haematopoietic tissue is less clear in ascidian adults. Regions with clusters of haemogenic cells ('lymphatic nodules') have been recognised both in the crossbars of the branchial and connective tissue surrounding the intestine of *Ciona intestinalis* and *Styela clava* and in the blood vessels in *Polyclinum planum* and *Pyura austor* [52–54]. A lymphatic or haematopoietic nodule is formed of undifferentiated cells in the centre, represented by haemoblasts, which are externally surrounded by differentiating blood cells [52].

The question of the origin of differentiation pathways during embryo-larval development was the main purpose of the present paper. It was dealt with using the cosmopolitan colonial ascidian *B. schlosseri*, and was concerned with the embryo, larva and oozoid, with the latter considered the result of metamorphosis and the founding individual of the colony. The choice of this species was made because it is ovoviviparous and easily reared in the laboratory. Entire embryonic development can be studied within parental individuals after obtaining colonies with embryos at well-determined stages of development [55–57]. Moreover, the metamorphosis stages of the hatched tadpole larva can be easily recognised [58]. In particular, the following stages of embryo–larval development have been considered: (i) early tailbud embryo, with the tail forming an initial turn around the cephalenteron; (ii) late tailbud embryo, with the tail at its maximum extension (1.5 turns around the cephalenteron); (iii) newly hatched tadpole larva (early metamorphosis); (iv) late metamorphosis larva; and (v) oozoid. The occurrence of various types of haemocytes was studied by means of ultrastructural observations, including characterisation of the enzymatic content, taking into account the functional peculiarities of the differentiation pathways of the immunocytes. The activity of lysosomal acid phosphatase has been assayed as a marker of the phagocytic line, and that of phenoloxidase has been assayed as a marker of the cytotoxic line. Finally, since the adult of *B. schlosseri*, despite belonging to the family Styelidae, does not possess haematopoietic or lymphatic nodules, the presence and localisation of the accumulation and differentiation zones of haemoblasts have been investigated inside the oozoid.

2. Materials and Methods

2.1. Animals

The animals used in this study were colonies and larvae of *B. schlosseri* from our laboratory cultures and wild colonies from the Lagoon of Venice. Colonies were kept in aerated aquaria attached to 5 × 5 cm glass slides filled with filtered sea water (salinity of 35 ± 1 psu, temperature of 19 ± 0.5 °C, pH of 8.1), which was renewed every other day. They were fed with microalgae mixtures. Stages of embryonic development from fragmented colonies and larvae were easily identified under a dissecting binocular stereomicroscope (Wild Heerbrugg Ltd., Heerbrugg, Switzerland) with 50× maximum magnification.

2.2. Transmission Electron Microscopy

Selected colonies and larvae were fixed for 2 h at 4 °C in 1% glutaraldehyde, buffered with 0.2 M sodium cacodylate buffer (CAB) containing 1.7% NaCl at a pH of 7.4 plus 1% caffeine to prevent the leakage of phenols contained inside morula cell vacuoles [59]. Since single zooids are 1–2 mm long in the colony, after washing in buffer with 1.7% NaCl, small fragments (approximately 5–6 mm) of the colonies were cut and inserted into glass vials, whereas entire larvae were inserted inside tubes with a plankton filter at the bottom. After postfixation in 1% OsO_4 in CAB, all specimens were dehydrated and embedded in Epon 812 (Fluka Chemie, Buchs, Switzerland). Thick sections (1 μm), obtained using an LKB ultrotome, were counterstained with 1% toluidine blue and observed under an Olympus CX31 light microscope equipped with a DV Lumenera Infinity 2 and Infinity Capture

Application software version 5.0.0 (Lumenera Co. 2002–2009, Ottawa, ON, Canada) for image capturing. Thin sections (80 nm) were briefly stained with uranyl acetate and lead citrate. Micrographs were taken with an FEI Tecnai G^2 transmission electron microscope operating at 75 kV.

2.3. *Histoenzymatic Assays*

Both intact larvae and lacerated colonies were preincubated in 0.1% Triton X-100 (Sigma-Aldrich, St. Louis, MO, USA) in CAB for 10 min. For each enzymatic assay, control specimens were prepared by omitting the specific enzyme substrate. The following assays are reported in brief according to the methods described in detail by Cima [60].

2.3.1. Assay for Acid Phosphatase

Among the hydrolytic enzymes, acid phosphatase was chosen for its presence in the lysosomes and phagosomes of the phagocytic line. Fixed specimens were incubated for 15 h at 30 °C in the following reaction mixture [61]: 5 mg of naphtol AS-BI phosphate (Sigma-Aldrich) previously dissolved in 200 μL of dimethylformamide (DMF); 200 μL of solution A (0.4 g of hexazonium p-rosanilin, Fluka dissolved in 2 mL of 36% HCl and 8 mL of distilled water); 200 μL of solution B (4% $NaNO_2$ in distilled water); and 10 mL of 0.1 M sodium acetate buffer plus 1.7% NaCl and 1% saccharose at a pH of 5.2. Positive sites were stained red under light microscopy and revealed an electron-dense granular precipitate under electron microscopy.

2.3.2. Assay for Phenoloxidase

Among the oxidative enzyme activities, phenoloxidase was chosen as the enzyme typical of the cytotoxic line. The enzyme substrate was β-(3,4-dihydroxyphenyl)-L-alanine (L-DOPA), purchased from Sigma-Aldrich. Fixed specimens were incubated for 15 h at 4 °C with L-DOPA-saturated 0.01 M CAB plus 5 mM $CaCl_2$ (CAB-Ca) at a pH of 7.0 [62], and then for 1 h at 37 °C in the same renewed medium before dehydration and embedding. In this case, postfixation in OsO_4 was omitted because of the high reducing activity inside morula cell vacuoles, which causes osmium precipitation and impedes the detection of electron-dense positive sites under electron microscopy.

3. Results

3.1. *Early Tailbud Embryo*

This stage immediately follows that of the neurula with the closure of the neuropore. The embryo appears elongated, and the cephalenteron region and tail can be distinguished, the latter appearing short and bending at 90° to the right side (Figure 2A). The chordamesodermal cells differentiate into cells of the notochord, whereas those of the mesoderm differentiate into myocytes in the tail and into a group of mesenchymal cells that will later give rise to the pericardium, heart and mantle mesenchyme. The archenteron cavity is also recognisable [56,63].

This is the first stage of embryonic development in which differentiating haemocytes are detected and distinguished with respect to the germ layers. They are mainly observed in the haemocoel, the space between the epidermis and the organ primordia (Figure 2B,C), and they are represented by haemoblasts and putative precursors of morula cells.

Haemoblasts appear as small cells (2–5 μm in diameter) resembling undifferentiated cells. The large nucleus occupies almost the entire volume of the cell and is placed in a central position, with chromatin mainly scattered or organised in small clusters close to the nuclear envelope. Within the nucleus, a nucleolus is recognisable with a granular structure. The plasmalemma shows an irregular contour, and the cytoplasm, which is represented by a thin peripheral layer, has no details of differentiation. Only some granules of the yolk, mitochondria and ribosomes, the latter both organised into polysomes and adherent to small cisternae of the rough endoplasmic reticulum (RER), can be found (Figure 2D).

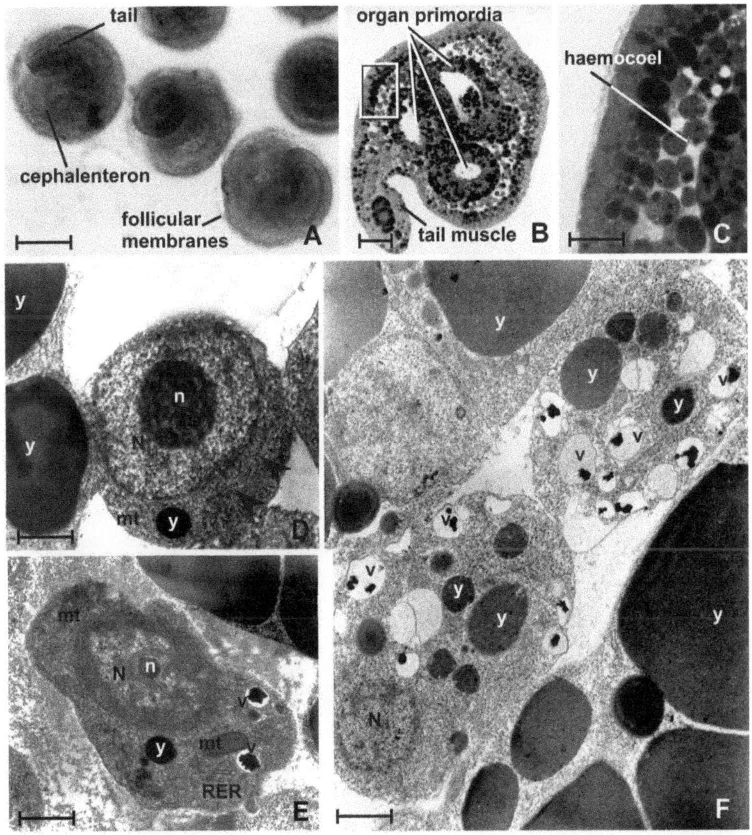

Figure 2. Early tailbud embryos: (**A**) embryos collected from parental zooids of the colony and viewed in toto; (**B**) longitudinal thick section of an embryo stained with toluidine blue; (**C**) detail of (**B**) showing numerous haemocytes in the haemocoel; (**D**) haemoblast showing scant cisternae of RER (*arrowheads*), mitochondria and yolk granules in the cytoplasm; and (**E**,**F**) putative precursors of morula cells with small vacuoles containing electron-dense aggregates. *mt*—mitochondrion, *N*—nucleus, *n*—nucleolus, *v*—vacuole, and *y*—yolk granule. Bar length: 150 μm in (**A**); 30 μm in (**B**); 10 μm in (**C**); 1.4 μm in (**D**–**F**); and 2 μm in (**E**).

Morula cells appear as vacuolated cells and have shared features with haemoblasts, such as a large spherical nucleus with a nucleolus (Figure 2E) and yolk granules in the cytoplasm. Vacuoles are small and scant and contain the typical material of this cell type, and form strongly electron-dense aggregates due to their high osmiophilicity (Figure 2F). In addition, long RER profiles and mitochondria with parallel cristae are present.

3.2. Late Tailbud Embryo

The tail is completely turned around the cephalenteron of the embryo. The first muscular movements of the tail begin [56,63]. Both the larval organs (such as the larval nervous system and photolith) and primordia of future organs of the adult (such as the great expansion of the peribranchial chambers that converge medially and an evident intestine) can be recognised (Figure 3A,B). Haemocytes are very numerous in the defined spaces of the haemocoel (Figure 3C). At this stage, haemoblasts, morula cells (Figure 3D) and, for the first time, hyaline amoebocytes and storage cells can be found.

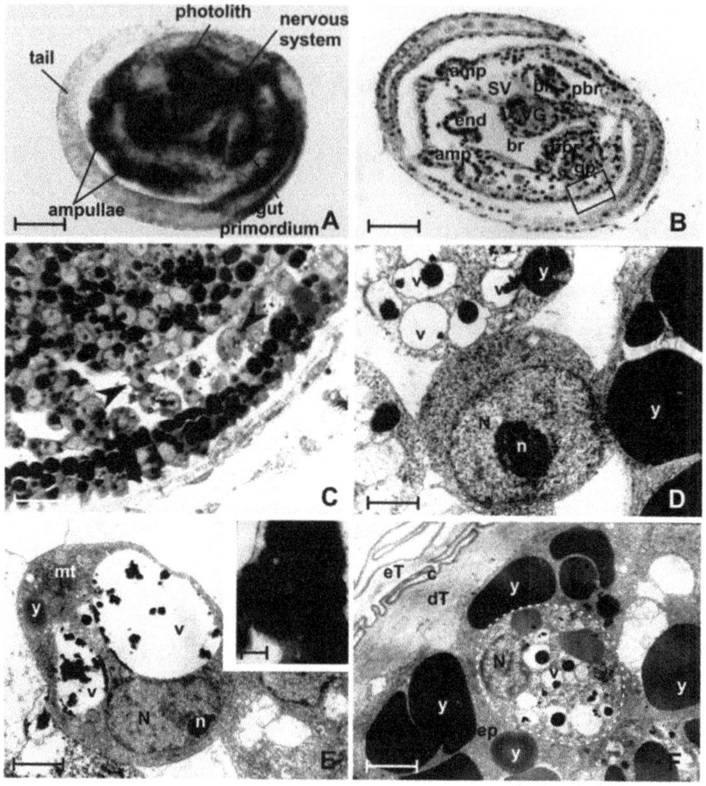

Figure 3. Late tailbud embryos: (**A**) embryo collected from a parental zooid of the colony and viewed in toto; (**B**) longitudinal thick section of an embryo showing both larval and adult organ primordia; (**C**) detail of (**B**) showing various types of haemocytes (*arrowheads*) in the haemocoel; (**D**) haemoblast (*in the centre*) and morula cell (*on the top*); (**E**) morula cell with a few vacuoles containing electron-dense aggregates occasionally clustered and merged (*inset*); and (**F**) morula cell (*dotted circle*) migrating from the haemocoel across the epidermis towards the definitive tunic, which can be distinguished from the embryonic tunic above for the presence of an electron-dense papillose cuticle. *amp*—ampulla, *br*—branchial chamber primordium, *c*—tunic cuticle, *dT*—definitive tunic, *end*—endostyle primordium, *ep*—epidermis, *eT*—embryonic tunic, *gp*—gut primordium, *mt*—mitochondrion, *N*—nucleus, *n*—nucleolus, *pbr*—peribranchial chamber primordium, *SV*—sensory vesicle, *v*—vacuole, *VG*—visceral ganglion, and *y*—yolk granule. Bar length: 60 μm in (**A**) and (**B**); 12 μm in (**C**); 1.7 μm in (**D**,**E**); 0.3 μm in inset of (**E**); and 2.7 μm in (**F**).

Morula cells occur with their typical features, i.e., large vacuoles containing strongly electron-dense granules (Figure 3E). The cytoplasm and nucleus are displaced into the spaces among the vacuoles. However, as a feature of young cells, the nucleus is still large and has a nucleolus; moreover, in the cytoplasm, mitochondria and cisternae of the RER, as well as some granules of yolk can still be observed. Beginning from this developmental stage, some morula cells are observed to cross the epidermis and migrate into the definitive tunic. The latter is secreted by the epidermis and is placed below the embryonic tunic, which is secreted by the egg envelopes. Moreover, the definitive tunic displays a cuticle, which is densely covered by small electron-dense spherical papillae (Figure 3F).

Hyaline amoebocytes are cells characterised by a large nucleus, occasionally reniform with clustered chromatin (Figure 4A). The cytoplasm contains scant granules of yolk and small granules of approximately 0.3–0.7 μm in diameter bound by an undulated membrane, as well as containing homogeneous and moderately electron-dense material (inset in Figure 4A). Small, clear vesicles of approximately 0.2 μm in diameter and many parallel cisternae of the RER, mitochondria and some multivesicular bodies are also present.

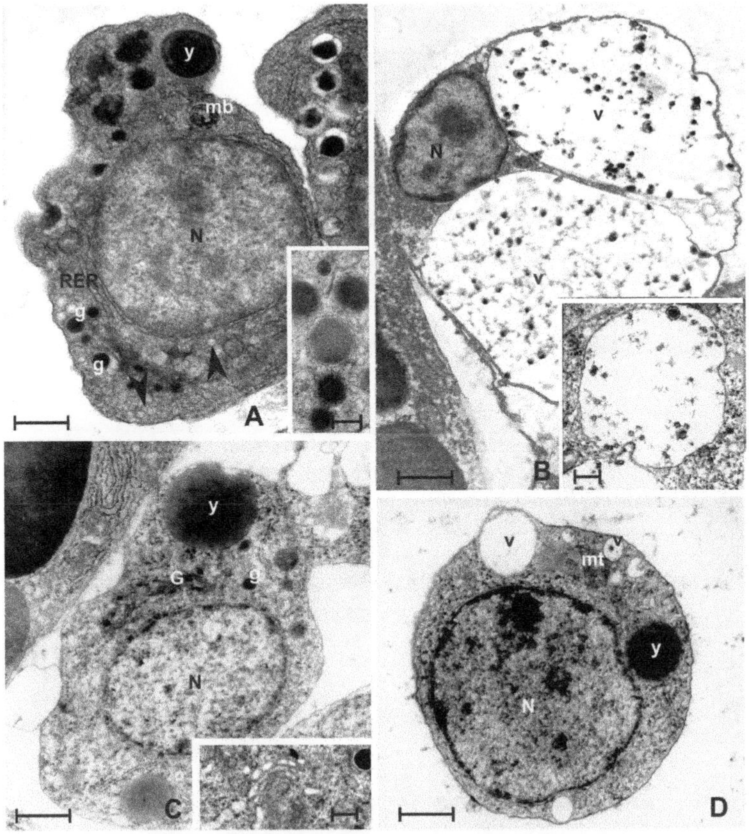

Figure 4. Late tailbud embryos: (**A**) hyaline amoebocyte with a large, reniform or horseshoe-shaped nucleus. Membrane-bound granules contain homogeneous and electron-dense material (*inset*). Numerous small, clear vesicles (*arrowheads*) are present in the cytoplasm; (**B**) pigment cell with large vacuoles containing few stratified pigment granules of 'blue phenotype' [1,37] (*inset*); (**C**) putative precursor of hyaline amoebocyte for the occurrence of scant, membrane-bound granules in the cytoplasm of a small cell (see text for details); and (**D**) putative precursor of morula cell, in which small vacuoles with highly electron-dense aggregates are forming. *g*—membrane-bound granule, *G*—Golgi apparatus, *mb*—multivesicular body, *mt*—mitochondrion, *N*—nucleus, *RER*—cisternae of rough endoplasmic reticulum, *v*—vacuole, and *y*—yolk granule. Bar length: 1 μm in (**A**,**C**,**D**); 1.7 μm in (**B**); 0.3 μm in inset of (**A**); 0.5 μm in inset of (**B**); and 0.2 μm in inset of (**C**).

Storage cells (Figure 4B) are only represented by pigment cells. They have large vacuoles of 3.5 μm in diameter containing small pigment crystals (approximately 180 nm in length) formed of overlapping electron-dense layers (inset in Figure 4B). In the cytoplasm, numerous mitochondria, free ribosomes and scant yolk granules are present.

Various putative precursors of hyaline amoebocytes (Figure 4C) and morula cells (Figure 4D) are also frequent. Both have characteristics similar to haemoblasts, that is, 5 μm in size, a high ratio of nucleus/cytoplasm, and the presence of a nucleolus in the nucleus. Moreover, the putative precursors of hyaline amoebocytes show, in the cytoplasm, small electron-dense membrane-bound granules and scattered clear vesicles, the latter budding from the Golgi apparatus (inset in Figure 4C). The putative precursors of morula cells have vacuoles with strongly electron-dense material inside (Figure 4D).

3.3. *Newly Hatched Tadpole Larva (Early Metamorphosis)*

The swimming larva represents the dispersal phase of the species (Figure 5A). It is anteriorly equipped with three transient sensory adhesive organs, namely 'papillae', for substratum detection and larval settlement, before the beginning of metamorphosis to the filter-feeding adult, i.e., the 'oozoid'. In the anterior part of cephalenteron, the larva has eight compacted ampullae, i.e., the primordium of the colonial vascular network. Ampullae are blind-sac structures that are, afterwards, involved both in the process of settlement and in the formation of the circulatory system of the colony [64–66]. They arise, before larval hatching, from a haemocoelic lacuna ventrally placed with respect to the branchial basket primordium. Subsequently, the ampullae spread into the newly formed definitive tunic and, during metamorphosis, protrude away from the area of origin, remaining connected to it through thin vascular peduncles. At the end of metamorphosis, the peduncles merge into a single vessel that is arranged to form an arch around the oozoid, namely, the marginal vessel, which communicates with the eight peripheral ampullae that are spreading onto the substratum [64]. As a clear sign of the beginning of metamorphosis, the epidermis thickens, and the cells of the notochord and musculature of the proximal area of the tail, are disrupted and begin to accumulate into the cephalenteron [67].

At this stage, haemoblasts, morula cells, hyaline amoebocytes and pigment cells continue to be present (Figure 5B). In the cytoplasm of these cells, the yolk granules are small and very scarce. Morula cells still have extensive RER and many mitochondria in the cytoplasm. Beginning from this stage, intense enzymatic activity of phenoloxidase in the vacuoles of some morula cells is revealed (Figure 5C). The specific histochemical labelling highlights the strong electron-density of the material inside the vacuoles. In the tunic, in addition to morula cells (Figure 5D), some granulocytes are detected (Figure 5E). The latter are large cells with a spherical nucleus in the peripheral position. In the cytoplasm, numerous large membrane-bound granules of 1.5 μm in diameter are recognised, the content of which shows various degrees of homogeneity and electron-density, occasionally with a more dense core. The salient feature of this cell type is the presence of a Golgi apparatus, which is enormously developed and displaced in the central area of the cytoplasm. Numerous cisternae of the RER and scant small mitochondria are also visible. This cell strongly resembles the cuboidal cells of the apical epithelium of the ampullae (Figure 5F).

3.4. *Late Larval Metamorphosis*

Tail resorption occurs for two thirds of its length, involving dismantlement of the notochord, posterior neural tube and musculature. Anteriorly, ampullae begin to expand on the substratum, contributing to the formation of a bell-shaped cephalenteron (Figure 6A). At this stage, heart contractions and haemolymph circulation begin. Almost all cell types of the haemolymph characteristic of the species are present and have a more advanced degree of differentiation in comparison with the previous stages. For example, morula cells (Figure 6B) show numerous large vacuoles with electron-dense content; however, a nucleolus is still present, and storage cells (Figure 6C)—still represented only by pigment cells—become remarkable in size (10 μm in diameter), although they are still smaller than the pigment cells observed in adults (maximum 20 μm in diameter). The nucleus is pushed to the periphery because the cell volume is almost entirely occupied by vacuoles variable in size, which often protrude towards the plasmalemma, deform the surface of the cell and contain small, multilayered granules.

In addition, the first occurrence of macrophage-like cells can be detected. They are large scavenger cells (about 13 μm in diameter) mainly involved in tail resorption by engulfing the dead cells detaching from the tissues. In this cell type, the large phagosomes displace the nucleus at the cell periphery and appear to mainly contain both necrotic masses and membrane systems, arranged to form myelin figures. In these heterophagic vacuoles, an intense activity of acid phosphatase is found inside their content, which appears to be labelled by an electron-dense product of the reaction (Figure 6D). Moreover, numerous cells with intermediated characteristics between those of hyaline amoebocytes and macrophage-like cells are observed (Figure 6E). They are represented by a medium-sized cell type (about 6 µm in diameter) and are variable in shape for their amoeboid activity and phagocytic function. The features shared with hyaline amoebocytes are a central nucleus, scattered clear vesicles, cisternae of the RER and moderately electron-dense membrane-bound granules in the cytoplasm (inset in Figure 6E). The features shared with macrophage-like cells are the presence of many digestive vacuoles with heterogeneous content (phagosomes) in the cytoplasm, which are positive to acid phosphatase activity (data not shown) and smaller than those of the macrophage-like cells.

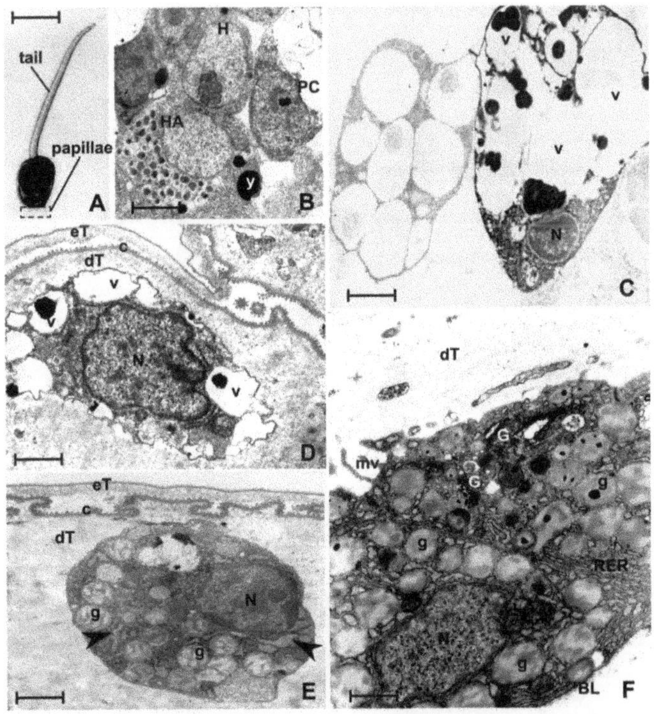

Figure 5. Larvae at early metamorphosis: (**A**) swimming larva viewed in toto. Note the presence of the anterior papillae, i.e., the organs involved in the settlement on the substratum; (**B**) haemoblasts (H), hyaline amoebocytes (HA) and pigment cells (PC) crowding in the haemocoel. Scant yolk granules are still recognisable in their cytoplasm; (**C**) two morula cells after the histochemical assay for phenoloxidase activity. The one on the left shows clear, i.e., not labelled, floccular granules and the one on the right reveals the vacuolar content positive to enzyme activity, which appears as highly electron-dense aggregates; (**D**) morula cell resident in the definitive tunic; and (**E**,**F**) granulocyte resident in the tunic (**E**) and cuboidal cells of the apical epithelium of the ampullae (**F**). Both share numerous large, membrane-bound granules showing a content with various degrees of homogeneity and electron-density, occasionally with a denser core, a well-developed Golgi apparatus and widened RER cisternae (*arrowheads*). *BL*—basal lamina, *c*—tunic cuticle, *dT*—definitive tunic, *eT*—embryonic tunic, *g*—granule, *G*—Golgi apparatus, *mv*—microvilli, *N*—nucleus, *RER*—cisternae of rough endoplasmic reticulum, *v*—vacuole, and *y*—yolk granule. Bar length: 350 µm in (**A**); 2.5 µm in (**B**); 1.5 µm (**C**) and (**D**); 1.6 µm in (**E**); and 1 µm in (**F**).

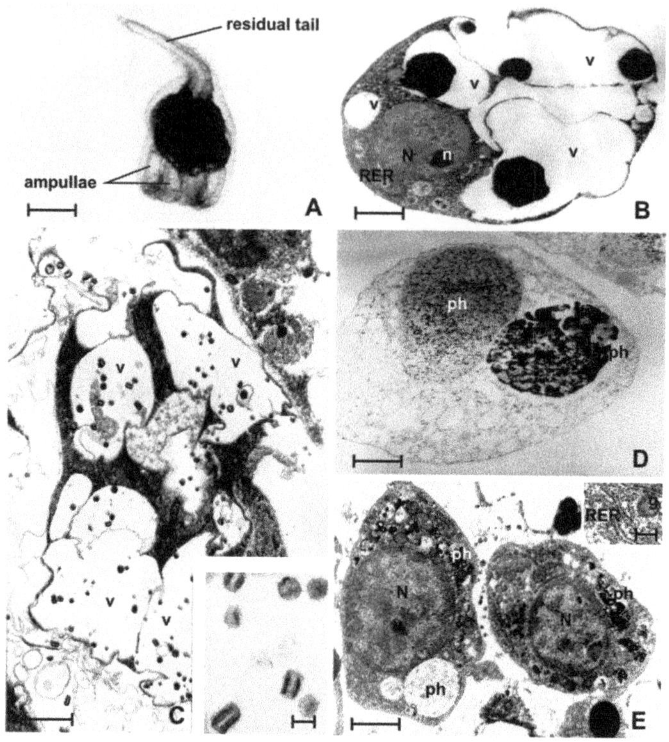

Figure 6. Late larval metamorphosis; (**A**) metamorphosing bell-shaped larva viewed in toto, showing partial resorption of the tail and protrusion of the ampullae; (**B**) morula cell; (**C**) pigment cell with evident pigment crystals of 'blue phenotype' inside vacuoles (*inset*); (**D**) macrophage-like cell. Phagosomes contain heterogeneous engulfed material and cell debris, and are labelled by the histochemical assay for the hydrolytic enzyme, acid phosphatase; and (**E**) two cells with intermediate characteristics between hyaline amoebocyte and macrophage-like cell (see text for details). *g*—membrane-bound granule, *N*—nucleus, *n*—nucleolus, *ph*—phagosome, *RER*—cisternae of rough endoplasmic reticulum, and *v*—vacuole. Bar length: 200 µm in (**A**); 1.2 µm in (**B**); 1 µm in (**C**) and (**E**); 1.7 µm in (**D**); 0.1 µm in inset of (**C**); and 0.2 µm in inset of (**E**).

3.5. Oozoid

The tail is not yet recognisable due to its complete resorption. The eight ampullae continue to protrude onto the substratum with the formation of the marginal vessel (Figure 7A). A blastogenetic bud is already visible on the right side that will later form a new zooid. On the left side, only a primordium of a bud appears, which will be reabsorbed. Siphons open, and the activity of filtration begins. The oozoid is the founding individual of the new colony and possesses all cell types of the adult tissues, as well as those of the haemolymph.

In particular, haemoblasts tend to accumulate in some areas at the level of the subendostylar sinus (Figure 7B,C) and along the lateral epithelium (i.e., the epidermis) of the peripheral ampullae (Figure 7D). Inside the ampullae, haemoblasts appear to closely adhere to each other (Figure 7E,F) and to the basal lamina of the epithelium (Figure 7F), forming small clusters. Some haemoblasts facing the lumen show specialisations, such as minute electron-dense granules in the cytoplasm and short pseudopodia (inset in Figure 7G). In the lumen of the ampullae, various free cells can be recognised. They are young cell types (Figure 8A) that include hyaline amoebocytes, small morula cells with vacuolar material in the process of aggregation, and small pigment cells. In the latter, the nucleus still shows organised chromatin and is not pycnotic as in the large forms of pigment cells at the end differentiation.

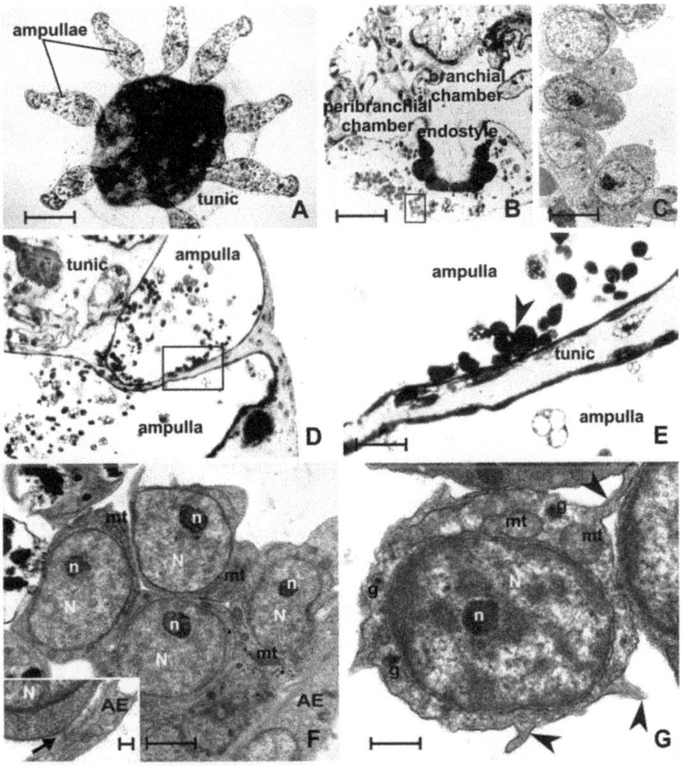

Figure 7. Oozoid: (**A**) individual at the end of metamorphosis viewed in toto, showing the maximum protrusion of the eight ampullae onto the substratum. Ampullae join with the vessels of the colonial circulatory system that are growing and arranging in the tunic; (**B**) transversal thick section at the level of the endostyle stained with toluidine blue. Note clusters of haemocytes in the subendostylar sinus; (**C**) ultrastructural detail of the selected area in (**B**) showing the presence of crowding haemoblasts; (**D**) longitudinal thick section at the level of ampullae; (**E**) detail of the selected area in (**D**) showing clusters of haemoblasts (*arrowheads*); and (**F**,**G**) ultrastructural images of these areas confirm the presence of groups of haemoblasts along the lateral epithelium of the ampullae (**F**), in which these cells adhere tightly to each other and with the basal lamina of the epithelium (*arrow* in *inset*). Specialisations such as granules and pseudopodia (*arrowheads*) are recognisable in haemoblasts facing the ampullar lumen (**G**). *AE*—ampullar epithelium, *g*—membrane-bound granule, *mt*—mitochondrion, *N*—nucleus, and *n*—nucleolus. Bar length: 65 μm in (**A**); 130 μm in (**B**); 5 μm in (**C**); 120 μm in (**D**); 8 μm in (**E**); 1.7 μm in (**F**); 0.7 μm in (**G**); and 0.3 μm in inset of (**F**).

In the haemolymph, circulating macrophage-like cells (Figure 8B) become larger (15 μm in diameter). Phagosomes appear of various sizes, some of which are so large that the nucleus is not always visible because it is pushed to the periphery by their presence. They contain entire cells, cell fragments, heterogenous and electron-dense material and lipid droplets. In the scant cytoplasm, RER, free ribosomes, some mitochondria and scattered clear vesicles are still present.

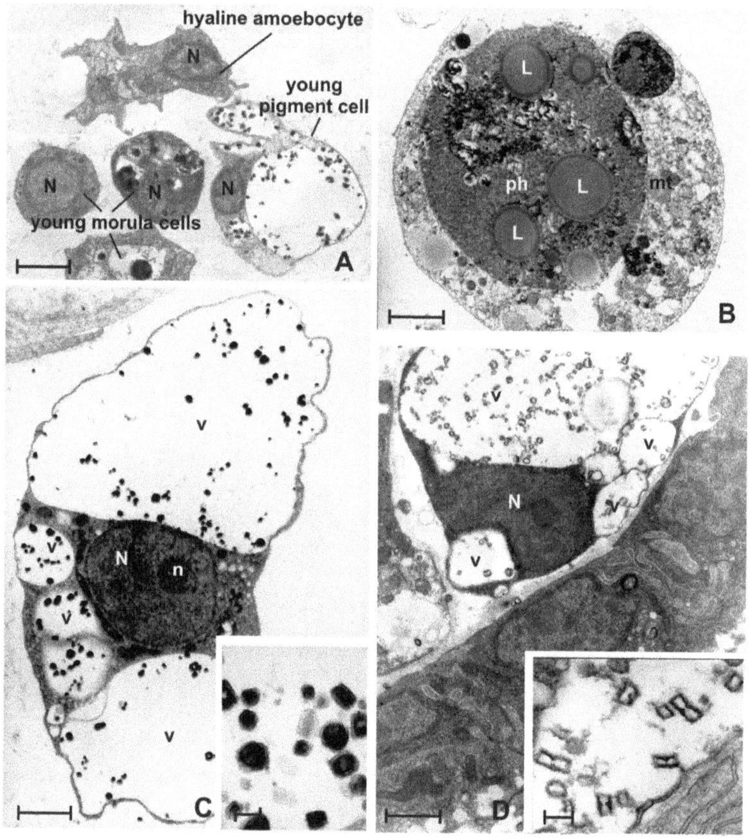

Figure 8. Oozoid: (**A**) young haemocyte types free in the lumen of the ampullae; (**B**) macrophage-like cell with a large phagosome containing digested cell debris and lipid droplets; (**C**) pigment cell with prismatic granules of 'blue phenotype' showing electron-dense layers (*inset*); and (**D**) nephrocyte showing the typical hourglass-shaped granules of uric acid in the large vacuoles. *L*—lipid droplet, *N*—nucleus, *n*—nucleolus, *ph*—phagosome, and *v*—vacuole. Bar length: 4 μm in (**A**); 3 μm in (**B**); 1.5 μm in (**C**); 1.6 μm in (**D**); 0.2 μm in inset of (**C**); and 0.3 μm in inset of (**D**).

Storage cells are represented by pigment cells and a new cell type, the nephrocyte, the latter appearing from this stage for the first time. Pigment cells (Figure 8C) become large (average size of 15 μm in diameter), with a few vacuoles containing a large number of crystalline granules, which reveal a prismatic structure (approximately 500 nm in length) with electron-dense layers (inset of Figure 8C). The nucleus is peripheral and often has a nucleolus. The cytoplasm is scarce and confined around the nucleus and vacuoles. The morphology of the nephrocyte is very similar to that of the pigment cell (Figure 8D). The most important difference lies in the structure of the crystalline granule (inset of Figure 8D), which appears with a typical hourglass shape and is transparent to electrons because it is composed of uric acid crystals.

4. Discussion

The problem of the origin of haemocytes in *B. schlosseri* still presents important questions, as specific haematopoietic tissues are unknown in adults of this species. During embryonic development, haemocytes become identifiable when the tail begins to form in the embryo, that is, at an advanced stage of organogenesis. They appear among the embryonic germ layers, and contain large yolk granules in the cytoplasm. The origin of the yolk is mainly exogenous and occurs during the vitellogenesis stage of the oocyte. Before ovulation, the outer follicle cells of the egg envelopes are engaged in the intense synthesis of proteins, which may be transferred and taken, via endocytosis, into the oocyte for yolk formation. The oocyte can also receive a protein contribution from the haemolymph via an intercellular pathway, overcoming the egg envelopes [68]. For the whole period of the embryo-larval development, the yolk granules inside the cytoplasm of all cell tissues, haemocytes included, progressively decrease in number and volume until they disappear for total consumption during metamorphosis.

As summarised in Table 1, the various haemocyte morphotypes appear progressively in the various stages of development, and at each stage, are always recognisable putative precursors, i.e., cell types with intermediate characteristics between the haemoblast and other cell types belonging to the three haemocyte lines.

Table 1. Occurrence of the various haemocyte types and enzymatic activities of phenoloxidase (PO) and acid phosphatase (AP) during embryonic development and metamorphosis.

	Early Tailbud Embryo	Late Tailbud Embryo	Early Metamorphosis	Late Metamorphosis	Oozoid
Haemoblasts					
Morula cells			PO	PO	PO
Hyaline amoebocytes				AP	AP
Pigment cells					
Macrophage-like cells				AP	AP
Nephrocytes					

The haemoblast can be considered a stem cell, a multipotent, self-renewing entity that generates one or more differentiated daughter cell types and is able to contribute to the germline and various somatic tissues of the zooid, the haemolymph included [69–71]. It possesses the typical characteristics of an undifferentiated cell with a large nucleus occupying the entire cytoplasm, the latter always showing scant organelles represented by a few cisternae of the RER and small mitochondria. In deuterostomes, this cell type differentiates into other haemocytes, as has been demonstrated in echinoderms [72,73] and has been hypothesised to occur in ascidians by various authors [8,36,38,39,74–77]. Differentiation towards other cell types is characterised by the development of long cisternae of the RER, elongated mitochondria, large Golgi apparatus, and the occurrence of granules, vacuoles and small vesicles, the latter appearing optically empty or filled with variously electron-dense material.

In the oozoid, which represents the first adult individual of the colony, we observed and documented, for the first time, the occurrence of clusters of undifferentiated cells of the haemoblast type, leaning against the flattened cells of the lateral epithelium of the ampullae, the basal lamina of which acquires close relationships through the intercalation of extracellular material. Mitotic figures that could demonstrate haematopoietic activities were not observed in the present study, although they could not be excluded. The presence of some haemoblasts showing incipient specialisations would indicate a centrifugal differentiation from the clusters of haemoblasts towards the ampullar lumen, where numerous young cell types represented by putative precursors of the main haemocyte cell lines (phagocytic, cytotoxic and storage lines) are recognisable. Beside the heart contractions, which give rise to haemolymph oscillating (advisceral and abvisceral) circulation, the ampullae play a role in the circulatory flow. They are associated with the colonial vascular network and exhibit coordinated swelling and contraction in regular cycles. Bundles of microfilaments in the ampullar cells are involved in contraction of the epithelium. During swelling, a decrease in blood pressure occurs in the lumen of the ampullae, causing haemolymph stagnation inside it, whereas during contraction, the blood pressure increases and haemolymph flows from the lumen of the ampullae within the connecting vessels [78]. The swelling phase could be useful for the storage of haemocytes inside the ampullae, and the contraction phase could cause the inflow of haemocytes in the bloodstream, modulating the release of newly formed haemocytes. From these preliminary observations, it is evident that further experimental and molecular data are required to support the hypothesis that ampullae could host a microenvironment of stem cells of the haemolymph. The origin of haemoblasts could be elsewhere, for example, in the endostylar niche. A 'niche' is a local tissue microenvironment that maintains and regulates stem cells in their ability to self-renew and to generate differentiated cell populations. In *B. schlosseri*, a highly coordinated asexual budding process, which occurs every week, generates new zooids. In this cyclic process, adult stem cells are involved in the generation of all somatic organs and the germline. In particular, the anterior subendostylar sinus, as reported for this species by Voskoboynik et al. [79], would represent a niche of populations of somatic stem cells, in which cells morphologically similar to haemoblasts would be able to proliferate and migrate to regenerate organs in the growing buds. Recently, molecular studies demonstrated the expression of genes of haematopoietic stem cells in the endostylar niche [70]. Moreover, it cannot be excluded that haemoblasts could also proliferate and differentiate towards functional immunocytes in the bloodstream after an immune stimulus, as has been shown for other invertebrates lacking haematopoietic organs, such as the bivalve mollusc *Ruditapes philippinarum* [80,81].

Morula cells, which belong to the cytotoxic line and are the most frequent circulating cells, occur very early during embryonic development, i.e., soon after the neurula stage. However, these precursors are different from the typical ones in the haemolymph of the adult called 'granular amoebocytes'. The latter have numerous cytoplasmic vacuoles with strongly electron-dense content and a central ovoidal nucleus without a nucleolus. As an important difference from the adult, the embryonic differentiation of this cell type probably starts very early, directly from the haemoblast. In the stages that precede the completion of larval formation, numerous transitional forms of haemoblasts were observed showing small vacuoles containing strongly electron-dense granules. Subsequently, in the swimming larva, morula cells lost yolk granules, and some revealed an intense activity of the enzyme phenoloxidase within their vacuoles. This enzyme is involved in the inflammatory responses of many invertebrates [82] and, with regard to ascidians, has been shown in the haemocytes of many species, such as *Goniocarpa rustica* and *Halocynthia aurantium* [83], *C. intestinalis* [84] and *B. schlosseri* [85]. In *B. schlosseri*, morula cells are the first cells to sense nonself molecules. As a consequence of recognition, they produce and release, via degranulation, the following: (i) cytokines able to recruit and activate phagocytes and (ii) the enzyme phenoloxidase and its polyphenol substrates, which are responsible for the formation of quinones and reactive oxygen species. The latter cause cytotoxicity against bacteria and nonfusion (rejection) reactions with melanin formation between two contacting colonies that are genetically incompatible [15,86–89].

At the late tailbud embryo stage, both embryonic and definitive tunics wrap the embryo. The definitive tunic is progressively populated by peculiar cells, some of which migrate from the haemocoel [90,91]. The classification of this cell population is even more difficult than that of circulating haemocytes because, once settled in the matrix, the cells significantly change their morphology and often appear at a more-or-less advanced stage of degeneration. Therefore, some tunic cells are very different from any haemocyte, and they most likely differentiate within the tunic [92]. They may also originate from epidermal cells [93,94] or from the mantle epithelium [95]. In *B. schlosseri*, according to the observations of histological sections stained via several methods, tunic cells have been classified into three types: (i) the 'vacuolated cell' [91] or 'vacuo-granular tunic cell' [92], which has many vacuoles each containing a round granule and long filopodia radiating from the cell body; (ii) the 'fibrocyte' [91] or 'amoeboid tunic cell' [92], which has small pseudopodia and a large nucleus with a nucleolus; and the 'fusiform cell' [91], which is found along the vessel wall of the colonial haemolymph circulation. In the present paper, morula cells were observed to cross the epidermis. This behaviour has been observed throughout the colonial life cycle. In adults, morula cells continuously undergo transmigration from the haemolymph to the tunic, where they play a role in immune defence as sentinel cells [96–98] responding to some reactions, such as lesions, bacterial invasion or allorecognition, after contacting genetically incompatible colonies of the same species [99–102]. The consequent rejection reaction is preceded by partial fusion of the contacting tunics, in front of the opposite, facing marginal ampullae, after the local disappearance of the respective cuticles. This enables the diffusion of soluble factors from the circulation of one colony to the other, which triggers the following events: (i) the selective crowding of morula cells inside the facing ampullae; (ii) the morula cell-crossing of the ampullar epithelium; and (iii) the degranulation of morula cells and release of their vacuolar contents, which are directly responsible for the formation of the cytotoxic foci (points of rejection) and consequent tissue necrosis in front of opposite contacting ampullae [31,32,103,104]. The early transmigrating activity observed in the present study could be related to the preparation of a defensive function for spatial competition in the larva during contact, exploration and settlement on the substratum. A mobilisation of phenoloxidase-containing morula cells via spatial competition by heterospecifics was demonstrated as a defensive mechanism, involving the recruitment of these sentinel cells in the tunic near competitor contact zones in colonies of the ascidian *Didemnum perlucidum* [96].

On the other hand, the other cell type observed in the tunic, i.e., the granulocyte, seems to derive from the cuboidal epithelial cells of the apical cap of the ampullae. This is supported by the fact that during vascular formation in this species, some epidermal cells detach from the apex of the tunic vessels and migrate into the tunic, possibly guiding vessel growth [105]. In this process, each migrating epithelial cell protrudes towards the tunic, but the basal lamina maintains its continuity. Tunic granulocytes, epithelial cells of both the ampullae and growing vessels, share many ultrastructural features. These cells show a basal nucleus, a developed rough endoplasmic reticulum (RER) and Golgi apparatus, and many membrane-bound secretory granules with similar electron-dense content. They could play an important role in the production of the constituents of the tunic since these ultrastructural aspects support the hypothesis of protein synthesis and secretion. In *Botryllus primigenus*, the granules in the ampullar epidermis were described as 'adhesive vesicles' and were assumed to be involved in the attachment to the substratum [65]. Therefore, the content of the granules will require further morpho-functional and molecular analyses to better characterise this cell type and understand not only its origin, but also its function; the latter concerns the immunity, formation and maintenance of the tunic, and adhesion to the substratum.

Hyaline amoebocytes, belonging to the phagocytic line, have been recognised starting from the late tailbud embryo. This cell type can be distinguished by the presence of a few membrane-bound granules and numerous small, clear vesicles, both characteristics observed even in adult individuals. The appearance of macrophage-like cells, derived from activated hyaline amoebocytes [38], was observed during the metamorphosis of the larva, confirming the observations in other ascidian species of the massive presence of this type of cell as soon as resorption of the tail tissues begins [106,107]. These giant cells also appear in the haemolymph when the regression of old zooids occurs in the colony. The latter are periodically replaced altogether from the next generation of zooids. During colonial regression, namely take-over, as well as during larval metamorphosis, the progressive phenomenon of apoptosis occurs, in which scavenger phagocytes infiltrate between degenerating organs for tissue resorption [108,109]. During the colonial take-over, the increase in macrophage-like cells is counterbalanced by a corresponding decrease in hyaline amoebocytes, confirming the role of the latter as precursors within the phagocytic line [110,111]. Hyaline amoebocytes can phagocytise various target particles in vitro. As they engulf foreign material, they increase in size, withdraw pseudopodia and acquire a spherical shape with large phagosomes containing hydrolytic enzymes [112]. To confirm what has been previously observed in this cell type in the colonies [38], even the phagosomes of the hyaline amoebocytes and macrophage-like cells of the larva in metamorphosis are rich in hydrolases such as acid phosphatase, supporting early functional activity.

Pigment cells are the first cells of the storage line to have been observed during embryonic development, just before larval hatching. They can be identified by the presence of giant vacuoles containing multi-layered, small granules of pigment, the phenotype of which—red or blue—is genetically regulated by two alleles [113]. They are specialised haemocytes, and the pigment contributes to the pigmentation of the colony and protects it against radiation light [114]. The transitional forms from the haemoblast to pigment cells are not easily identifiable. These young cells always show characteristics peculiar to adults [37] but, as a difference, show a small volume of the entire cell and vacuoles, few stratified pigment granules, and a nucleus that is still well-represented morphologically and functionally. Nephrocytes appear only after metamorphosis in the oozoid, and their occurrence is related to the metabolism change with the beginning of filter-feeding activity. The granules have a typical hourglass shape with an electron-dense edge. Nephrocytes are cells specialised in the storage of nitrogenous catabolites such as uric acid and purines [115], and are localised in the connective tissue of the mantle lacunae and in the ampullae of both the oozoid and the blastozoids [36]. Even in this case, transitional forms from the haemoblast were not detected; there was only the occasional occurrence of circulating small, vacuolated cells with optically empty vacuoles, in which the germs of the future granules could be formed.

5. Conclusions

This study on the origin and differentiation of haemocytes by means of ultrastructural observations in both the embryo and larva of *B. schlosseri* made it possible to establish a timescale of the appearance of the various cell types. This approach represents a starting point for future studies on haematopoiesis, which still remains to be clarified with regard to many aspects, distinguishing the possible roles of the haemoblast. Many questions still remain open about this stem cell, such as (i) the mode and time of separation of the somatic line from the germline in the embryo and in blastogenetic generations; (ii) the reduction of pluripotency in somatic stem cells towards the haemolymph stem cell line; (iii) the niches in which the formation of various cell types of the haemolymph occur in both the embryos and adults, with a particular focus on their locations inside the individuals during the blastogenetic stages of the colonial life-cycle; (iv) the presence of differentiation sites different from those of formation; and (v) the capacity of the haemoblasts to proliferate and differentiate in the bloodstream after an immune stimulus. Therefore, further studies are necessary to understand and characterise these aspects from a functional and molecular point of view.

The early transcellular migration of certain cell types of both the haemolymph and epidermis into the tunic for immune defence and tunic repair is another aspect that requires in-depth analysis to understand the mechanism behind this behaviour, which appears to be very similar to that of the motile cells of the connective tissues of vertebrates. Confocal microscopy with fluorescent molecular probes specific to the various haemocyte morphotypes, and 3D-reconstructions, could be useful to show the location of these cells in the individuals during embryonic development, metamorphosis and blastogenesis.

Finally, the presence of immunocytes with active enzymes in the larva demonstrates the great potential of innate immunity, beginning from the settlement on the substratum. This is probably the basis of the ecological success of this species, which is competitive and widespread along the coasts in all the world's seas and is able to cope with the challenges of climate change, pollutants and the settlement of other fouling competitors.

Funding: This research was supported by grants from the Italian MIUR (DOR 2021) to F.C.

Acknowledgments: The author wishes to thank Andrea Sambo, the technician in charge of the staff at the 'Umberto D'Ancona' Hydrobiological Station of Chioggia (Venice, Italy), for their assistance with ascidian collection and boat driving.

Conflicts of Interest: The author declares no conflict of interest. The funder had no role in the design of the study; in the collection, analyses, or interpretation of data; in the writing of the manuscript; or in the decision to publish the results.

References

1. Cima, F.; Ballarin, L.; Caicci, F.; Franchi, N.; Gasparini, F.; Rigon, F.; Schiavon, F.; Manni, L. Life history and ecological genetics of the colonial ascidian *Botryllus schlosseri*. *Zool. Anz.* **2015**, *257*, 54–70. [CrossRef]
2. Reem, E.; Douek, J.; Rinkevich, B. A critical deliberation of the 'species complex' status of the globally spread colonial ascidian *Botryllus schlosseri*. *J. Mar. Biol. Assoc. UK* **2021**, *101*, 1047–1060. [CrossRef]
3. Rosengarten, R.D.; Nicotra, M.L. Model systems of invertebrate allorecognition. *Curr. Biol.* **2011**, *21*, R82–R92. [CrossRef] [PubMed]
4. Ballarin, L.; Cima, F.; Franchi, N. Origin and functions of tunicate hemocytes. In *The Evolution of the Immune System: Conservation and Diversification*, 1st ed.; Malagoli, D., Ed.; Elsevier Academic Press: Amsterdam, The Netherlands, 2016; Chapter 2; pp. 29–49.
5. Delsuc, F.; Tsagkogeorga, G.; Lartillot, N.; Philippe, H. Additional molecular support for the new chordate phylogeny. *Genesis* **2008**, *46*, 592–604. [CrossRef]
6. Putnam, N.; Butts, T.; Ferrier, D.; Furlong, R.F.; Hellsten, U.; Kawashima, T.; Robinson-Rechavi, M.; Shoguchi, E.; Terry, A.; Dubchak, I.; et al. The amphioxus genome and the evolution of the chordate karyotype. *Nature* **2008**, *453*, 1064–1071. [CrossRef]
7. Fujimoto, H.; Watanabe, H. The characterization of granular amebocytes and their possible roles in the asexual reproduction of the plystyelid ascidian, *Polyzoa vesiculiphora*. *J. Morphol.* **1976**, *150*, 623–638. [CrossRef]
8. Goodbody, I. The physiology of ascidians. *Adv. Mar. Biol.* **1974**, *12*, 1–149.
9. Oka, H.; Watanabe, H. Vascular budding, a new type of budding in *Botryllus*. *Biol. Bull.* **1957**, *112*, 225–240. [CrossRef]
10. Freeman, G. The role of blood cells in the process of asexual reproduction in the tunicate *Perophora viridis*. *J. Exp. Zool.* **1964**, *156*, 157–183. [CrossRef]
11. Sabbadin, A.; Zaniolo, G. Sexual differentiation and germ cell transfer in the colonial ascidian *Botryllus schlosseri*. *J. Exp. Zool.* **1979**, *207*, 289–304. [CrossRef]
12. Kassmer, S.H.; Langenbacher, A.D.; De Tomaso, A.W. Integrin-alpha-6$^+$ candidate stem cells are responsible for whole body regeneration in the invertebrate chordate *Botrylloides diegensis*. *Nat. Commun.* **2020**, *11*, 4435. [CrossRef] [PubMed]
13. Manni, L.; Anselmi, C.; Cima, F.; Gasparini, F.; Voskoboynik, A.; Martini, M.; Peronato, A.; Burighel, P.; Zaniolo, G.; Ballarin, L. Sixty years of experimental studies on the blastogenesis of the colonial tunicate *Botryllus schlosseri*. *Dev. Biol.* **2019**, *448*, 293–308. [CrossRef] [PubMed]
14. Hirose, E.; Taneda, Y.; Ishii, T. Two modes of tunic cuticle formation in a colonial ascidian *Aplidium yamazii*, responding to wounding. *Dev. Comp. Immunol.* **1997**, *21*, 25–34. [CrossRef]
15. Franchi, N.; Ballarin, L. Immunity in protochordates: The tunicate perspective. *Front. Immunol.* **2017**, *8*, 674. [CrossRef] [PubMed]
16. Anderson, R.S. Cellular responses to foreign bodies in the tunicate *Molgula manhattensis* (DeKay). *Biol. Bull.* **1971**, *141*, 91–98. [CrossRef]
17. Wright, R.K.; Cooper, E.L. Immunological maturation in the tunicate *Ciona intestinalis*. *Am. Zool.* **1975**, *15*, 21–27. [CrossRef]
18. Wright, R.K.; Cooper, E.L. Inflammatory reactions of protochordata. *Am. Zool.* **1983**, *23*, 205–211. [CrossRef]
19. Ballarin, L.; Cima, F.; Sabbadin, A. Phagocytosis in the colonial ascidian *Botryllus schlosseri*. *Dev. Comp. Immunol.* **1994**, *18*, 467–481. [CrossRef]
20. Cima, F.; Ballarin, L.; Sabbadin, A. New data on phagocytes and phagocytosis in the compound ascidian *Botryllus schlosseri* (Tunicata: Ascidiacea). *Ital. J. Zool.* **1996**, *63*, 357–364. [CrossRef]
21. Kelly, K.L.; Cooper, E.L.; Raftos, D.A. Purification and characterization of a humoral opsonin from the solitary urochordate *Styela clava*. *Comp. Biochem. Physiol. B Biochem. Mol. Biol.* **1992**, *103*, 749–753. [CrossRef]
22. Raftos, D.A.; Cooper, E.L.; Habicht, G.S.; Beck, G. Invertebrate cytokines: Tunicate cell proliferation stimulated by an interleukin-1-like molecule. *Proc. Natl. Acad. Sci. USA* **1991**, *88*, 9518–9522. [CrossRef] [PubMed]
23. Menin, A.; Ballarin, L. Immunomodulatory molecules in the compound ascidian *Botryllus schlosseri*: Evidence from conditioned media. *J. Invertebr. Pathol.* **2008**, *99*, 275–280. [CrossRef]
24. Melillo, D.; Marino, R.; Della Camera, G.; Italiani, P.; Boraschi, D. Assessing immunological memory in the solitary ascidian *Ciona robusta*. *Front. Immunol.* **2019**, *10*, 1977. [CrossRef] [PubMed]
25. Gasparini, F.; Franchi, N.; Spolaore, B.; Ballarin, L. Novel rhamnose-binding lectins from the colonial ascidian *Botryllus schlosseri*. *Dev. Comp. Immunol.* **2008**, *32*, 1177–1191. [CrossRef] [PubMed]
26. Franchi, N.; Schiavon, F.; Carletto, M.; Gasparini, F.; Bertoloni, G.; Tosatto, S.C.E.; Ballarin, L. Immune roles of a rhamnose-binding lectin in the colonial ascidian *Botryllus schlosseri*. *Immunobiology* **2011**, *216*, 725–736. [CrossRef] [PubMed]
27. Franchi, N.; Ballarin, L. Preliminary characterization of complement in a colonial tunicate: C3, Bf and inhibition of C3 opsonic activity by compstatin. *Dev. Comp. Immunol.* **2014**, *46*, 430–438. [CrossRef]
28. Peronato, A.; Drago, L.; Rothbächer, U.; Macor, P.; Ballarin, L.; Franchi, N. Complement system and phagocytosis in a colonial protochordate. *Dev. Comp. Immunol.* **2020**, *103*, 103530. [CrossRef] [PubMed]

29. Lehrer, R.I.; Lee, I.H.; Menzel, L.; Waring, A.; Zhao, C. Clavanins and styelins, α-helical antimicrobial peptides from the hemocytes of *Styela clava*. In *Phylogenetic Perspectives on the Vertebrate Immune System. Advances in Experimental Medicine and Biology*; Beck, G., Sugumaran, M., Cooper, E.L., Eds.; Springer: Boston, MA, USA, 2001; Volume 484, pp. 71–76.
30. Parrinello, N.; Cammarata, M.; Parrinello, D. The inflammatory response of Urochordata: The basic process of the ascidians' innate immunity. In *Advances in Comparative Immunology*; Cooper, E.L., Ed.; Springer: Cham, Switzerland, 2018; pp. 521–590.
31. Sabbadin, A.; Zaniolo, G.; Ballarin, L. Genetic and cytological aspects of histocompatibility in ascidians. *Ital. J. Zool.* **1992**, *59*, 167–173. [CrossRef]
32. Cima, F.; Sabbadin, A.; Ballarin, L. Cellular aspects of allorecognition in the compound ascidian *Botryllus schlosseri*. *Dev. Comp. Immunol.* **2004**, *28*, 881–889. [CrossRef]
33. De Tomaso, A.W.; Nyholm, S.V.; Palmeri, K.J.; Ishizuka, K.J.; Ludington, W.B.; Mitchel, K.; Weissman, I.L. Isolation and characterization of a protochordate histocompatibility locus. *Nature* **2005**, *438*, 454–459. [CrossRef]
34. Nyholm, S.V.; Passegue, E.; Ludington, W.B.; Voskoboynik, A.; Mitchel, K.; Weissman, I.L.; De Tomaso, A.W. *fester*, a candidate allorecognition receptor from a primitive chordate. *Immunity* **2006**, *25*, 163–173. [CrossRef] [PubMed]
35. McKitrick, T.R.; Muscat, C.C.; Pierce, J.D.; Bhattacharya, D.; De Tomaso, A.W. Allorecognition in a basal chordate consists of independent activating and inhibitory pathways. *Immunity* **2011**, *34*, 616–626. [CrossRef] [PubMed]
36. Milanesi, C.; Burighel, P. Blood cell ultrastructure of the ascidian *Botryllus schlosseri*. I. Hemoblast, granulocytes, macrophage, morula cell and nephrocyte. *Acta Zool.* **1978**, *59*, 135–147. [CrossRef]
37. Burighel, P.; Milanesi, C.; Sabbadin, A. Blood cell ultrastructure of the ascidian *Botryllus schlosseri*. II. Pigment cells. *Acta Zool.* **1983**, *64*, 15–23. [CrossRef]
38. Ballarin, L.; Cima, F. Cytochemical properties of *Botryllus schlosseri* haemocytes: Indications for morpho-functional characterisation. *Eur. J. Histochem.* **2005**, *49*, 255–264. [PubMed]
39. Wright, R.K. Urochordates. In *Invertebrate Blood Cells*; Ratcliffe, N.A., Rowley, A.F., Eds.; Academic Press: New York, NY, USA, 1981; Volume 2, pp. 565–626.
40. De Leo, G. Ascidian hemocytes and their involvement in defense reactions. *Boll. Zool.* **1992**, *59*, 195–213. [CrossRef]
41. Burighel, P.; Cloney, R.A. Urochordata: Ascidiacea. In *Microscopic Anatomy of Invertebrates: Hemichordata, Chaetognatha and the Invertebrate Chordates*; Harrison, F.W., Ruppert, E.E., Eds.; Wiley-Liss, Inc.: New York, NY, USA, 1997; Volume 15, pp. 221–347.
42. Jiménez-Merino, J.; de Abreu, I.S.; Hiebert, L.S.; Allodi, S.; Tiozzo, S.; De Barros, C.M.; Brown, F.D. Putative stem cells in the hemolymph and in the intestinal submucosa of the solitary ascidian *Styela plicata*. *Evo Devo* **2019**, *10*, 31. [CrossRef]
43. Hirano, T.; Nishida, H. Developmental fates of larval tissues after metamorphosis in ascidian *Halocynthia roretzi*. I. Origin of mesodermal tissues of the juvenile. *Dev. Biol.* **1997**, *192*, 199–210. [CrossRef]
44. Cowden, R.R. The embryonic origin of blood cells in the tunicate *Clavelina*. *Trans. Am. Microsc. Sci.* **1968**, *87*, 521–524. [CrossRef]
45. Andrew, W. Phase microscope studies of living blood cells of the tunicates under normal and experimental conditions with a description of a new type of motile cell appendage. *Q. J. Microsc. Sci.* **1961**, *102*, 89–105. [CrossRef]
46. Davidson, B.; Swalla, B.J. A molecular analysis of ascidian metamorphosis reveals activation of an innate immune response. *Development* **2002**, *129*, 4739–4751. [CrossRef] [PubMed]
47. Roberts, B.; Davidson, B.; MacMaster, G.; Lockhart, V.; Ma, E.; Wallace, S.S.; Swalla, B.J. A complement response may activate metamorphosis in the ascidian *Boltenia villosa*. *Dev. Genes Evol.* **2007**, *217*, 449–458. [CrossRef] [PubMed]
48. Wieczorek, S.; Todd, C. Inhibition and facilitation of bryozoan and ascidian settlement by natural multi-species biofilms: Effects of film age and the roles of active and passive larval attachment. *Mar. Biol.* **1997**, *128*, 463–473. [CrossRef]
49. Chase, A.L.; Dijkstra, J.A.; Harris, L.G. The influence of substrate material on ascidian larval settlement. *Mar. Pollut. Bull.* **2016**, *106*, 35–42. [CrossRef]
50. Freckelton, M.; Nedved, B.; Hadfield, M. Induction of invertebrate larval settlement; different bacteria, different mechanisms? *Sci. Rep.* **2017**, *7*, 42557. [CrossRef]
51. Dobretsov, S.; Rittschof, D. Love at first taste: Induction of larval settlement by marine microbes. *Int. J. Mol. Sci.* **2020**, *21*, 731. [CrossRef]
52. Ermak, T.H. An autoradiographic demonstration of blood cell renewal in *Styela clava* (Urochordata: Ascidiacea). *Experientia* **1975**, *31*, 837–838. [CrossRef]
53. Ermak, T.H. The hematogenic tissues of tunicates. In *Phylogeny of Thymus and Bone Marrow-Bursa Cells*; Wright, R.K., Cooper, E.L., Eds.; North-Holland Publishing Company: Amsterdam, The Netherlands, 1976; pp. 45–56.
54. Wright, R.K.; Ermak, T.H. Cellular defense systems of the protochordata. In *Phylogeny and Ontogeny*; Cohen, N., Sigel, M.M., Eds.; Springer: Boston, MA, USA, 1982; pp. 283–320.
55. Zaniolo, G.; Burighel, P.; Martinucci, G.B. Ovulation and placentation in *Botryllus schlosseri* (Ascidiacea): An ultrastructural study. *Can. J. Zool.* **1987**, *65*, 1181–1190. [CrossRef]
56. Manni, L.; Lane, N.J.; Sorrentino, M.; Zaniolo, G.; Burighel, P. Mechanism of neurogenesis during the embryonic development of a tunicate. *J. Comp. Neurol.* **1999**, *412*, 527–541. [CrossRef]
57. Manni, L.; Burighel, P. Common and divergent pathways in alternative developmental processes of ascidians. *Bioessays* **2006**, *28*, 902–912. [CrossRef]
58. Cima, F.; Varello, R. Effects of exposure to trade antifouling paints and biocides on larval settlement and metamorphosis of the compound ascidian *Botryllus schlosseri*. *J. Mar. Sci. Eng.* **2022**, *10*, 123. [CrossRef]

59. Müller, W.C.; Greenwood, A.D. The ultrastructure of phenolic-storing cells fixed with caffeine. *J. Exp. Bot.* **1978**, *29*, 757–764. [CrossRef]
60. Cima, F. Enzyme histochemistry for functional histology in invertebrates. In *Single Molecule Histochemistry: Methods and Protocols*; Pellicciari, C., Biggiogera, M., Eds.; Methods in Molecular Biology; Springer Science: New York, NY, USA, 2017; Volume 1560, pp. 69–90.
61. Barka, T.; Anderson, P.J. Histochemical methods for acid phosphatase using hexazonium pararosanilin as coupler. *J. Histochem. Cytochem.* **1962**, *10*, 741–753. [CrossRef]
62. Hose, J.E.; Martin, G.G.; Nguyen, V.A.; Lucas, J.; Rosenstein, A.T. Cytochemical features of shrimp hemocytes. *Biol. Bull.* **1987**, *173*, 178–187. [CrossRef]
63. Kowarsky, M.; Anselmi, C.; Hotta, K.; Burighel, P.; Zaniolo, G.; Caicci, F.; Rosental, B.; Neff, N.F.; Ishizuka, K.J.; Palmeri, K.J.; et al. Sexual and asexual development: Two distinct programs producing the same tunicate. *Cell Rep.* **2021**, *34*, 108681. [CrossRef]
64. Burighel, P.; Brunetti, R. The circulatory system in the blastozooid of the colonial ascidian *Botryllus schlosseri* (Pallas). *Boll. Zool.* **1971**, *38*, 273–289. [CrossRef]
65. Katow, H.; Watanabe, H. Fine structure and possible role of ampullae on tunic supply and attachment in a compound ascidian, *Botryllus primigenus* Oka. *J. Ultrastruct. Res.* **1978**, *64*, 23–34. [CrossRef]
66. Rodriguez, D.; Nourizadeh, S.; De Tomaso, A.W. The biology of the extracorporeal vasculature of *Botryllus schlosseri*. *Dev. Biol.* **2019**, *448*, 309–319. [CrossRef]
67. Schiaffino, S.; Burighel, P.; Nunzi, M.G. Involution of the caudal musculature during metamorphosis in the ascidian, *Botryllus schlosseri*. *Cell Tissue Res.* **1974**, *153*, 293–305. [CrossRef]
68. Manni, L.; Zaniolo, G.; Burighel, P. 1993 Egg envelope cytodifferentiation in the colonial ascidian *Botryllus schlosseri* (Tunicata). *Acta Zool.* **1993**, *74*, 103–113. [CrossRef]
69. Rinkevich, Y.; Voskoboynik, A.; Rosner, A.; Rabinowitz, C.; Paz, G.; Oren, M.; Douek, J.; Alfassi, G.; Moiseeva, E.; Ishizuka, K.J.; et al. Repeated, long-term cycling of putative stem cells between niches in a basal chordate. *Dev. Cell.* **2013**, *24*, 76–88. [CrossRef]
70. Rosental, B.; Kowarsky, M.; Seita, J.; Corey, D.M.; Ishizuka, J.K.; Palmeri, K.J.; Chen, S.Y.; Sinha, R.; Okamoto, J.; Mantalas, G.; et al. Complex mammalian-like haematopoietic system found in a colonial chordate. *Nature* **2018**, *564*, 425–429. [CrossRef] [PubMed]
71. Ballarin, L.; Karahan, A.; Salvetti, A.; Rossi, L.; Manni, L.; Rinkevich, B.; Rosner, A.; Voskoboynik, A.; Rosental, B.; Canesi, L.; et al. Stem cells and innate immunity in aquatic invertebrates: Bridging two seemingly disparate disciplines for new discoveries in biology. *Front. Immunol.* **2021**, *12*, 688106. [CrossRef] [PubMed]
72. Fontane, A.R.; Lambert, P. The fine structure of the leucocytes of the holothurian *Cucumaria miniata*. *Can. J. Zool.* **1977**, *55*, 1530–1544. [CrossRef] [PubMed]
73. Smith, U.Z. The echinoderms. In *Invertebrate Blood Cells*; Ratcliffe, N.A., Rowley, A.F., Eds.; Academic Press: New York, NY, USA, 1981; Volume 2, pp. 513–562.
74. Azéma, M. Le sang des Botrylles. *Comptes Rendus Séances Soc. Biol. Paris* **1929**, *192*, 823–825.
75. Azéma, M. Recherches sur le sang et l'excrétion chez les ascidies. *Ann. Inst. Océan. Monaco* **1937**, *17*, 1–150.
76. Endean, R. The blood cells of the ascidian *Phallusia mammillata*. *Q. J. Microsc. Sci.* **1960**, *101*, 177–197.
77. George, W.C. A comparative study of the blood of the tunicates. *Q. J. Microsc. Sci.* **1939**, *81*, 391–428.
78. Mackie, G.O.; Singla, C.L. Coordination of compound ascidians by epithelial conduction in the colonial blood vessels. *Biol. Bull.* **1983**, *165*, 209–220. [CrossRef]
79. Voskoboynik, A.; Soen, Y.; Rinkevich, Y.; Rosner, A.; Ueno, H.; Reshef, R.; Ishizuka, K.J.; Palmeri, K.J.; Moiseeva, E.; Rinkevich, B.; et al. Identification of the endostyle as a stem cell niche in a colonial chordate. *Cell Stem Cell* **2008**, *3*, 456–464. [CrossRef]
80. Matozzo, V.; Marin, M.G.; Cima, F.; Ballarin, L. First evidence of cell division in circulating haemocytes from the Manila clam *Tapes philippinarum*. *Cell Biol. Int.* **2008**, *32*, 865–868. [CrossRef] [PubMed]
81. Cima, F.; Matozzo, V. Proliferation and differentiation of circulating haemocytes of *Ruditapes philippinarum* as a response to bacterial challenge. *Fish Shellfish Immunol.* **2018**, *81*, 73–82. [CrossRef] [PubMed]
82. Cerenius, L.; Söderhäll, K. Immune properties of invertebrate phenoloxidases. *Dev. Comp. Immunol.* **2021**, *122*, 104098. [CrossRef] [PubMed]
83. Chaga, O.Y. Ortho-diphenoloxidase system of ascidians. *Tsitologia* **1980**, *22*, 619–625.
84. Smith, V.J.; Söderhäll, K. A comparison of phenoloxidase activity in the blood of marine invertebrates. *Dev. Comp. Immunol.* **1991**, *15*, 251–262. [CrossRef]
85. Ballarin, L.; Cima, F.; Sabbadin, A. Phenoloxidase and cytotoxicity in the compound ascidian *Botryllus schlosseri*. *Dev. Comp. Immunol.* **1998**, *22*, 479–492. [CrossRef]
86. Ballarin, L.; Franchini, A.; Ottaviani, E.; Sabbadin, A. Morula cells as the major immunomodulatory hemocytes in ascidians: Evidences from the colonial species *Botryllus schlosseri*. *Biol. Bull.* **2001**, *201*, 59–64. [CrossRef]
87. Ballarin, L.; Cima, F.; Sabbadin, A. Morula cells and histocompatibility in the colonial ascidian *Botryllus schlosseri*. *Zool. Sci.* **2005**, *12*, 757–764. [CrossRef]

88. Ballarin, L.; Franchi, N.; Schiavon, F.; Tosatto, S.C.E.; Mičetić, I.; Kawamura, K. Looking for putative phenoloxidases of compound ascidians: Haemocyanin-like proteins in *Polyandrocarpa misakiensis* and *Botryllus schlosseri*. *Dev. Comp. Immunol.* **2012**, *38*, 232–242. [CrossRef]
89. Taketa, D.A.; De Tomaso, A.W. *Botryllus schlosseri* allorecognition: Tackling the enigma. *Dev. Comp. Immunol.* **2014**, *48*, 254–265. [CrossRef]
90. Izzard, C.S. Contractile filopodia and in vivo cell movement in the tunic of the ascidian, *Botryllus schlosseri*. *J. Cell Sci.* **1974**, *15*, 513–535. [CrossRef]
91. Zaniolo, G. Histology of the ascidian *Botryllus schlosseri*: In particular, the test cells. *Boll. Zool.* **1981**, *48*, 169–178. [CrossRef]
92. Hirose, E. Ascidian tunic cells: Morphology and functional diversity of free cells outside the epidermis. *Invertebr. Biol.* **2009**, *128*, 83–96. [CrossRef]
93. Hirose, E.; Mukai, H. An ultrastructural study on the origin of glomerulocytes, a type of blood cells in a styelid ascidian, *Polyandrocarpa misakiensis*. *J. Morphol.* **1992**, *211*, 269–273. [CrossRef] [PubMed]
94. Kimura, S.; Itoh, T. Evidence for the role of glomerulocyte in cellulose synthesis in the tunicate, *Metandrocarpa uedai*. *Protoplasma* **1995**, *186*, 24–33. [CrossRef]
95. Di Bella, M.A.; Carbone, M.C.; De Leo, G. Aspects of cell proliferation in mantle tissue of *Ciona intestinalis* L. (Tunicata, Ascidiacea). *Micron* **2005**, *36*, 477–481. [CrossRef] [PubMed]
96. Dias, G.M.; Yokoyama, L.Q. Spatial competition induces the mobilization of morula cells in the colonial ascidian *Didemnum perlucidum* (Tunicata: Didemnidae). *Invertebr. Biol.* **2011**, *130*, 186–192. [CrossRef]
97. Ballarin, L. Ascidian cytotoxic cells: State of the art and research perspectives. *Invertebr. Surv. J.* **2012**, *9*, 1–6.
98. Melillo, D.; Marino, R.; Italiani, P.; Boraschi, D. Innate immune memory in invertebrate metazoans: A critical appraisal. *Front. Immunol.* **2018**, *9*, 1915. [CrossRef]
99. Franchi, N.; Ballarin, L.; Cima, F. Insights on cytotoxic cells of the colonial ascidian *Botryllus schlosseri*. *Invertebr. Surv. J.* **2015**, *12*, 109–117.
100. Franchi, N.; Ballarin, L.; Peronato, A.; Cima, F.; Grimaldi, A.; Girardello, R.; de Eguileor, M. Functional amyloidogenesis in immunocytes from the colonial ascidian *Botryllus schlosseri*: Evolutionary perspective. *Dev. Comp. Immunol.* **2019**, *90*, 108–120. [CrossRef] [PubMed]
101. Taneda, Y.; Watanabe, H. Studies on colony specificity in the compound ascidian, *Botryllus primigenus* Oka. 1. Blood cells infiltration. *Dev. Comp. Immunol.* **1982**, *6*, 43–52. [CrossRef]
102. Hirose, E.; Saito, Y.; Watanabe, H. Subcuticular rejection: An advanced mode of the allogeneic rejection in the compound ascidians *Botrylloides simodensis* and *B. fuscus*. *Biol. Bull.* **1997**, *192*, 53–61. [CrossRef] [PubMed]
103. Ballarin, L.; Cima, F.; Floreani, M.; Sabbadin, A. Oxidative stress induces cytotoxicity during rejection reaction in the compound ascidian *Botryllus schlosseri*. *Comp. Biochem. Physiol. C Toxicol. Pharmacol.* **2002**, *133*, 411–418. [CrossRef]
104. Cima, F.; Sabbadin, A.; Zaniolo, G.; Ballarin, L. Colony specificity and chemotaxis in the compound ascidian *Botryllus schlosseri*. *Comp. Biochem. Physiol. A Mol. Integr. Physiol.* **2006**, *145*, 376–382. [CrossRef]
105. Gasparini, F.; Longo, F.; Manni, L.; Burighel, P.; Zaniolo, G. Tubular sprouting as a mode of vascular formation in a colonial ascidian (Tunicata). *Dev. Dyn.* **2007**, *236*, 719–731. [CrossRef]
106. Grimm, L.M.; Cloney, R.A. Ultrastructural study of ascidian metamorphosis: Blood cell migration across the epidermis. *Am. Zool.* **1965**, *5*, 644.
107. Cloney, R.A. Ascidian larvae and the events of metamorphosis. *Am. Zool.* **1982**, *22*, 817–826. [CrossRef]
108. Burighel, P.; Schiavinato, A. Degenerative regression of the digestive tract in the colonial ascidian *Botryllus schlosseri* (Pallas). *Cell Tissue Res.* **1984**, *235*, 309–318. [CrossRef]
109. Cima, F.; Ballarin, L. Apoptosis and pattern of *Bcl-2* and *Bax* expression in the alimentary tract during the colonial blastogenetic cycle of *Botryllus schlosseri* (Urochordata, Ascidiacea). *Ital. J. Zool.* **2009**, *76*, 28–42. [CrossRef]
110. Cima, F.; Basso, G.; Ballarin, L. Apoptosis and phosphatidylserine-mediated recognition during the take-over of the colonial life-cycle in the ascidian *Botryllus schlosseri*. *Cell Tissue Res.* **2003**, *312*, 369–376. [CrossRef] [PubMed]
111. Cima, F.; Manni, L.; Basso, G.; Fortunato, E.; Accordi, B.; Schiavon, F.; Ballarin, L. Hovering between death and life: Natural apoptosis and phagocytes in the blastogenetic cycle of the colonial ascidian *Botryllus schlosseri*. *Dev. Comp. Immunol.* **2010**, *34*, 272–285. [CrossRef] [PubMed]
112. Ballarin, L.; Cima, F.; Sabbadin, A. Histoenzymatic staining and characterization of the colonial ascidian *Botryllus schlosseri* hemocytes. *Boll. Zool.* **1993**, *60*, 19–24. [CrossRef]
113. Sabbadin, A. Formal genetics of ascidians. *Am. Zool.* **1982**, *22*, 765–773. [CrossRef]
114. Mukai, H. Photo-induced accumulation of pigment cells in a compound ascidian, *Botryllus primigenus*. *Ann. Zool. Jpn.* **1974**, *47*, 43–47.
115. Sabbadin, A.; Tontodonati, A. Nitrogenous excretion in the compound ascidians *Botryllus schlosseri* (Pallas) and *Botrylloides leachi* (Savigny). *Monit. Zool. Ital.* **1967**, *1*, 185–190.

Journal of
Marine Science and Engineering

MDPI

Article

Naturally Occurring Rock Type Influences the Settlement of *Fucus spiralis* L. zygotes

William G. Ambrose Jr. [1,*], Paul E. Renaud [2,3], David C. Adler [4] and Robert L. Vadas [5]

1 School of the Coastal Environment, Coastal Carolina University, Conway, SC 29528, USA
2 Akvaplan-niva, 9007 Tromsø, Norway; per@akvaplan.niva.no
3 University Centre in Svalbard, 9170 Longyearbyen, Norway
4 East Coast Outfitters, 2017 Lower Prospect Rd., Halifax, NS B3T 1Y8, Canada; dave@hookedinc.ca
5 Department of Biological Science, University of Maine, Orono, ME 04469, USA; vadas@maine.edu
* Correspondence: wambrose@coastal.edu

Abstract: The settlement of spores and larvae on hard substrates has been shown to be influenced by many factors, but few studies have evaluated how underlying bedrock may influence recruitment. The characteristics of coastal rock types such as color, heat capacity, mineral size, and free energy have all been implicated in settlement success. We examined the influence of naturally occurring rock types on the initial attachment of zygotes of the brown alga *Fucus spiralis* Linnaeus 1753. We also assessed the dislodgment of zygotes on four bedrock types after initial attachment in laboratory experiments using wave tanks. Settling plates were prepared from limestone, basalt, schist, and granite, found in the region of Orrs Island, Maine, USA. The plate surfaces tested were either naturally rough or smooth-cut surfaces. We measured the density of attached zygotes after 1.5 h of settlement and subsequently after a wave treatment, in both winter and summer. The pattern of initial attachment was the same on natural and smooth surfaces regardless of season: highest on limestone (range 7.0–13.4 zygotes/cm^2), intermediate on schist (1.8–8.5 zygotes/cm^2) and basalt (3.5–14.0 zygotes/cm^2), and lowest on granite (0.8–7.8 zygotes/cm^2). Patterns of survivorship following the wave treatment were similar to those of initial settlement with the mean survivorship varying from 60.1% (SE = 3.8) (limestone, smooth substrate) to 31.8% (SE = 0.59) (granite, natural substrate), and with the highest mean survival on limestone, basalt, and schist, and the lowest on granite. Our results suggest that rock type has a significant effect on zygote attachment and persistence. Patterns of attachment were the same on smooth and rough surfaces, indicating that surface roughness is not the predominant factor controlling the difference in successful attachment among rock types. Other properties of bedrock, possibly grain size, surface free energy, or chemical interaction with the adhesives used by the zygotes, directly affect the attachment of these algal propagules. These results suggest that patterns of benthic community structure could be determined in part by the distribution of bedrock types.

Keywords: *Fucus*; rock type; settlement; attachment; mineral composition

Citation: Ambrose, W.G., Jr.; Renaud, P.E.; Adler, D.C.; Vadas, R.L. Naturally Occurring Rock Type Influences the Settlement of *Fucus spiralis* L. zygotes. *J. Mar. Sci. Eng.* **2021**, *9*, 927. https://doi.org/10.3390/jmse9090927

Academic Editor: Francesca Cima

Received: 18 July 2021
Accepted: 24 August 2021
Published: 26 August 2021

1. Introduction

Explaining mechanisms for the variation in community structure on multiple spatial scales is one of the fundamental problems in marine ecology. Physical factors such as water flow [1–4], larval supply [5–9], substratum inclination [10–13], wave exposure [1,14–16], disturbance [17], upwelling [18,19], and salinity [20,21] have all been shown to affect the distribution and abundance of organisms on rocky shores, singly or in combination. One variable that has received relatively little attention, however, is the direct effect of bedrock type on the settlement and development of epibenthic organisms [22–25]. Rock types vary in physical and chemical characteristics that might influence the settlement and survival of sessile organisms and thereby influence community structure. For example, the amount of quartz in substrates can influence the settlement of epibenthic and infaunal organisms [26,27]. If substrate minerology is important in the recruitment of marine biota, it would offer considerable insight into explaining the patterns in spatial distribution in these communities.

Physical characteristics of substrates such as substrate roughness, microtopography, refs. [28–41] and mineralogy [23,27] are known to affect the settlement and attachment of marine organisms. Bedrock type can certainly influence these characteristics. It is difficult, however, to compare the results of these studies because the scale of substratum heterogeneity in each study varies widely. Further, little distinction is made between surface heterogeneity on a scale smaller than the size of the propagule ('texture'), and that on a scale larger than the size of the propagule ('contour') [2,35]. The size of features relative to the size of the settling propagule is known to be important in determining the success of settlement [42].

The chemical influence of natural substrates on the adhesion of marine propagules has been open to debate [36,43]. There is increasing evidence that the attachment success of macroalgae is directly linked to the chemical characteristics of the adhesives involved and their interaction with the physical–chemical characteristics of the surface to which they bond [43–46]. The presence of biofilms, which develop rapidly on immersed material, however, can make the identification of causal mechanisms challenging.

Studies of the adhesion of zygotes and larvae to artificial and natural surfaces are not only of academic interest. The need to develop nontoxic coatings to prevent the adhesion of marine fouling organisms has revived this area of research (see reviews by [46,47]). This knowledge has implications for developing materials with anti-fouling properties and building artificial reefs and eco-friendly structures. Furthermore, in areas where coastal geologic formations are highly variable, understanding the effect of rock types on the settlement and attachment of sessile organisms may help explain the variability in successful invasion by nonnative species.

Fucoids are common members of intertidal and subtidal hard substrate communities worldwide, ranging from the Arctic to the tropics [48]. Studies have addressed the importance of surface roughness [49,50], water flow and wave action [16,49,51], and substrate type and surface properties [43,44,52] on the settlement and early development of fucalean algae. No studies that we are aware of, however, have addressed differences in settlement among different bedrock types where these algae regularly occur.

We tested the effects of four naturally occurring rock types and their surface contour on the initial attachment of zygotes of the brown alga *Fucus spiralis* Linnaeus 1753. To determine the effect of contour (surface heterogeneity on a scale larger than the size of the settling zygote), adhesion was tested on rock plates that were prepared with both natural surfaces and smooth-cut surfaces. We addressed the following questions in a series of laboratory settling experiments: (1) Are there differences in the primary adhesion of *Fucus spiralis* zygotes on limestone, schist, basalt, and granite substrates? (2) Do differences in the surface contour of these rock types affect the primary adhesion of zygotes? (3) Does exposure to a wave alter the initial settlement patterns? As there were differences in attachment success to different substrata that were independent of surface contour, we explored other physical and chemical characteristics of natural substrata to explain our results.

2. Materials and Methods

2.1. Rock Types

The four rock types we used were selected based on their varying physical appearance and geological composition, and because of their close juxtaposition in the intertidal zone (Supplemental Figure S1). All rock types occurred within 4 km of each other in mid-coast Maine, USA. Basalt and schist were collected from the southeast coast of Bailey Island (43°43′30″ N, 69°59′40″ W); limestone was collected from the southwest shore of South Harpswell at an outcrop across from Bar Island (43°44′25″ N, 70°00′05″ W); and granite was collected from the southernmost tip of Bethel Point (43°47′30″ N, 69°54′40″ W). For a detailed lithologic description of these rock units, see [53].

Basalt: This was the youngest rock type used in our study (Triassic age, 195–230 myo). Basalt was collected from an intrusive dike located at Bailey Island, which fills a 5–7 m fissure in heavily folded metasedimentary schist. The mineralogy of the basalt is quite uniform (because parent magma was uniform), and the rock is undeformed and unmetamorphosed (because of its young age). The grain size of this rock is very fine, but not as fine as the Spurwink Limestone used in this study. The homogeneity of basalt and its dark color result in rapid desiccation and quick thermal regulation to the environment [7].

Granite: The granite we used has been identified as two-mica granite on the Orrs Island 7 1/2′ United States Geographic Society quadrangle [53]. This is an intrusive rock of middle to late Devonian age (about 345–370 myo). Minerals include garnet, biotite and muscovite mica flakes, potassium feldspar (which gives the rock its yellow appearance), and significant quartz content. The two-mica granite has a large grain size (relative to the other rocks in this study), and a heterogeneous composition.

Schist: The schist used in this study is a metasedimentary rock of the Cape Elizabeth formation, and is estimated to be of Ordovician, Silurian, or Devonian age (345–500 myo) [53]. Such stratified metasedimentary rocks were laid down in a deep basin environment as fine clays. These layers were then compressed and deformed by the formation of the Atlantic Ocean, which resulted in their metamorphism. The clay component of this and the granite has come from the similar chemical weathering of crustal rocks, and thus, the schist and two-mica granite have similar mineralogy. The main chemical difference between the granite and schist is the presence of aluminum (from the fine clays) in schist. Because of the parallel foliation of the dominant mica flakes in this schist, the microheterogeneity (<500 μ) of the surface can vary significantly as a result of the exposure of different minerals. Large-scale surface topography (on the scale of meters) is also extremely variable in this rock formation because of the intense folding that the unit has undergone.

Limestone: The limestone we used belongs to the Spurwink Formation [53] and has a very similar tectonic history as the schist (Creasy, pers. comm.). This unit crops out as a thin exposure in the high intertidal zone of a sandflat on the west side of South Harpswell. Due to the thin exposure of this limestone unit (15–30 m), it was difficult to locate in the local study area (future workers are directed to limestone outcrops near Rockport Harbor, Maine). The Spurwink Limestone is a laminated rock that has undergone severe deformation. The dark component resembles the schist in composition and derivation, containing biotite, muscovite, quartz, and feldspar. The lighter colored member is very fine-grained gray limestone that contains 90% calcium carbonate ($CaCO_3$) by definition. Both members were equally represented in the experimental plates due to the highly mixed nature of the rock. The most significant difference between this rock type and others used in this study is the extremely small grain size, and the high levels of $CaCO_3$.

2.2. Fucus Spiralis

Zygotes of *Fucus spiralis*, a brown alga common to the high intertidal zone, were selected for study because of their relatively simple fertilization process and well documented attachment strategy. The *Fucus* zygote is an excellent model system for bioadhesion studies because it is unicellular, develops synchronously, and adheres rapidly to the substrate after fertilization [54]. *F. spiralis* was specifically chosen because it is monoecious (male and female reproductive structures on the same thallus), has a wide temporal range of reproductive activity, and is well studied [55]. Attachment of the *Fucus* zygote to the substrate is a multistage process. Shortly after fertilization, the negatively buoyant zygote drops to the substrate surface [3], and then adheres initially via an extracellular mucilaginous material of unknown chemical composition. After approximately 9 to 12 h, primary rhizoid development is initiated, and the zygote becomes permanently attached to the substrate [54].

2.3. Preparation of Settling Plates

Settlement plates were prepared from samples of basalt, granite, schist, and limestone collected from bedrock using a sledgehammer. No cobbles or boulders were used. Plates with natural settling surfaces were prepared by first cutting a 1 cm slab from the surface of a bedrock sample using a slab saw. Smooth surfaces were prepared by cutting 1 cm slabs from the interior of the bedrock samples. The angle of cuts on all samples was dictated by the foliation, cleavage, and mineral veins in the rocks, which affected the structural integrity of the samples. As the blade of the slab saw was lubricated with oil, cut slabs were vigorously washed to eliminate any possible effect on the adhesion of zygotes (cf. [56]). In order to remove microorganisms and eliminate the formation of a biofilm, the slabs were scrubbed in a hot solution of water, dilute hydrochloric acid, and detergent (Ajax), and then rinsed with fresh water. Slabs were then cut into 5 cm × 5 cm (+/−1 mm plates) with a rock saw (blade water-lubricated). All plates were rinsed in running fresh water for 24 h. Slab thicknesses of between 7 and 12 mm for the natural plates and 8 and 10 mm for the smooth plates were maintained to ensure minimal differences in surface properties (see [4]). Ten plates of each surface type of each rock type were prepared, resulting in a total of 80 plates (40 natural and 40 smooth).

2.4. Preparation of Zygote Solution

We conducted two settlement trials, one in February 1995 ('winter') and one in August 1995 ('summer'). We collected receptacles of *Fucus spiralis* near Gun Point, Maine (69°56′55″ W 43°45′55″ N) at low tide on 14 February 1995, and 13 August 1995, respectively. We only collected receptacles that were visibly producing gametes, and these were sealed in a plastic bag with absorbent tissue as a desiccant. Receptacles were stored at 5 °C for ten days after the February 14 collection, and two days after the August 13 collection until gamete release was induced in the laboratory. To obtain gametes, we placed receptacles in a freshwater-ice bath for five minutes. The receptacles were then desiccated in direct sunlight until they had swelled and emitted gametes (ca. 10 min), and then soaked in seawater until a sufficient gamete release had been observed (ca. 1 h in winter and 45 min in summer). After release, we manually agitated receptacles to shake off any remaining gametes. This solution was then placed in a growth chamber at 20 °C for 15 min on a magnetic stirring plate to keep the gametes in suspension and promote fertilization.

2.5. Settlement of Zygotes on Plates

Settlement plates were randomly arranged in two shallow metal trays (natural vs. smooth surfaces) in a 4 × 4 grid consisting of 40 plates (10 of each rock type). Natural and smooth plates were placed in different trays to minimize the variation in profile heights of the plates and thereby reduce the possibility that uneven plate heights would modify the water flow and influence settlement. There were no gaps left between plates, and the trays were larger than the 4 × 10 plate grid, leaving a 5 cm perimeter around settling plates. Fifteen minutes prior to zygote addition, artificial seawater (Instant Ocean, 32 psu, 5 °C) was added to the trays so that all plates were covered by ca. 2 cm of water. After fertilization (15 min), we poured 250 mL of the zygote solution (in suspension) over the settlement plates in a constant flow in a grid pattern to homogenize zygote densities over each plate. The settlement trays were left undisturbed under diffuse fluorescent light at room temperature (14–16 °C) for 1.5 h. The surface water was then carefully siphoned out of the trays with a plastic tube (6 mm inside diameter) placed at least 5 cm from any plate. The flow caused by siphoning, estimated by recording the time required for zygotes on the bottom surface of the tray to pass by 2 plates (10 cm), was approximately 1 cm/s.

Five plates of each rock type were chosen at random, using a table of random numbers, from the settlement-plate array for the wave treatment, and the remaining 'control' plates were sprayed with a fine mist of Instant Ocean (32 psu, 5 °C). These control plates were transferred to individual plastic containers with airtight lids and moist towels underneath and maintained in a horizontal position at 5 °C until zygotes were counted under a dissecting scope. From the remaining plates, one plate of each rock type to be tested for wave-treatment effects was placed in random order in the track of a wave tank (see [16,57] for description). The direction of water flow was recorded for each plate and the surface types (natural vs. smooth) were tested separately. A single 3 L wave was then released and allowed to wash over the plates and drain through holes behind the plates. We allowed the water to drain off the plates (ca. 30 s), and then the plates were carefully transferred to plastic containers with air-tight lids and handled as the control plates described above. The wave treatment experiment was repeated 4 times with rock plates randomly assigned to positions in the flume each time, leading to 5 individuals of each rock type exposed to a wave ('survival' after a single wave treatment), and 5 replicate plates of each rock type serving as controls (initial settlement only).

2.6. Counting Methodology

We counted zygotes using a dissecting microscope at 40× magnification. The 25 cm^2 surface of the plates was divided into nine evenly spaced 1 cm^2 quadrats. The peripheral 1 cm of the plates was avoided (to reduce edge effects). A humidity chamber to prevent desiccation was created by placing a moist towel beneath the plate and covering the plate with a gridded petri dish during counting. In order to reduce processing time, we randomly selected six of the nine 1 cm^2 areas. We counted the zygotes in each quadrat and calculated a mean number of zygotes per cm^2.

2.7. Statistical Analyses

Mean zygote densities per square centimeter from the initial attachment trial were compared using a three-way ANOVA with season, rock type, and wave exposure as main factors. Variances were not homogenous until the data were $\log_{10}$-transformed (F-max test, [58]). Separate 3-way ANOVAs were conducted for natural and smooth plates because surface types were tested independently. We used Tukey's HSD post hoc test to compare differences among rock types because there were no significant interactions in either of the ANOVAs.

Zygote densities at initial attachment were significantly different among rock types (Figure 1). It was not possible, therefore, to simply compare zygote densities after the wave treatment plates to reveal how successfully zygotes remained attached to different rock types. In order to examine the success of the initial attachment of zygotes to plates of both natural and smooth surfaces of different rock types following the wave treatment, we calculated zygote 'survivorship' [16]. This was calculated by dividing the mean zygote density found on each surface type and rock type following the wave treatment by the mean zygote density for each plate in each control. The resultant dataset consisted of 8 groups (4 rock types x 2 wave treatments) of 5 (number of replicates) percentages for each experiment (summer and winter). These percentages were then arcsine-transformed to make the variances homogenous [59]. The transformed data were analyzed using a two-way ANOVA with season and rock type as main factors. As with the analysis of patterns of initial attachment, we analyzed survivorship data for surface type (natural, smooth) using separate ANOVAs. The interaction term in these ANOVAs was not significant, so as with the 3-way ANOVAs, we used Tukey's HSD post hoc test to compare differences among rock types. We used the terms 'survival' and 'survivorship' to represent the percentage of attached zygotes persisting after the wave treatment, not in the biological sense of surviving for days beyond the treatment.

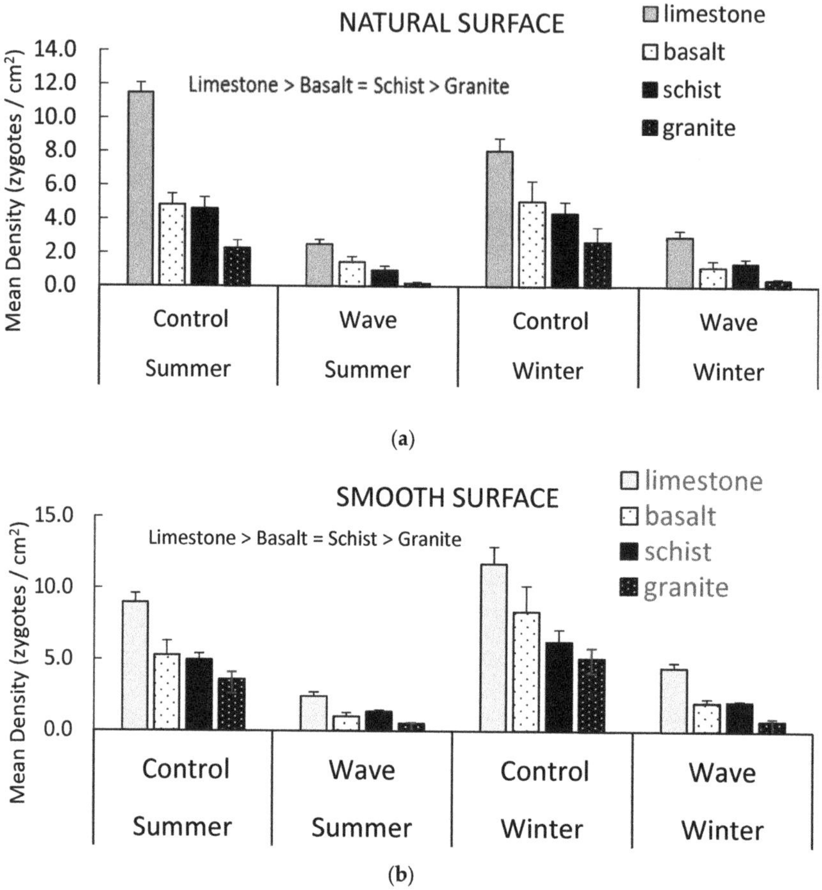

Figure 1. Mean density (zygotes per cm^2 + 1SE, N = 5) of *Fucus spiralis* zygotes on 4 rock types after 1.5 h of settlement (control) and 1.5 h of settlement followed by one wave in summer and winter experiments. A three-way ANOVA compared zygote density as a function of season, rock type, and wave treatment (wave or no wave) for each surface type: (**a**) natural and (**b**) smooth. The zygote density was higher in winter than summer on smooth plates ($p < 0.001$), but there was no significant difference ($p > 0.05$) between seasons on natural plates. The density was always significantly lower ($p < 0.0001$) regardless of rock type on plates subjected to a wave compared to control plates. The mean density among rock types was compared using Tukey's HSD post hoc test. For both natural and smooth plates, the density was significantly higher ($p < 0.0001$) on limestone than on basalt and schist, which were not significantly different from each other ($p > 0.05$); granite had a significantly ($p < 0.0001$) lower density than all other rock types.

3. Results

3.1. Initial Attachment

The mean number of zygotes per square centimeter that settled on plates in the winter ranged from 0.67 (natural granite receiving a wave treatment) to 14.2 (smooth limestone control) and in summer from 0.4 (natural granite receiving a wave treatment) to 13.4 (smooth and natural limestone control). There was a significant difference in mean zygote density among rock types and between wave treatments for both natural and smooth surfaces, but there was only a significant difference between seasons on the smooth surface substrate (Table 1). None of the interactions between factors were significant for either surface. Plates subjected to a wave always had lower zygote densities, between 61% (limestone, smooth, winter) and 91% (granite, natural, summer), than the control plates. On the smooth surface, the mean initial attachment was significantly greater ($p < 0.0001$) in the winter (5.1 zygotes per cm^2, SE = 1.3) than in the summer experiment (3.5 zygotes per cm^2, SE = 1.0) (Figure 1). For both natural and smooth plates, the density was significantly higher ($p < 0.0001$) on limestone than on basalt and schist, which were not significantly different from each other ($p > 0.05$); granite had a significantly ($p < 0.0001$) lower density than all other rock types. The pattern was the same before and after the wave treatment.

Table 1. Results of 3-way ANOVA analyzing the effect of season (summer and winter), rock type (limestone, schist, basalt, and granite), and wave or no wave on the density of zygotes recorded on settlement plates of natural and smooth surfaces. There were 5 replicate plates of each rock type and flow combination. Data were $\log_{10}$-transformed before analysis.

Natural Surface					
Source	*df*	*SS*	*MS*	*F*	*p*
Season	1	0.0001	0.0001	0.01	0.92
Rock type	3	2.3207	0.7735	53.9	0.0001
Wave	1	3.4149	3.4149	237.9	0.0001
Interaction (season x rock)	3	0.0374	0.0125	0.86	0.46
Interaction (season x wave)	1	0.0261	0.0261	1.82	0.18
Interaction (rock x wave)	3	0.0131	0.0044	0.30	0.82
Interaction (season x rock) x wave	3	0.0433	0.0144	1.01	0.39
Error	64	0.9185	0.0144		
Total	79	6.7742	0.0858		
Smooth Surface					
Source	*df*	*SS*	*MS*	*F*	*p*
Season	1	0.328	0.328	23.29	0.0001
Rock type	3	1.433	0.478	33.91	0.0001
Wave	1	4.065	4.065	288.68	0.0001
Interaction (season x rock)	3	0.025	0.008	0.6	0.62
Interaction (season x wave)	1	0.0004	0.0005	0.03	0.85
Interaction (rock x wave)	3	0.044	0.014	1.05	0.37
Interaction (season x rock) x wave	3	0.016	0.005	0.38	0.77
Error	64	0.901	0.014		
Total	79	6.812	0.086		

3.2. *Survivorship*

Across all treatments, survivorship averaged 27.3% (SE = 3.2) in the winter and 21.4% (SE = 2.5) in the summer. The results of the two-way ANOVAs comparing the effects of season and rock type on the survivorship of zygotes after a wave treatment were the same for natural and smooth surfaces (Table 2). The mean survivorship of zygotes was significantly affected by rock type, but not season, and there were no significant interactions in the ANOVAs. On natural rock, there was no significant difference ($p > 0.05$) in mean percent survival of zygotes among limestone, basalt, and schist treatments, but survival was significantly ($p < 0.01$) higher on these rock types than on granite (Figure 2). The survival of zygotes was similar on smooth and natural rock surfaces except that there was no significant difference ($p > 0.05$) in mean percent survival between basalt and granite.

(a)

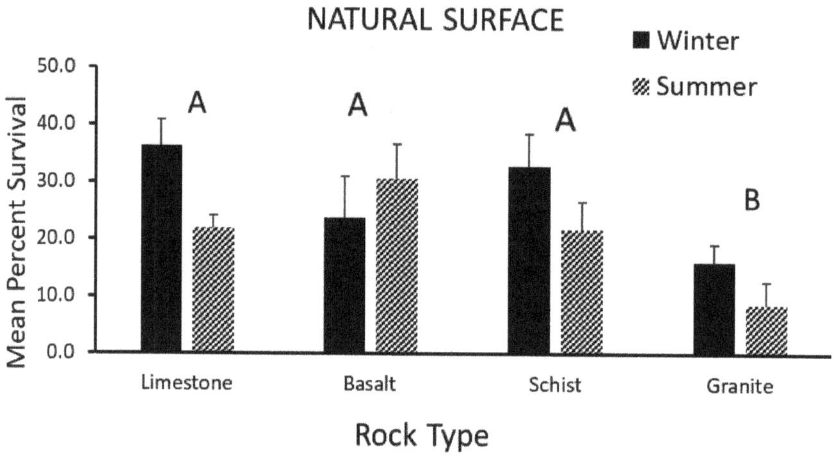

(b)

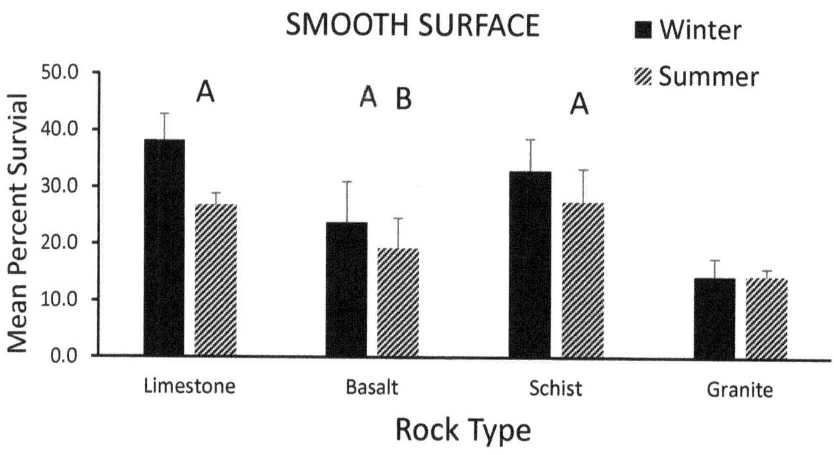

Figure 2. Mean percent survival (+1SE, N = 5) of *Fucus spiralis* zygotes after one-wave treatment in experiments with natural (**a**) and smooth (**b**) rock surfaces. Percentages are based on mean densities of no-wave treatment for each rock type. Arcsine-transformed proportions were compared between seasons and among rock types for each experiment using a two-way ANOVA. The season x rock type interaction was not significant in either ANOVA ($p > 0.05$). Means (by rock type) with a common letter over the bar are not significantly different from each other (Tukey's HSD post hoc test, $p < 0.05$).

Table 2. Results of 2-way ANOVA analyzing the effect of season (summer and winter) and rock type (limestone, schist, basalt, and granite) on the mean percent survivorship of zygotes after treatment by a wave. Survivorship is calculated as a percentage based on the mean zygote density of control plates in each rock and surface type. Data were arcsine-transformed before analysis. There were 5 replicates of each rock and surface type combination.

Natural Surface					
Source	*df*	*SS*	*MS*	*F*	*p*
Season	1	0.0908	0.0908	4.11	0.051
Rock type	3	0.3953	0.1318	5.96	0.002
Interaction (season x type)	3	0.1133	0.0377	1.71	0.181
Error	32	0.7076	0.0221		
Total	39	1.3071			
Smooth Surface					
Source	*df*	*SS*	*MS*	*F*	*p*
Season	1	0.0218	0.218	1.66	0.207
Rock type	3	0.2698	0.089	6.83	0.001
Interaction (season x type)	3	0.0258	0.0086	0.65	0.587
Error	32	0.4216	0.0132		
Total	39	0.7391			

4. Discussion

Zygote attachment and survivorship following a wave were greatest on limestone, least on granite, and intermediate on schist and basalt. A wave dislodged some initial settlers, but it did not change the general patterns of initial attachment we recorded. While we did not statistically compare natural and smooth surfaces (experimental design prohibits this), initial attachment patterns were the same between the two surfaces, and patterns of survival were very similar. These results suggest that minerology, fine-scale characteristics of the rock types, or some other characteristic of rock type, *and not contour* (natural vs. smooth), accounted for the differences we observed.

We examined the initial attachment phase, which may not indicate the effect of bedrock type on zygotes attached for longer periods [16]. The exact timing of the transition from initial mucilaginous adhesion to permanent rhizoidal adhesion is not clear. Initial attachment may occur about one hour after fertilization [43], while another study [54] reported that initial attachment occurs from 3 to 6 h after fertilization and refers to the time after fertilization as the 'pre-adhesive' stage. The zygote undergoes an adhesive maturation phase from 7 to 9 h after fertilization [54], which involves the hardening or gelling of the polysaccharide [3,44]. The chemical composition of the adhesives involved in the attachment of *Fucus* zygotes is largely unknown [54,60]. There is evidence, however, that polysaccharide–protein complexes are involved [44], and that two polysaccharides, an alginic acid complex and sulfated fucan, and polyphenols make up the adhesive secretion [54,61]. In our study, we had an initial attachment of up to 120,000 per m^2, and approximately 25% were able to maintain adherence after wave exposure. This indicates that even if 3–6 h may be termed a 'pre-adhesion' stage, the shorter time used in our experiment is certainly enough for good settlement and persistence following a wave.

4.1. Surface Roughness

The initial attachment of the nonmotile zygote is, in part, dependent on the physical characteristics of the substrate. The most implicated substrate characteristic in earlier studies of settlement and attachment of marine organisms is surface roughness, or contour [34–37,62]. Surface roughness has been quantified with the development of the engineering roughness index, a dimensionless parameter that relates the proportion of the surface that is recessed and the amount of freedom a spore has to move based on surface topography [42,63]. Zygotes are likely to find refuge from turbulent flow in rock substrates that are rough enough to offer cracks and crevices that protect germlings [35,44]. Such crevices may be present within a rock type, or between juxtaposed rock types. Most studies on the effects of surface roughness on settling behavior have used artificial substrates (but see [42,64]). Typically, roughness is examined by cutting or drilling grooves or pits of varying size in otherwise homogenous substrates [16,34,62,64,65], by attaching silica grains of different sizes to a homogenous surface [31,49], or by producing a range of rough surfaces using varying grades of abrasive materials on otherwise smooth materials [66]. These results are helpful in indicating what is possible and, therefore, what can and cannot be expected of larvae (as in these studies), or algal propagules [67]. Although Caffey [68] argues that artificial substrates provide little insight as to the factors influencing attachment in the natural environment, they can, if consistent with field observations, provide valuable corroborative data on control mechanisms.

We did not statistically compare natural and smooth surfaces for the initial attachment or survivorship, because plates of different surface types were held in separate trays (i.e., there was a lack of interspersion of treatments). Settling trays were identical and settlement occurred simultaneously under identical conditions, so it is unreasonable to expect a significant tray effect. Furthermore, analyzing survivorship instead of settler density removes much of the possible variability in the concentrations of the zygote solutions added to each tray. Nevertheless, comparisons between surface types are qualitative.

There were few differences in our study in patterns of initial attachment or survivorship following a wave between plates with natural and smooth surfaces (Figures 1 and 2). With both surface types, limestone, which has the finest grain size and least inherent surface roughness, proved to be most suitable, and granite, which has the largest grain size of the rock types tested, the least suitable for zygote attachment. As this pattern persisted for both natural and smooth surfaces, a factor other than the contour of the surface must explain the variation among rock types.

Although topography on the scale of centimeters to tens of meters is important in assessing intertidal environments and community structure, the effect of substratum microheterogeneity, or texture must also be addressed to investigate settlement processes. For a range of taxa, a microtopography smaller than the length of an organism's attachment point generally reduces settlement [42]. Algal spores and zygotes are typically about 5 μm in size, so the most appropriate scale to measure the surface texture of substrates in algal settling experiments may be the grain size of minerals [64]. Mineral grains are the units that are cleaved from rocks as a result of physical weathering. This suggests that grain size can dictate the surface texture in natural environments because the weathering of rocks with large grains will expose larger mineral surfaces than those with small grain size, although it may be difficult to be so conclusive about this parameter for rock types with heterogeneous mineralogic composition (e.g., two-mica granite). Surface heterogeneity on this scale has been shown to affect community structure in many species [3,35,43,44]. Although on a large scale, the natural surface of the rocks used in this study can be classified in order of increasing roughness (limestone-basalt-schist-granite), this may not be the appropriate scale for our study. Surface roughness as it applies to zygotes has been defined as the number of surface planes of the substrate encountered by the zygote [44,62]. Zygotes may experience more surface planes on a rock with small grain size, and fewer on a rock with large grain size. Thus, the scale of roughness as it applies to zygotes may be the converse from that determined on a large scale. Future studies should investigate differences in roughness among natural substrata on scales relevant to zygotes.

The limestone used in this study was very fine grained [53], with grains smaller than the typical size of the zygotes used. Such substrata would present a greater number of surface planes for mucilaginous adhesion than would a substrate with large grain size (i.e., granite). This classification would also be relevant for surfaces cut with a diamond blade. The large mineral grains of granite (due to slow cooling of parent magma) and schist (due to recrystallization of marine sediments) were large enough so that individual mineral grains were actually cut, leaving a smooth surface, while the small grains of basalt and limestone remained intact. Thus, any advantage in attachment strength due to grain size would persist even if the natural rock surface were cut, which is what we observed in documenting a few differences in the pattern of survival between natural and smooth surfaces (Figure 2). This hypothesis is consistent with the results of one study [64] in which the settlement of barnacle cyprids was enhanced on fine-grained natural rock plates and inhibited on course plates, even though all settlement plates were machine-cut and polished. Our result, however, does not agree with more recent work that suggests that features the same size or smaller than the settling organisms inhibit settlement [69]. Scardino et al. [69], however, found that for very small motile propagules on the order of 7 μm, the effect of attachment points was weak, so the relationship between surface features and sizes of settling propagules may not apply to *Fucus* spores.

It is important to note that while we found no differences between smooth and rough plates in the pattern of initial attachment of zygotes, our results may not be easily extrapolated to the field. In the field, roughness exists on many scales that we did not test in our study. These roughness differences are due in part to rock minerology, wave energy, crystal size, geological processes juxtaposing rock types, and erosional history, and can well influence very local flow patterns and potential attachment angles—and thus, settlement/recruitment. This should be considered in field experiments that may follow from our study.

4.2. Chemical Interactions between Adhesives and Substrata

As discussed above, there is evidence that a polysaccharide chain of alginic acid comprises one of the initial attachment adhesives in the Phaeophyta. Alginic acid is a linear 1,4-linked block copolymer comprised of beta-D-mannuronic acid (M) and alpha-L-guluronic acid (G) residues [3]. The strength of polysaccharide gels such as alginate are increased by the binding of Ca^{2+} [3]. Due to differences in structure, polyguluronic acid has a greater affinity for Ca^{2+} than for polymannuronic acid. The structure of polyguluronic acid better accommodates insertion of the calcium ion, resulting in cross-linking of the polymer chain and a stiffer gel. Higher concentrations of Ca^{2+} ions at the surface interface of specific bedrock types may result in more cross-linking of the alginic acid polymer chain, thereby producing a more rigid gel and a stickier adhesive. It is probable that such a difference in Ca^{2+} concentration exists between limestone and granite because the limestone consists of at least 90% $CaCO_3$, which readily dissociates. The availability of free calcium ions may partially explain a higher initial attachment and survivorship of zygotes on limestone compared to some other rock types.

4.3. Free Energy of Rock Surfaces

Applied research into fouling by marine organisms has focused on the alteration of potential settling surfaces to prevent adhesion. The most implicated physical–chemical characteristic of substrates is the surface free energy, also referred to as wettability [44]. This is defined as an unsatisfied bonding potential at the surface of a material that results in the greater propensity of that surface to bond to dissimilar particles in the surrounding water [45]. Increased surface free energies have been shown to increase the adhesive strength of algal spores [41,44], change the shape of rhizoids that are produced [43], and enhance the attachment of barnacle and bryozoan larvae [70]. The early settlement of meiospores is best described by water contact angle [41], which is related to surface free energy [41]. As this research is generally applied to the development of nontoxic coatings for use in marine industries, artificial substrates such as synthetic polymers, glass, ceramic tiles, and Teflon have been studied in adhesive comparisons [41,43,44]. The homogenous composition of these surfaces and their predictable matrices make it possible to measure the free energy of these surfaces and then to compare these measurements with strength of attachment.

The modeling of natural substrates such as rock, however, is much more complex than for artificial surfaces due to the heterogenous nature of natural surfaces. The heterogenous mineralogy of natural rock makes surface free energy difficult to measure (Berry, pers. com.). Even the lattice of a simple salt such as NaCl is heterogenous (Na is not the same as Cl), so in introducing the highly variable chemical composition of rocks, it is probable that the surface energy of the substrate will vary over small scales within one sample of any given rock type [71]. Furthermore, there will likely be differences in the free energy of surfaces based on the cleavage of the crystals. Adamson [72] indicated that a clean cleavage of a crystal (i.e., through natural weathering) will have a different and probably lower surface energy than would a ground or abraded surface of the same material.

Differences in surface free energy among substrates are also expected to decrease with prolonged immersion in sea water [70,73]. The adsorption of organic and possibly inorganic molecules can occur on clean surfaces, creating a 'conditioning' film within minutes of immersion in sea water [45,74]. This film can alter physio-chemical properties such as the surface energy of the original surface and the effect of the settlement of algae (see [75] for review). This effect is important to note when comparing short-term laboratory settling experiments using fresh substrata to possible long-term effects of substrates on community composition in the field. Holm et al. [73] concluded that while surface energy may be important in determining initial settlement patterns in some fouling communities, it is probably not a major influence on long-term community development. Conversely, Callow and Fletcher [45] reported that, although surfaces with different original surface energies acquired similar films, differences in attachment persisted after immersion.

No direct measurements of surface free energy were available for comparison in our study. Nonetheless, it is possible that differences in surface charges contributed to the differences in survivorship among our substrates, and this possibility should be explored in future experiments. The extreme contrast between the chemical composition of limestone and granite, for example, make it probable that differences in surface energies of these substrates exist, and that limestone is the more highly charged of the two (Berry, pers. com.). The surface free energy of limestone aggregate is greater than that for granite aggregate [76], which is consistent with this speculation. While the chemical characteristics of aggregates and natural rocks in the field are not expected to be the same, the relative differences might be. The higher settlement on limestone compared to granite we measured would agree with Callow and Fletcher [45] who found that adhesion for a wide range of organisms is higher on surfaces with higher surface energies.

4.4. Other Factors

Other factors that have been suggested to influence algal and larval attachment are salinity gradients [77], color differences [7], pH gradients [3], substrate hardness [43], and the presence of microbial films [78,79]. The effects of color on attachment have been attributed to differences in thermal properties of dark vs. light substrates [7]. As our experiments were run indoors under diffuse light, color is unlikely to be important (but may certainly be an important contrast between our experiments and settlement under field conditions). A pilot study indicated that there was no detectable difference (to +/− 0.01) in pH at the substrate–water interface after 1.5 h among the four rock types. Such an effect cannot be completely discounted, however, because zygotes may be able to detect much smaller differences in pH than the instrument used could measure [80]. Substrate hardness has been implicated in affecting the adult communities found on rocks because softer rocks (such as limestone) will erode more rapidly than granite or basalt. Hardy and Moss [43] concluded that ephemeral species tend to grow on soft substrates, while perennials occur on hard substrates. Variations in microbial films affect the surface tension of substrates and, hence, attachment [70,81,82]. Microbial films were not likely a factor in our experiments, because settling plates were rigorously cleaned and sterilized before testing, experimental trials were short, and the artificial seawater used contained far fewer micro-organisms than did natural seawater. Finally, Amsler et al. [80] reported that surfaces concentrate several nutrients, which stimulate kelp spore chemotaxis or settlement. It is plausible that a similar effect may occur in *Fucus*, but the measurement of this effect was beyond the scope of our study.

5. Conclusions

Our results indicate that the adhesion of *F. spiralis* zygotes is influenced by characteristics of natural rocks other than surface contour and that the general pattern of initial attachment to different rock types we observed persists after the zygotes are exposed to a wave. The microheterogeneity (texture) of substrates based on grain size, the effect of variations in surface charge of natural substrates, and the chemical interaction of the initial adhesives at the substrate interface are suggested as possible factors affecting initial attachment success, consistent with our results of contrasting attachment and survivorship of the zygote on different rock types. The difficulty in assessing characteristics such as free energy in natural substrates and the lack of data on the precise chemical composition of the adhesives produced by settling propagules leave some of these questions open for further study. It is clear, however, that natural substrates present a higher degree of variation in factors affecting settlement than simply different degrees of roughness and that these factors operate within the first few hours of zygote attachment. A further study of this interaction and patterns of settlement among naturally occurring rock types may contribute to a better understanding of attachment of algal spores and to the distribution and dynamics of natural communities on rocky shores. Finally, 26 years have elapsed since our study and, while we do not opine that zygote affinity for different rock surfaces changed as a result of a circa 0.5–1 °C increase in water temperature [83], many factors that affect settlement could have changed over this period, warranting repeating these experiments.

Supplementary Materials: The following are available online at https://www.mdpi.com/article/10.3390/jmse9090927/s1, Figure S1: Images of the bedrock types used in the settlement experiments.

Author Contributions: All authors contributed to the research idea and experimental design. The experiments were performed by D.C.A. in the laboratory of and with assistance of R.L.V. The first draft and initial data analysis were completed by D.C.A. Subsequent drafts and data analysis were performed by W.G.A.J. and P.E.R. Funding for the research was acquired by R.L.V. R.L.V. read and agreed to publish an earlier version of this manuscript, but poor health prevented him from reading the current version. All authors have read and agreed to the published version of the manuscript.

Funding: The research was funded by the NOAA Sea Grant program of the University of Maine (#NA89AADSG020 and #NA16RGO157) and the USDA Hatch funds to Maine Agricultural and Forestry Experiment Station.

Institutional Review Board Statement: Not applicable.

Data Availability Statement: Data are available upon request from the first author.

Acknowledgments: We appreciate the use of the Bates College Geology Department rock saw, the use of the Fisheries and Aquaculture Research Group (FARG) facilities, and the assistance of FARG personnel with the seawater system. Comments by an anonymous reviewed significantly improved the manuscript.

Conflicts of Interest: The authors declare no conflict of interest.

References

1. Lewis, J.R. Water movements and their role in rocky shore ecology. *Sarsia* **1968**, *34*, 13–36. [CrossRef]
2. Crisp, D.J. Factors influencing the settlement of marine invertebrate larvae. In *Chemoreception in Marine Organisms*; Grant, P.T., Mackie, A., Eds.; Academic Press: London, UK, 1974; pp. 177–265.
3. Lobban, C.S.; Harrison, P.; Duncan, A.J. *The Physiological Ecology of Seaweeds*; Cambridge University Press: New York, NY, USA, 1985.
4. Mullineux, L.S.; Butnam, C.A. Initial contact, exploration, and attachment of barnacle (*Balanus amphitrite*) cyprids settling in flow. *Mar. Biol.* **1991**, *110*, 93–103. [CrossRef]
5. Smith, C.M. Diversity in intertidal habitats: An assessment of the marine algae of select islands in the Hawaiian Archipelago. *Pac. Sci.* **1992**, *46*, 466–479.

6. Gaines, S.; Brown, S.; Roughgarden, J. Spatial variation in larval concentrations as a cause of spatial variation in settlement of the barnacle *Balanus glandula*. *Oecologia* **1985**, *67*, 267–272. [CrossRef]
7. Raimondi, P. Rock type affects settlement, recruitment, and zonation of the barnacle *Chthamalus anispoma* Pilsbury. *J. Exp. Mar. Biol. Ecol.* **1988**, *123*, 253–267. [CrossRef]
8. Menge, B.A. Relative importance of recruitment and other causes of variation in rocky intertidal community structure. *J. Exp. Mar. Biol. Ecol.* **1991**, *146*, 69–100. [CrossRef]
9. Valdivia, H.; Aguilera, M.A.; Navarrete, S.A.; Broitman, B.R. Disentangling the effects of propagule supply and environmental filtering on the spatial structure of a rocky shore metacommunity. *Mar. Ecol. Prog. Ser.* **2015**, *538*, 67–79. [CrossRef]
10. Pornerat, C.M.; Reiner, E.R. The influence of surface angle and light on the attachment of barnacles and other sedentary organisms. *Biol. Bull.* **1942**, *82*, 14–25.
11. Barnes, H.; Crisp, D.J.; Powell, H.T. Observations on the orientation of some species of barnacles. *J. Anim. Ecol.* **1951**, *20*, 227–241. [CrossRef]
12. Vaselli, S.; Bertocci, I.; Maggi, E.; Benedetti-Cecchi, L. Assessing the consequences of sea level rise: Effects of changes in the slope of the substratum on sessile assemblages of rocky shorelines. *Mar. Ecol. Prog. Ser.* **2008**, *368*, 9–22. [CrossRef]
13. Firth, L.B.; White, F.J.; Schofield, M.; Hanley, M.E.; Burrows, M.T.; Thompson, R.C.; Skov, M.W.; Evans, A.J.; Moore, P.J.; Hawkins, S.J. Facing the future: The importance of substratum features for ecological engineering of artificial habitats in the rocky intertidal. *Mar. Fresh. Res.* **2014**, *67*, 131–143. [CrossRef]
14. Paine, R.T. Disaster, catastrophe, and local persistence of the sea palm *Postelsia palmaeformis*. *Science* **1979**, *17*, 685–687. [CrossRef] [PubMed]
15. Ricketts, E.F.; Calvin, J.; Hedgpeth, J.W.; Phillips, D.W. *Between Pacific Tides*, 5th ed.; Stanford Univ. Press: Palo Alto, CA, USA, 1985.
16. Vadas, R.L.; Wright, W.A.; Miller, S.L. Recruitment of *Ascophyllum nodosum*: Wave action as a source of mortality. *Mar. Ecol. Prog. Ser.* **1990**, *61*, 263–272. [CrossRef]
17. Petraitis, P.S.; Methratta, E.T.; Rhile, E.C.; Vidargas, N.A.; Dudgeon, S.R. Experimental confirmation of multiple community states in a marine ecosystem. *Oecologia* **2009**, *161*, 139–148. [CrossRef] [PubMed]
18. Menge, B.A.; Menge, D.N.L. Dynamics of coastal meta-ecosystems: The intermittent upwelling hypothesis and a test in rocky intertidal regions. *Ecol. Monogr.* **2013**, *83*, 283–310. [CrossRef]
19. Ibanez-Erquiaga, B.; Pacheco, A.S.; Rivadeneira, M.M.; Tejada, C.L. Biogeographical zonation of rocky intertidal communities along the coast of Peru (3.5–13.5° S Southeast Pacific). *PLoS ONE* **2018**, *13*, e0208244. [CrossRef] [PubMed]
20. Fenberg, P.B.; Menge, B.A.; Raimondi, P.T.; Rivadeneira, M.M. Biogeographic structure of the northeastern Pacific rocky intertidal: The role of upwelling and dispersal drive patterns. *Ecography* **2015**, *38*, 83–95. [CrossRef]
21. Grabowska, M.; Grzelak, K.; Kuklińsk, P. Rock encrusting assemblages: Structure and distribution along the Baltic Sea. *J. Sea Res.* **2015**, *103*, 24–31. [CrossRef]
22. Malm, T.; Kautsky, L.; Claesson, T. The density and survival vesiculous L. (Fucales, Phaeophyta) on different bedrock types on a Baltic Sea moaraine coast. *Botanica Mar.* **2003**, *46*, 256–262. [CrossRef]
23. Guidetti, P.; Bianchi, C.N.; Chiantore, M.; Schiaparelli, S.; Morri, C.; Cattaneo-Vietti, R. Living on the rocks: Substrate mineralogy and the structure of subtidal rocky substrate communities in the Mediterranean Sea. *Mar. Ecol. Prog. Ser.* **2004**, *274*, 57–68. [CrossRef]
24. Sempere-Valverde, J.; Ostalé-Valriberas, E.; Farfán, G.M.; Espinosa, F. Substratum type affects recruitment and development of marine assemblages over artificial substrata: A case study in the Albaron Sea. *Est. Coast. Shelf Sci.* **2018**, *204*, 56–65. [CrossRef]
25. Liversage, K.; Janetzki, N.; Benkendorff, K. Association of benthic fauna with different rock types, and evidence of changing effects during succession. *Mar. Ecol. Prog. Ser.* **2014**, *505*, 131–143. [CrossRef]
26. Cerrano, C.; Arillo, A.; Bavestrello, G.; Benatti, U.; Calcinai, B.; Cattaneo-Vietti, R.; Cortesogno, L.; Gaggero, L.; Giovine, M.; Puce, S.; et al. Organism-quartz interactions in structuring benthic communities. Towards a marine bio-mineralogy? *Ecol. Lett.* **1999**, *2*, 1–3. [CrossRef]
27. Bavestrello, G.; Bianchi, C.N.; Calcinai, B.; Cattaneo-Vietti, R.; Cerrano, C.; Morri, C.; Puce, S.; Sara, M. Bio-mineralogy as a structuring factor for marine epibenthic communities. *Mar. Ecol. Prog. Ser.* **2000**, *193*, 241–249. [CrossRef]
28. Moore, H.B.; Kitching, J.A. The biology of *Chthamalus stellatus* (Poli). *J. Mar. Biol. Assoc. UK* **1939**, *23*, 521–541. [CrossRef]
29. Crisp, D.J.; Barnes, H. The orientation and distribution of barnacles and settlement with particular reference to surface contour. *J. Anim. Ecol.* **1954**, *23*, 142–162. [CrossRef]
30. Crisp, D.J.; Ryland, J.S. Influence of filming and surface texture on the settlement of marine organisms. *Nature* **1960**, *185*, 119. [CrossRef]
31. Harlin, M.M.; Lindbergh, J. Selection of substrata by seaweeds: Optimal surface relief. *Mar. Biol.* **1977**, *40*, 33–40. [CrossRef]
32. Wethey, D.S. Ranking of settlement cues by barnacle larvae: Influence of surface contour. *Bull. Mar. Sci.* **1986**, *39*, 393–400.
33. Barkai, A.; Branch, G.B. The influence of predation and substrate complexity on recruitment to settlement plates: A test of the theory of alternative states. *J. Exp. Mar. Biol. Ecol.* **1988**, *124*, 215–237. [CrossRef]
34. Chabot, R.; Bourget, E. Influence of substratum heterogeneity and settled barnacle density on the settlement of Cypris larvae. *Mar. Biol.* **1988**, *97*, 45–56. [CrossRef]

35. Le Tourneux, F.; Bourget, E. Importance of physical and biological settlement cues used at different spatial scales by the larvae of *Semibalanus balanoides*. *Mar. Biol.* **1988**, *97*, 57–66. [CrossRef]
36. Wells, J.; Moll, E.J.; Bolton, J.J. Substrate as a determinant of marine intertidal algal communities at Smitswinkle Bay, False Cay, Cape. *Mar. Bot.* **1989**, *32*, 499–502. [CrossRef]
37. Anderson, M.J.; Underwood, A.J. Effects of substratum on the recruitment and development of an intertidal estuarine fouling assemblage. *J. Exp. Mar. Biol. Ecol.* **1994**, *184*, 217–236. [CrossRef]
38. Gersun, L.; Anderson, R.J.; Hart, J.R.; Maneveldt, G.W.; Bolton, J.J. Sublittoral seaweed communities on natural and artificial substrata in a high-latitude coral community in South Africa. *Afr. J. Mar. Sci.* **2016**, *38*, 303–316. [CrossRef]
39. Chase, A.L.; Dijkstra, J.A.; Harris, L.G. The influence of substrate material on ascidian larval settlement. *Mar. Poll. Bull.* **2016**, *106*, 35–42. [CrossRef] [PubMed]
40. Brzozowska, A.M.; Maassen, S.; Rong, R.G.Z.; Benke, P.I.; Lim, C.-S.; Marzinelli, E.M.; Janczewski, D.; Teo, S.L.-M.; Vancso, G.J. Effect of variations in micropatterns and surface modulus on marine fouling of engineering polymers. *Appl. Mater. Interfaces* **2017**, *9*, 17508–17516. [CrossRef]
41. Kerrison, P.; Stanley, M.; Mitchell, E.; Cunningham, L.; Hughes, A. A life-stage conflict of interest in kelp: Higher meiospore settlement where sporophyte attachment is weak. *Algal Res.* **2018**, *35*, 309–318. [CrossRef]
42. Erramilli, S.; Genzer, J. Influences of surface topography attributes on settlement and adhesion of natural and synthetic species. *Soft Matter* **2019**, *15*, 4045–4067. [CrossRef]
43. Hardy, F.G.; Moss, B.L. The effects of the substratum on the morphology of the rhizoids of *Fucus* germlings: *Est. Coast. Mar. Sci.* **1979**, *9*, 577–584. [CrossRef]
44. Fletcher, R.L.; Callow, M. The settlement, attachment, and establishment of marine algal spores. *Brit. Phycol. J.* **1992**, *27*, 303–329. [CrossRef]
45. Callow, M.E.; Fletcher, R.L. The influence of low surface energy materials on bioadhesion—A review. *J. Int. Biodeterior. Biodegrad.* **1994**, *34*, 333–348. [CrossRef]
46. Pradhan, S.; Kumar, S.; Mohanty, S.; Nayak, S.K. Environmentally Benign Fouling-Resistant Marine Coatings: A Review. *Polym.-Plast. Technol. Eng.* **2019**, *58*, 498–518. [CrossRef]
47. Leonardi, A.K.; Ober, C.K. Polymer-based marine antifouling and fouling release surfaces: Strategies for synthesis and modification. *Ann. Rev. Chem. Biomol. Engin.* **2019**, *10*, 241–264. [CrossRef]
48. Chapman, A.R.O. Functional ecology of fucoid algae: Twenty-three years of progress. *Phycologia* **1995**, *34*, 1–32. [CrossRef]
49. Norton, T.A.; Fetter, R. The settlement of Sargassum muticum propagules in stationary and flowing water. *J. Mar. Biol. Assoc. UK* **1981**, *61*, 929–940. [CrossRef]
50. Le, H.N.; Hughes, A.D.; Kerrison, P.D. Early development and substrate twice selection for the cultivation of *Sargassum muticum* (Yendo) Fensholt under laboratory conditions. *J. Appl. Phycol.* **2018**, *30*, 2475–2483. [CrossRef]
51. Taylor, D.I.; Schiel, D.R. Wave-related mortality in zygotes of habit-forming algae from different exposures in southern New Zealand: The importance of 'stickability'. *J. Exper. Mar. Biol. Ecol.* **2003**, *290*, 229–245. [CrossRef]
52. Dimartino, S.; Mather, A.V.; Nowell-Usticke, J.S.; Fischer, B.; Nock, V. Investigation of the adhesive from Hormosira banksia germlings and its performance over different material surfaces and topographies. *Inter. J. Adhes. Adhes.* **2017**, *75*, 114–123. [CrossRef]
53. Hussey, A.M., II; Berry, H.N., IV. *Bedrock Geology of the Bath 1:100,000 Map Sheet, Coastal Maine*; Map 02–152; Maine Geological Survey: Augusta, ME, USA, 2002.
54. Vreeland, V.; Grotkopp, S.; Espinosa, S.; Quiroz, D.; Laetsch, W.M.; West, J. The pattern of cell wall adhesive formation by Fucus zygotes. *Hydrobiologia* **1993**, *260/261*, 485–491. [CrossRef]
55. Niemeck, R.A.; Mathieson, A.C. An ecological study of *Fucus spiralis* (L.). *J. Exp. Mar. Biol. Ecol.* **1976**, *24*, 33–48. [CrossRef]
56. Thelin, I. Effects, en culture, de deux petroles bruts et d'un petrolier sur les zygotes et les plantules, de *Fucus serratus* Linnaeus (Fucales, Phaeophyceae). *Bot. Mar.* **1981**, *24*, 515–519. [CrossRef]
57. Vadas, R.L.; Johnson, S.; Norton, T.A. Recruitment and mortality of early post-settlement stages of benthic algae. *Brit. Phycol. J.* **1992**, *27*, 331–351. [CrossRef]
58. Rohlf, F.J.; Sokal, R.R. *Statistical Tables*; W. H. Freeman and Company: San Francisco, CA, USA, 1969.
59. Zar, J.H. *Biostatistical Analysis*, 2nd ed.; Prentice Hall: Englewood Cliffs, NJ, USA, 1984.
60. Potin, P.; Leblanc, C. Phenolic-based adhesives of marine brown algae. In *Biological Adhesives*; Smith, A.M., Callow, J.A., Eds.; Springer: Berlin/Heidelberg, Germany, 2006; pp. 105–124.
61. Britton, R.; Ben-Yehuda, M.; Davidovich, M.; Balazs, Y.; Potin, P.; Delage, L.; Colin, C.; Bianco-Peled, H. Structure of algal-born phenolic polymeric adhesives. *Macromol. Biosci.* **2006**, *6*, 737–746. [CrossRef] [PubMed]
62. Bourget, E.; DeGuise, J.; Daigle, G. Scale of substratum heterogeneity structural complexity, and the early establishment of a marine epibenthic community. *J. Exp. Mar. Biol. Ecol.* **1994**, *181*, 31–51. [CrossRef]
63. Schumacher, J.F.; Carman, M.L.; Estes, T.G.; Feinberg, A.W.; Wilson, L.H.; Callow, M.E.; Callow, J.A.; Finlay, J.A.; Brennan, A.G. Engineered Antifouling Microtopographies—Effect of Feature Size, Geometry, and Roughness on Settlement of Zoospores of the Green Alga *Ulva*. *Biofouling* **2007**, *23*, 55–62. [CrossRef] [PubMed]
64. Holmes, S.P.; Sturgess, C.J.; Davies, M.S. The effect of rock-type on the settlement of *Balanus balanoides* (L.) cyprids. *Biofouling* **1997**, *11*, 137–147. [CrossRef]

65. McGuinness, K.A.; Underwood, A.J. Habitat structure of communities on intertidal boulders. *J. Exp. Mar. Biol. Ecol.* **1986**, *104*, 97–123. [CrossRef]
66. Barnes, H. Surface roughness and the settlement of Balanus balanoides. *Arch. Soc. Zool.-Bot. Fenn. Vanamo* **1956**, *10*, 164–168.
67. Caffey, H.M. No effect of naturally occurring rock types on settlement or survival of the intertidal barnacle *Tessoropora rosea*. *J. Exp. Mar. Biol. Ecol.* **1982**, *63*, 119–132. [CrossRef]
68. Caffey, H.M. Spatial and temporal variation in settlement and recruitment of intertidal barnacles. *Ecol. Monogr.* **1994**, *55*, 313–332. [CrossRef]
69. Scardino, A.J.; Guenther, J.; de Nys, R. Attachment point theory revisited: The fouling response to a microtextured matrix. *Biofouling* **2008**, *24*, 45–53. [CrossRef] [PubMed]
70. Roberts, D.; Rittschof, D.; Holm, E.; Schmidt, A.R. Factors influencing initial larval settlement temporal, spatial and surface molecular components. *J. Exp. Mar. Biol. Ecol.* **1991**, *150*, 203–211. [CrossRef]
71. Jaycock, M.J.; Parfitt, G.D. *Chemistry of Interfaces*; John Wiley and Sons: New York, NY, USA, 1981.
72. Adamson, A.W. *Physical Chemistry of Surfaces*; John Wiley and Sons: New York, NY, USA, 1982.
73. Holm, E.R.; Cannon, G.; Roberts, D.; Schmidt, A.R.; Sutherland, J.P.; Rittschof, D. The influence of initial surface chemistry on development of the fouling community at Beaufort North Carolina. *J. Exp. Mar. Biol. Ecol.* **1997**, *215*, 189–203. [CrossRef]
74. Grzegorczyk, M.; Pogorzelski, S.J.; Pospiech, A.; Boniewicz-Szmyt, B. Monitoring of marine biofilm formation dynamics at submerged solid surfaced with miltitechnique sensors. *Front. Mar. Sci.* **2018**, *10*, 363. [CrossRef]
75. Mieszkin, S.; Callow, M.E.; Callow, J.A. Interactions between microbial biofilms and marine fouling algae: A mini review. *Biofouling* **2013**, *29*, 1097–1113. [CrossRef]
76. Hou, Y.; Xiaoping, J.; Li, J.; Xianghang, L. Adhesion between asphalt and recycled concrete aggregate and its impact on properties of asphalt mixture. *Materials* **2018**, *11*, 2528. [CrossRef]
77. Wright, J.; Reed, R.H. Effects of osmotic stress on gamete size, rhizoid initiation and germling growth in fucoid algae. *Brit. Phycol. J.* **1990**, *25*, 149–155. [CrossRef]
78. Maki, J.S.; Rittschof, D.; Schmidt, A.R.; Snyder, A.G.; Mitchell, R. Factors controlling attachment of bryozoan larvae: A comparison of bacterial films and unfilmed surfaces. *Biol. Bull.* **1989**, *177*, 295–302. [CrossRef]
79. Becker, K. Attachment strength and colonization patterns of two macrofouling species on substrata with different surface tension (in situ studies). *Mar. Biol.* **1993**, *117*, 301–309. [CrossRef]
80. Allen, T.F.H. Scale in microscopic algal ecology: A neglected dimension. *Phycologia* **1977**, *16*, 253–257. [CrossRef]
81. Amsler, C.D.; Reed, D.C.; Neushul, M. The microclimate inhabited by macroalgal propagules. *Brit. Phycol. J.* **1992**, *27*, 253–270. [CrossRef]
82. Maki, J.S.; Rittschof, D.; Mitchell, R. Inhibition of larval barnacle attachment to bacterial films: An investigation of physical properties. *Microb. Ecol.* **1992**, *23*, 97–106. [CrossRef] [PubMed]
83. Pershing, A.J.; Alexander, M.A.; Brady, D.C.; Brickman, D.; Curchitser, E.N.; Diamond, A.W.; McClenachan, L.M.; Mills, K.E.; Nichols, O.C.; Pendleton, D.E.; et al. Climate impacts of the Gulf of Maine ecosystem: A review of observed and expected changes in 2050 from rising temperatures. *Elem. Sci. Anth.* **2021**, *9*, 76. [CrossRef]

Journal of
Marine Science and Engineering

MDPI

Article

The Application of UVC Used in Synergy with Surface Material to Prevent Marine Biofouling

Kailey N. Richard *, Kelli Z. Hunsucker, Harrison Gardner, Kris Hickman and Geoffrey Swain

Center for Corrosion and Biofouling Control, Florida Institute of Technology, Melbourne, FL 32901, USA; khunsucker@fit.edu (K.Z.H.); hgardner2009@my.fit.edu (H.G.); khickman2019@my.fit.edu (K.H.); swain@fit.edu (G.S.)

* Correspondence: krichard2018@my.fit.edu

Abstract: Biofouling is problematic for the shipping industry and can lead to functional and financial setbacks. One possible means of biofouling prevention is the use of ultraviolet-C (UVC) light. Previous studies have investigated UVC with marine coatings, but the synergistic effect with color and surface material, specifically reflectance, has yet to be determined. This study comprised three parts: UVC and color (red vs. white), UVC and reflectance (stainless steel vs. polycarbonate), and UVC and exposure intervals (weekly intervals and 10 min intervals). There was no variance in the biofouling communities for colored surfaces when exposed to 254 nm UVC. Reflectance studies demonstrated that the surface material plays a role in biofouling settlement. Stainless steel panels had significantly greater macrofouling settlement than polycarbonate, specifically among encrusting bryozoan, tubeworms, and tunicate communities. Panels of both surface materials exposed to indirect UVC significantly differed from controls and those exposed directly to UVC. Exposure intervals were also found to reduce biofouling settlement especially with long frequent intervals (i.e., 10 min/day). UVC can be utilized on various colored surfaces and different surface types, but the effectiveness in preventing biofouling is ultimately determined by the duration and frequency of UVC exposure.

Keywords: ultraviolet light; biofouling; color; surface material; exposure interval

Citation: Richard, K.N.; Hunsucker, K.Z.; Gardner, H.; Hickman, K.; Swain, G. The Application of UVC Used in Synergy with Surface Material to Prevent Marine Biofouling. *J. Mar. Sci. Eng.* **2021**, *9*, 662. https://doi.org/10.3390/jmse9060662

Academic Editor: Weicheng Cui

Received: 29 April 2021
Accepted: 4 June 2021
Published: 15 June 2021

1. Introduction

Biofouling is a significant problem for the shipping industry as it leads to many functional and financial setbacks. During idle periods, a ship can accumulate biofouling. Within minutes of immersion, a surface, such as a ship hull, accumulates a conditioning film, followed by bacteria and other unicellular organisms (i.e., diatoms). This accumulation is known as slime or microfouling [1]. Within hours to days, macrofouling communities such as barnacles, sponges, and tunicates may begin to grow, creating a complex biofouling community. This accumulation causes changes in surface roughness that lead to increased drag, lower fuel efficiency, and increased greenhouse gas emissions [2]. A ship with only biofilm (microfouling) and no macrofouling can cause up to $1.2 million in fuel consumption [2]. Significant growth on a ship hull may also require physical removal (e.g., brushing, power washing, scrapping) of the biofouling organisms [2]. Biofouling can also damage oceanographic equipment and is a known vector for the transfer of invasive species [3–5].

Currently, the most common approach to biofouling prevention on ship hulls is the application of marine coatings. These fall into two main categories: antifouling (AF) and fouling release (FR) [6]. Antifouling coatings are biocide-based, such as copper, which deters growth, making it hard for other organisms to attach and develop [1,6,7]. The drawback is these coatings emit copper and other chemicals into the water, causing environmental issues [8] and potentially resulting in nutrient loading [7]. Fouling release coatings have been developed as an alternative to biocide coatings. These coatings are often silicone-based and work by reducing the adhesion strength of organisms, allowing for easier removal either via cleaning or hydrodynamic forces imparted on the ship hull as it moves through water [1,6,9]. Other approaches include mechanical methods which either prevent the buildup of biofouling or remove a fully established biofouling community. Grooming, which uses remotely operated vehicles (ROVs) equipped with brushes to frequently and gently wipe the ship hull, can remove fouling before it has time to fully establish [10,11]. Grooming at a frequency of once a week has been effective in removing micro- and macrofouling on both AF and FR coatings [10–13]. Cleaning is another alternative, which is a reactive approach to fouling control, as it aims to remove growth via power washing either in or out of water. Currently, there are concerns that grooming and cleaning may release coating biocidal materials into the water; thus, there are methods in place to capture the effluent [11,14]. While there are many types of antifouling methods and strategies available, there is still a need for solutions which are environmentally friendly [12].

Ultraviolet-C (UVC) light is commonly used for the prevention of bacteria in the medical field. For medicinal purposes, it has proven to be 99% effective in eliminating bacterial biofilms growing on catheters. UVC light (254 nm) damages bacterial cells by attacking their DNA [15]. Recently, it has been applied in the marine field to prevent biofouling formation on multiple surfaces due to its increased affordability, and it is also considered an environmentally friendly method of biofouling removal or prevention [16–23]. UV radiation can disrupt the detection and settlement of coral larvae along with decreasing the biofilm formation on various surfaces [16–23]. Barnacles have also been shown to undergo periods of blindness when exposed to UVB light, preventing them from detecting surfaces and settling [21]. Salters and Piola (2017) [17] used embedded UVC LEDs as a way to prevent biofouling and identified no biofouling within a small vicinity of the light source.

Recent experiments have investigated the use of UVC as an externally applied source for biofouling prevention on ship hulls. Hunsucker et al. (2019) [22] and Braga et al. (2020) [23] determined that biofouling was prevented or significantly decreased in abundance when various frequencies of UVC were applied. Both experiments proved UVC can work in synergy with the fouling control coatings (biocidal and fouling release) mentioned above. On the basis of these prior experiments, a three-part study was designed to address how UVC can be used in synergy with different surfaces while considering the importance of exposure frequency. UVC has yet to be applied to surfaces with varying reflective properties or colors that are unattractive to biofouling organisms. The use of inert surfaces removes the use of antifouling or fouling release coatings, in turn removing the adverse effects that these coatings could potentially have on the environment. Additionally, the use of color is known to influence biofouling settlement and recruitment. Swain et al. (2006) [24] found that white surfaces contained less biofouling than black surfaces. Red-colored surfaces are a highly preferred color for settlement of barnacles including *Balanus amphitrite*, tubeworms, and some coral species [25–29]. Colors reflect light at different frequencies, and it is thought that color would influence settlement during periods when there is no UVC exposure.

The aim of this study was to investigate the interaction of UVC on surfaces with different reflective properties and color, as well as its effect on biofouling settlement. Additionally, only daily exposure intervals (1 min/day and 1 min/6 h) were tested in previous studies. Thus, study examines various exposure times and intervals (e.g., 5 min/week and 10 min twice a week) in order to prevent biofouling formation.

2. Materials and Methods

A three-part study was designed to address how UVC interacts with surfaces. Specifically, Section 2.1 looked at how surface color and UVC exposure work in synergy to prevent biofouling settlement, while Section 2.2 compared the interaction of UVC light with surfaces of different reflective properties (stainless steel and polycarbonate). In Sections 2.1 and 2.2, low doses of UVC known to have a limited effect on fouling [22,23], were applied to panels to allow for a comparison of fouling intensity and composition. Section 2.3 investigated the effectiveness of different time intervals to prevent biofouling formation on the surfaces used in Sections 2.1 and 2.2.

2.1. UVC and Color

To test the synergistic effect of UVC with color, red and white surfaces were selected according to prior research on biofouling settlement [25–29]. Both colors were printed on weatherproof computer paper. Each colored paper was sandwiched between 10 cm × 20 cm × 0.16 cm polycarbonate plates to create a uniform surface (herein referred to as a panel). The polycarbonate panels were 1.6 mm thick to allow for the color paper to be clearly visible. General Electric (GE) all-purpose silicone was used to seal the edges of panels to prevent water ingression which would alter the color and paper. Trilux-33 (Interlux), an antifouling spray paint, was used as a border on the panels to prevent the edge effect of unwanted biofouling.

Eight polycarbonate panels were housed on two sides of a 3D printed box (referred to as Box 1), which also allowed water and biofouling larvae to flow through [22] (Figure 1). A 25 W Aqua UVC (254 nm) lamp was placed in the center, at a distance of 25 mm from the panels [22,23]. In fresh water, the light intensity of the UVC lamp 25 mm from a surface was measured at 1.31 ± 0.88 μW/cm^2. Four replicates of red polycarbonate panels and four white polycarbonate panels were placed in Box 1 (Table 1). A PRIME-digital lighting timer was used to set to a time frequency of 1 min per day within the box [22,23]. Control panels were constructed in the same fashion and were hung on a PVC frame with no UVC exposure.

2.2. UVC and Reflectance

An experiment was designed to determine how fouling interacted with surfaces that reflect UVC or absorb it. Substrates of stainless steel and polycarbonate were used because they are both inert and have known specular reflectance at 254 nm. Stainless steel reflects 55% [30] and polycarbonate reflects 0% [31] of UVC light at near-normal angles of incidence. The reflectivity of the two substrates was measured in the lab using the same lamp as seen in Figure 1 and a Solar Light (model no. PMA2122-WP) UVC sensor. Laboratory testing confirmed that stainless steel panels reflected significantly more UVC light than the polycarbonate panels.

In order to test the interaction of UVC with surfaces of different reflectance, three different treatments were established (Table 1): all polycarbonate panels (Box 1), all stainless steel panels (Box 2), and a combination of polycarbonate panels and stainless steel panels (Box 3). Note that the box setup used to investigate the synergistic effect of UVC and color was also Box 1. Eight replicates of 316 stainless steel, polished to an ASTM 480 # 8 mirror finish [32], were cut into 10 cm × 20 cm panels and placed in Box 2. Box 3 contained four stainless steel panels on one side of the box and four red polycarbonate panels on the opposite side of the box, to test for the indirect effect of the reflective stainless steel on polycarbonate. Each box was set to a time frequency of 1 min per day using a PRIME-digital lighting timer based on prior experiments [22,23]. A set of control panels was constructed for each surface material, immersed during the same time period, and not exposed to UVC.

Figure 1. A visual representation of the UVC box set up. The sides of the box were composed of two epoxy backing plates (10 cm × 20 cm) with 3D printed caps that allowed for the lamp to be held in place. The photos in this figure reflect the box prior to the insertion of panels. Each box housed eight panels, four on either side, placed 25 mm from the lamp source.

Table 1. A summary of the experiments conducted with the number and type of substrate and the time frequency.

Experiment	Panels	Time Frequency
UVC and Color	4 Red and 4 White Polycarbonate (Box 1)	1 min/day
UVC and Reflectance	4 Red and White Polycarbonate (Box 1) 8 Stainless steel (Box 2) 4 Polycarbonate and 4 Stainless steel (Box 3)	1 min/day
UVC and Exposure Intervals	4 Polycarbonate and 4 Stainless steel	1 min/week
	4 Polycarbonate and 4 Stainless steel	5 min/week
	4 Polycarbonate and 4 Stainless steel	10 min/week
	4 Polycarbonate and 4 Stainless steel	10 min/MF
	4 Polycarbonate and 4 Stainless steel	10 min/MWF
	4 Polycarbonate and 4 Stainless steel	10 min/day

2.3. UVC and Exposure Intervals

According to the above results from Section 2.1, two separate time–frequency experiments were conducted to investigate the duration in which UVC would be the most effective on the inert surfaces. Color was not considered as a factor following the results from Section 2.1. The previous two experiments, Sections 2.1 and 2.2, tested UVC at daily intervals (Table 1), which was able to provide reduced macrofouling settlement. In order to determine if a weekly dose could also be effective, three weekly frequencies were tested for 1 month immersion (Table 1): 1 min a week, 5 min a week, and 10 min a week. Following these results, a second experiment was conducted using 10 min UVC intervals at a more frequent rate: 10 min twice a week (every Monday and Friday), 10 min three times a week (every Monday, Wednesday, and Friday), and 10 min every day.

The same box setup described above and shown in Figure 1 was used during these experiments, housing four stainless steel panels and four polycarbonate panels. Each box contained a 25 W Aqua UVC (254 nm) lamp placed in the center at a distance of 25 mm from the panels on each side of the box and connected to a PRIME-digital lighting timer to control the frequency. A set of control panels was constructed for each surface material, immersed, and not exposed to UVC.

2.4. Immersion

Testing was conducted at the Center for Corrosion and Biofouling Control's static immersion facility located at Port Canaveral, Florida (28°24′31.01″ N, 80°37′39.54″ W). The average salinity at this location is 34 ± 2 ppt, and the average water temperature is 27 ± 2 °C. Biofouling is high year-round, with seasonality observed with different fouling organisms. For example, in the warmer months (June, July, August), encrusting bryozoans, calcareous tubeworms, and barnacles dominate, while, in the cooler months (December, January, February), arborescent bryozoans and biofilms dominate. All panels were immersed 0.5 m below the surface water at Port Canaveral for a 1 month period [33].

Fouling coverage was visually assessed monthly. Only organisms that were directly attached to the surface were recorded [33]. For experiments that continued past 1 month of immersion, all panels were cleaned back and wiped down with vinegar, which is a mild acid that helps neutralize the surface and remove settlement cues. After cleaning, the timer was then switched to the next set of predetermined time frequencies.

2.5. Statistical Analysis

A permutational multivariate analysis of variance (PERMANOVA) was performed on data collected for each month, to compare the total biofouling communities based on UVC treatment. A PERMANOVA analysis was run on both UVC-exposed and nonexposed panels to compare color (red and white), surface material (polycarbonate and stainless steel), the indirect effect of UVC, and the exposure intervals. A nonmetric multidimensional scaling (MDS) plot was also conducted, when PERMANOVA results proved to be significant, to display differences in fouling communities based on surface material. A SIMPER analysis was used to indicate differences among fouling communities for the MDS plots. In addition, a multivariate analysis of variance (ANOVA) followed by a Tukey test was used to compare individual biofouling groups (i.e., barnacles, tubeworms) for different surface materials and between exposed samples and controls. All analyses were done using R statistical software (2019).

3. Results

3.1. UVC and Color

The synergistic effect of color and UVC exposure on biofouling was assessed using panels immersed in Box 1 (red and white polycarbonate) compared to control panels (no UVC). After 1 month exposure of UVC at a frequency of 1 min per day, all panels were 100% fouled. The red UVC-exposed panels were primarily fouled with biofilm and had an average of 31 ± 21.7% of macrofouling composed of tubeworms, tunicates, encrusting bryozoan, and intermediate bryozoan (Figure 2). White UVC-exposed panels had an average of 9 ± 4.7% of macrofouling, with the rest of the community driven by biofilm. Between the red and white exposed panels, no statistical difference was found with the overall biofouling community ($p > 0.05$). However, there was a statistical difference in tubeworm settlement between the two colors ($p < 0.05$), with red having an average of 7% and white having an average of 3% tubeworm abundance. While the intermediate bryozoan visually appeared to vary between the red and white UVC-exposed panels, statistically, there was no significance ($p > 0.05$). Control panels were both 100% fouled, but red panels had 67 ± 35.5% macrofouling, while white panels had 12 ± 4.6% macrofouling (Figure 2). Community composition of red controls was primarily characterized by tubeworms with a minimal amount of tunicate settlement. Macrofouling on the white controls was mainly tubeworm settlement, with a low settlement of many other organisms such as bryozoans, barnacles, and sponge. No statistical significance was found between UVC-exposed panels and controls ($p > 0.05$). Similar to the UVC-treated panels, the controls showed no significant variance between colors for total biofouling coverage ($p > 0.05$) but did show significant difference between the red and white for tubeworm settlement with 56 ± 35.5% and 6 ± 4.6%, respectively ($p < 0.05$).

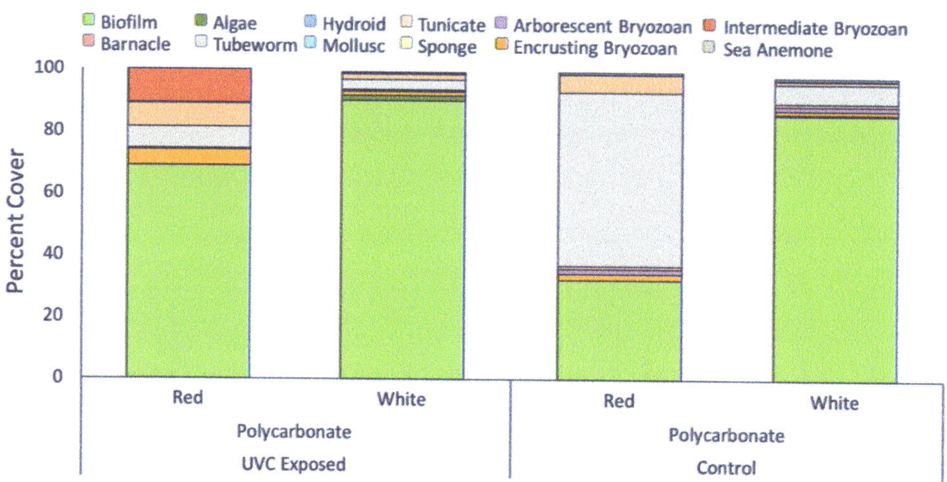

Figure 2. Biofouling composition on the UVC exposed red and white panels and their corresponding controls.

3.2. UVC and Reflectance

To address if surface material and UVC can be used in synergy to prevent biofouling, Box 1 (all polycarbonate) and Box 2 (all stainless steel) were utilized. After 1 min a day exposure, all the stainless steel and polycarbonate panels were 100% fouled (Figure 3). UVC polycarbonate panels had 20 ± 18.7% macrofouling coverage, which consisted primarily of encrusting bryozoan, tubeworm, tunicates, and intermediate bryozoan. UVC-exposed stainless steel panels had an average macrofouling coverage of 54 ± 20.7%. The community was driven by tubeworms, tunicates, and encrusting bryozoan. For both types of material, biofilm was also a major contributor to the overall fouling community (Figure 3). PERMANOVA and ANOVA results indicated a difference among stainless steel and polycarbonate biofouling communities, with significant differences for biofilm, encrusting bryozoan, and tubeworms ($p < 0.05$) (Figure 4).

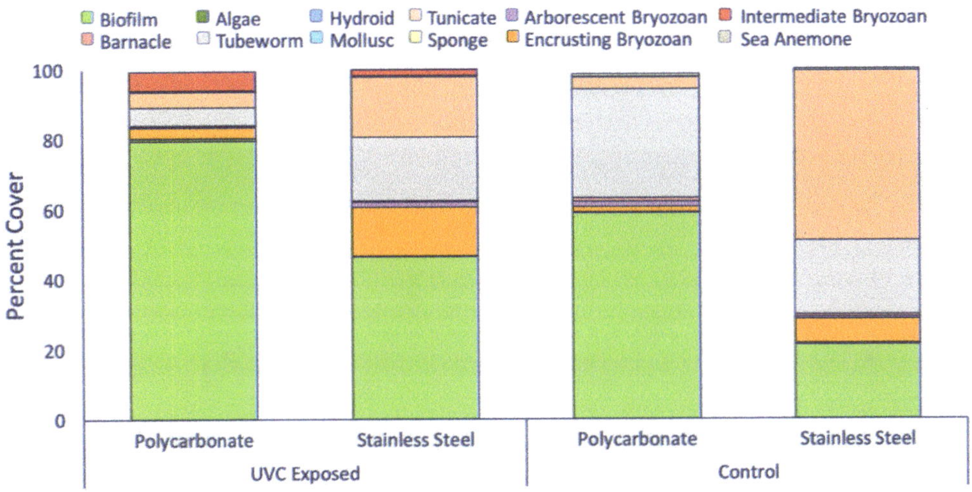

Figure 3. Biofouling composition on the UVC-exposed polycarbonate (average of red and white) and stainless steel panels and their corresponding controls.

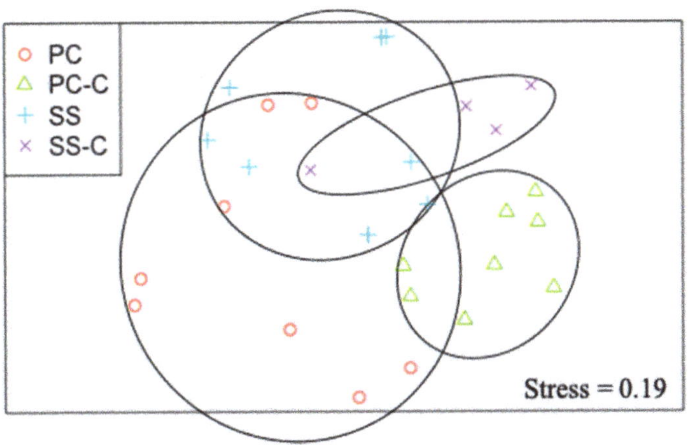

Figure 4. MDS of average biofouling communities on both UVC-exposed polycarbonate (PC) and stainless steel (SS) and nonexposed polycarbonate (PC-C) and stainless steel control surfaces (SS-C) following the 1 min/day exposure time.

Controls had parallel results to the UVC exposed panels, with significant biofouling community differences ($p < 0.05$) (Figure 4). Polycarbonate control panels were predominantly fouled with biofilm and 40 ± 37.5% macrofouling which consisted primarily of tubeworms. Stainless steel control panels had a greater macrofouling community presence (79 ± 17.9%) which was typified by tunicates, tubeworms, and encrusting bryozoan (Figure 3). UVC-exposed panels had significantly less biofouling abundance than untreated panels ($p < 0.001$), with greater tunicate settlement on the controls ($p < 0.001$). These trends can also be seen on the MDS plot in Figure 4.

To test if there was an indirect effect with surface material, a box (Box 3) was arranged to have both red polycarbonate and stainless steel panels placed directly across from one another. After 1 month, biofouling was similar among exposed surfaces ($p > 0.05$) (Figure 5). The macrofouling for control panels was significantly different between the stainless steel and polycarbonate panels, with less on the polycarbonate panels ($p < 0.001$). Macrofouling on control stainless steel panels differed significantly from the UVC stainless steel panels that were exposed to indirect UVC ($p < 0.05$), and similar results were found among polycarbonate panels ($p < 0.05$). Comparisons were also made with polycarbonate and stainless steel panels from Box 1 and Box 2, respectively, where the panels were not placed directly across from one another. As described above, the total macrofouling on the directly exposed UVC polycarbonate and stainless steel panels was 20 ± 18.7% and 54 ± 20.7%, respectively, comprising tubeworm, tunicates, and encrusting bryozoan. For the indirect panels, polycarbonate had 20 ± 9.4% macrofouling and stainless steel panels had 26 ± 6.3% macrofouling. ANOVA indicated that macrofouling recruitment was significantly different between those panels which received direct UVC and those which received indirect UVC, for both polycarbonate and stainless steel ($p < 0.05$). Specifically, tunicates were less abundant on the stainless steel panels in the indirect treatment, but tubeworms were greater on directly exposed stainless steel panels (Box 2).

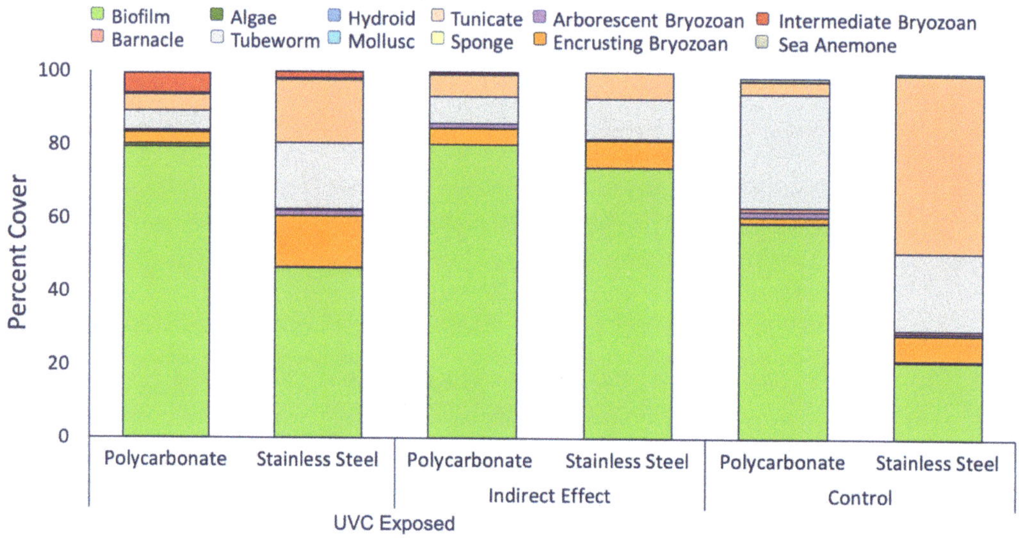

Figure 5. Biofouling composition on the UVC-exposed polycarbonate (red) and stainless steel panels and their corresponding controls.

3.3. Determination of UVC Exposure Intervals

Two exposure interval experiments were performed to evaluate the duration at which UVC may be effective in preventing biofouling, i.e., weekly and 10 min increments. During the first month, weekly time frequencies were tested (Table 1): 1 min/week, 5 min/week, and 10 min/week. During this exposure, stainless steel and polycarbonate panels were 100% fouled with biofilm and little macrofouling (Figure 6). The 1 min/week polycarbonate panels were predominantly fouled with biofilm and had 4 ± 1.3% macrofouling which was mainly tubeworms, while stainless steel panels had 23 ± 26.1% macrofouling which was primarily hydroids (~20%). When the 1 min/week panels were compared to the controls, there was a significant difference in the macrofouling community which had developed ($p < 0.05$). Both control polycarbonate and stainless steel panels had greater coverage of tubeworms (~9% and ~8%, respectively) and tunicates (~20% and 11%, respectively) than panels exposed to UVC.

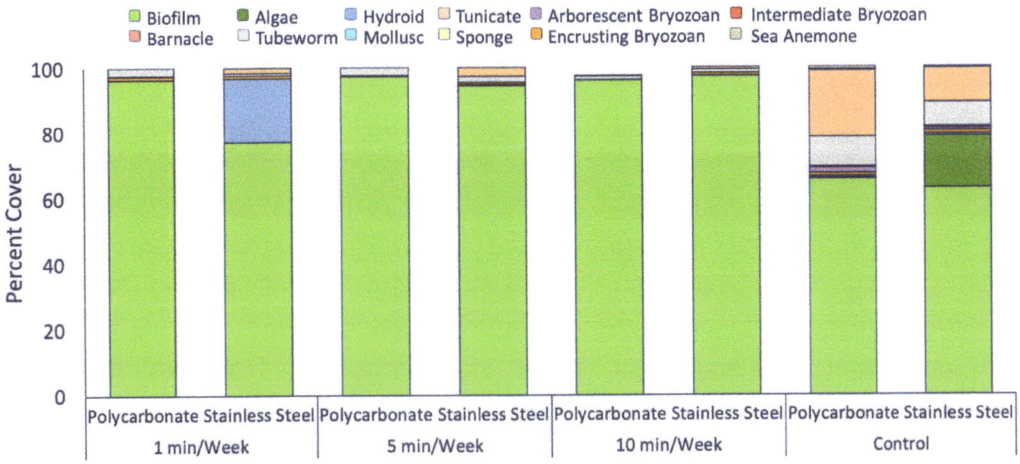

Figure 6. Biofouling composition on the UVC-exposed polycarbonate and stainless steel panels and their corresponding controls.

Biofouling communities did not differ between polycarbonate and stainless steel for the 5 min/week UVC-exposed panels ($p > 0.05$). This frequency had significantly less arborescent bryozoan, tubeworms, and sea anemones than the control panels ($p < 0.001$). The 10 min/week panels also had similar results to the 1 min and 5 min per week panels ($p > 0.05$). An ANOVA test was performed on all time frequencies (1 min/week, 5 min/week, and 10 min/week), but showed no statistical difference with regard to the total biofouling community ($p > 0.05$) among the three exposure treatments. UVC treatment for all frequencies reduced the macrofouling that occurred on both polycarbonate and stainless steel surfaces.

According to the first month of results, 10 min UVC exposure treatments were applied (Table 1): 10 min twice weekly (MF), 10 min three times weekly (MWF), and 10 min/day. All treatments significantly reduced the macrofouling compared to the controls, and the 10 min/day treatment prevented all macrofouling and significantly reduced biofilm (Figure 7). For the twice weekly exposed panels, polycarbonate had 45 ± 45.4% macrofouling, while the stainless steel had 74 ± 28.6% macrofouling. As for the panels exposed three times weekly, polycarbonate had 18 ± 10.4% macrofouling and stainless steel had 65 ± 37.9% macrofouling. Both the polycarbonate and the stainless steel panels exposed to UVC daily were approximately 40% covered in primarily biofilm. The stainless steel panels also had a very minimal amount macrofouling, 1 ± 0.6% (Figure 7).

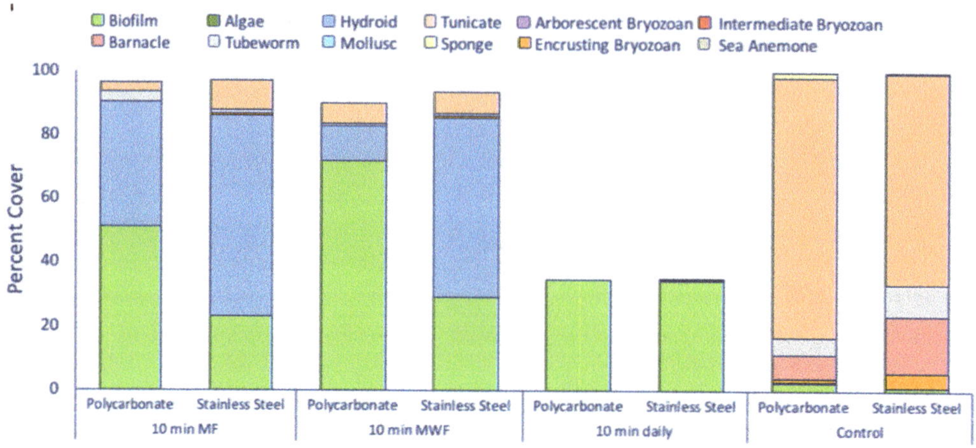

Figure 7. Biofouling composition on the UVC-exposed polycarbonate and stainless steel panels and their corresponding controls.

The total biofouling community and macrofouling community did not differ statistically among the polycarbonate and stainless steel panels exposed to UVC twice weekly ($p > 0.05$). However, when compared to the controls, they had significantly higher encrusting bryozoan ($p < 0.05$), barnacles ($p < 0.001$), tubeworms ($p < 0.05$), and tunicates ($p < 0.001$). The polycarbonate and stainless steel panels exposed to UVC three times a week presented with parallel results to the panels exposed twice weekly, where there was no significant difference among UVC-exposed panels ($p > 0.05$). When comparing time exposure treatments, there was a significant difference in total biofouling community and the macrofouling community with reference to surface material ($p < 0.05$) and time exposures ($p < 0.001$), which can be seen in Figure 8. SIMPER results indicated that the division between exposed surfaces and controls and among time frequencies was driven by biofilm formation.

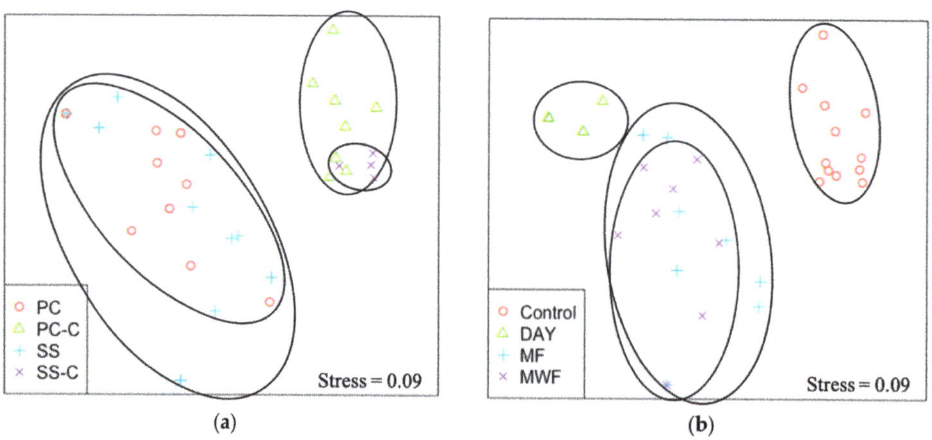

Figure 8. MDS of average biofouling communities among (**a**) polycarbonate (PC), stainless steel (SS), and nonexposed polycarbonate control (PC-C) and stainless steel control (SS-C), and (**b**) for the 10 min intervals: 10 min/day (Day), 10 min twice weekly (MF), 10 min three times weekly (MWF).

4. Discussion

UVC was applied to surfaces of different colors and two different materials, polycarbonate and stainless steel, to study the synergistic effect on biofouling prevention. This research demonstrated that UVC may be applied intermittently to various surfaces submerged in seawater as a means to control or to prevent fouling. It showed that the effectiveness is moderated by the type of surface that is being treated, and by the frequency and duration of the treatment. No significant differences in total biofouling recruitment were observed for the UVC-treated colors evaluated by these experiments. However, it was found that, in certain cases, calcareous tubeworm recruitment to the red surfaces was greater than that to the white surfaces. These results were similar to the results of Satheesh and Wesley (2010) [27], who looked at substrate color effects on the settlement of tubeworms, and Swain et al. (2006) [24], who looked at the effects of black and white surfaces on in situ fouling communities. Both studies found that fouling organisms settle on darker shades of color. Swain et al. (2006) [24] had increased settlement on black surfaces and Satheesh and Wesley (2010) [27] had greater settlement on red surfaces. Color preference is a contributing factor in biofouling recruitment and settlement for specific organisms, but it appears to be hindered with UVC exposure due to no statistical difference with respect to fouling. The UVC treatment significantly reduced the tubeworm abundance on treated panels versus the controls. As another possibility, UVC exposure to the surrounding water column could have potentially limited the settlement of tubeworms compared to the controls. However, it is unclear exactly what factor resulted in similar fouling communities.

Comparisons were made of direct UVC exposure on polycarbonate and stainless steel panels to those subjected to indirect influence. A lower abundance of biofouling was seen on both polycarbonate and stainless steel panels that were exposed indirectly. The low abundance of fouling may be due to the coupling of both surfaces reflecting and absorbing the UVC light. For example, polycarbonate reflects 0% of UVC [31], which indicates that it is absorbing the light. The absorption of light could potentially be heating the surfaces, preventing fouling from settling. The same statement could be made about stainless steel because it only reflects 50% of light [30]. The differences may also be the result of spatial variation within the test site, which could be teased out with additional replication and testing. Overall, throughout the different experiments, stainless steel had more macrofouling than the polycarbonate, which was seen specifically for tubeworms, tunicates, and encrusting bryozoan. Kim and Kang (2020), [34] saw similar results for stainless steel and polyvinyl chloride (PVC) when exposing UVC to foodborne pathogens. Bacterial colonies were less abundant on PVC (another commonly used inert surface) compared to stainless steel, indicating that reflectance may not play a role in biofouling prevention as much as the actual surface material.

While differences in color or surface material may play a small role, the major factor influencing settlement on stainless steel and polycarbonate was the duration and frequency of UVC exposure. Panels exposed for intervals of two (MF) and three times (MWF) a week resulted in biofouling communities that were similar. However, a significant decrease in fouling was seen with daily exposures of UVC. This indicated that, with a high enough dose of UVC, most if not all macrofouling can be prevented on both substrates. Hunsucker et al. (2019) [22] and Braga et al. (2020) [23] also found that longer exposure times to antifouling and fouling release coatings were effective at preventing both biofilms and macrofouling. While more work is needed to understand the impacts of UVC exposure at the different life stages of biofouling organisms, previous studies have also found veliger [35] and nematode larvae [36] to have higher rates of mortality after long exposures to UV irradiance. Both organisms were exposed to continuous UV light for longer than 24 h; however, high mortality was observed within the first 24 h. These studies correlate with what was observed when UVC was exposed to panels for 10 min a day herein. The panels exposed to UVC for 10 min a day had minimal fouling, indicating that a slight increase in time may be equivalent to continuous exposure and may display complete biofouling prevention or removal. For the current study, testing was conducted in warmer months of Florida which are typically the more aggressive fouling season, thus requiring a longer UVC exposure. During lower fouling months or environments, a lower time and frequency of UVC exposure would be required to prevent fouling [22]. Studies comparing fouling response in different seasons will need to be further investigated since the current study was conducted in the summer months where animal larvae are the dominate fouling organisms compared to the winter months, which are more algal-dominated. Along with exposure time and intensity, distance can play a role in UVC effectiveness. Braga et al. (2020) [23] applied UVC light to pre-existing adult barnacles attached to an inert surface and saw greater mortality on panels closer to the lamp (25 mm) than farther away (275 mm).

On both polycarbonate and stainless steel, biofouling was minimal when UVC was applied for long and frequent intervals, which was also observed by Hunsucker et al. (2019) [22] and Braga et al. (2020) [23]. According to the results herein, UVC can be considered a more effective environmentally friendly use for biofouling prevention compared to fouling control coatings. Applying UVC to inert coatings or substrates could reduce the use of antifouling coatings such as copper, which can leach, and other chemicals from antifouling coatings into water bodies. UVC exposure to copper-based coatings has demonstrated that UVC may accelerate coating release [22]. Further testing is still needed to understand how long-term UVC impacts both fouling control coatings and inert surfaces. Furthermore, material selection should be considered due to the possibility of surface degradation by the UVC exposure [37].

5. Summary

These experiments demonstrated that low doses of UVC reduce macrofouling coverage, and its effectiveness can be improved by accurately matching it with the appropriate substrate. Both the color and the surface material experiments demonstrated that material selection influences fouling organism settlement. While inert surfaces are not typically used in the construction of naval ships, they can be used in other oceanographic fields, such as with ROVs and instrumentation. Biofouling on marine sensors is problematic because it hinders the functionality of the instrument. Being able to apply UVC to the instruments, through external exposure or embedded into the device [38,39], to prevent fouling (instead of manually removing fouling) would save time and increase longevity of the instrument. Proper UVC exposure could also prove beneficial in invasive species reduction especially as a treatment in niche areas (i.e., ballast tanks and propellers) where antifouling coatings are not typically applied [38–41]. The versatility of UVC [17,22,23,38] makes it possible to use in confined spaces such as niche areas, while it can additionally be used to treat the waterbody to eliminate larvae before they are able to reproduce.

Author Contributions: Conceptualization, K.N.R., K.Z.H. and G.S.; methodology, K.N.R., K.H., K.Z.H., G.S. and H.G.; formal analysis, K.N.R.; investigation, K.N.R. and K.H.; data curation, K.N.R. and K.H.; writing—original draft preparation, K.N.R. and K.Z.H.; writing—review and editing, K.H., K.Z.H., G.S., and H.G.; supervision, K.Z.H. and G.S.; project administration, K.Z.H. and G.S.; funding acquisition, K.Z.H. and G.S. All authors have read and agreed to the published version of the manuscript.

Funding: This research was funded by The Office of Naval Research, grant numbers N00014-16-1-3123 and N00014-20-1-2214. Support for the publication fee was provided by the OA Subvention Fund at Florida Institute of Technology.

Institutional Review Board Statement: Not applicable.

Informed Consent Statement: Not applicable.

Data Availability Statement: Request to corresponding author of this article.

Acknowledgments: The authors would like to thank the members of the Center for Corrosion and Biofouling Control for their assistance on this project, Glenn Miller for his guidance for the statistical analysis, and the OA Subvention Fund at Florida Institute of Technology for funding this publication.

Conflicts of Interest: The authors declare no conflict of interest.

References

1. Callow, M.E.; Callow, J.E. Marine Biofouling: A Sticky Problem. *Biologist (London)* **2002**, *49*, 10–14.
2. Schultz, M.P.; Bendick, J.A.; Holm, E.R.; Hertel, W.M. Economic Impact of Biofouling on a Naval Surface Ship. *Biofouling* **2011**, *27*, 87–98. [CrossRef] [PubMed]
3. Bax, N.; Williamson, A.; Aguero, M.; Gonzalez, E.; Geeves, W. Marine Invasive Alien Species: A Threat to Global Biodiversity. *Mar. Policy* **2003**, *27*, 313–323. [CrossRef]
4. Coutts, A.D.M.; Taylor, M.D. A Preliminary Investigation of Biosecurity Risks Associated with Biofouling on Merchant Vessels in New Zealand. *N. Z. J. Mar. Freshw. Res.* **2004**, *38*, 215–229. [CrossRef]
5. Campbell, M.L.; Hewitt, C.L. Assessing the Port to Port Risk of Vessel Movements Vectoring Non-Indigenous Marine Species within and across Domestic Australian Borders. *Biofouling* **2011**, *27*, 631–644. [CrossRef]
6. Swain, G. Redefining antifouling coatings. *J. Prot. Coat. Linings* **1999**, *16*, 26–35.
7. Almeida, E.; Diamantino, T.C.; de Sousa, O. Marine Paints: The Particular Case of Antifouling Paints. *Prog. Org. Coat.* **2007**, *59*, 2–20. [CrossRef]
8. Srinivasan, M.; Swain, G.W. Managing the Use of Copper-Based Antifouling Paints. *Environ. Manag.* **2007**, *39*, 423–441. [CrossRef]
9. Swain, G.W.; Kovach, B.; Touzot, A.; Casse, F.; Kavanagh, C.J. Measuring the Performance of Today's Antifouling Coatings. *J. Ship Prod.* **2007**, *23*, 164–170. [CrossRef]
10. Tribou, M.; Swain, G. The Use of Proactive In-Water Grooming to Improve the Performance of Ship Hull Antifouling Coatings. *Biofouling* **2010**, *26*, 47–56. [CrossRef]
11. Swain, G.; Melissa, T.; Harrison, G.; Kelli, H. *In-Water Grooming of Fouling Control Coatings: From Research to Reality*; PortPIC: Hamburg, Germany, 2020; pp. 29–37.
12. Tribou, M.; Swain, G. The Effects of Grooming on a Copper Ablative Coating: A Six Year Study. *Biofouling* **2017**, *33*, 494–504. [CrossRef]
13. Hearin, J.; Hunsucker, K.Z.; Swain, G.; Stephens, A.; Gardner, H.; Lieberman, K.; Harper, M. Analysis of Long-Term Mechanical Grooming on Large-Scale Test Panels Coated with an Antifouling and a Fouling-Release Coating. *Biofouling* **2015**, *31*, 625–638. [CrossRef]
14. Woods, C.M.C.; Floerl, O.; Jones, L. Biosecurity Risks Associated with In-Water and Shore-Based Marine Vessel Hull Cleaning Operations. *Mar. Pollut. Bull.* **2012**, *64*, 1392–1401. [CrossRef]
15. Von Sonntag, C.; Kolch, A.; Gebel, J.; Oguma, K.; Sommer, R. The Photochemical Basis of UV Disinfection. In Proceedings of the European Conference on UV Radiation-Effects and Technologies, Karlsruhe, Germany, 22–24 September 2004; pp. 22–24.
16. Bak, J.; Ladefoged, S.D.; Tvede, M.; Begovic, T.; Gregersen, A. Dose Requirements for UVC Disinfection of Catheter Biofilms. *Biofouling* **2009**, *25*, 289–296. [CrossRef]
17. Salters, B.; Piola, R. UVC Light for Antifouling. *Mar. Technol. Soc. J.* **2017**, *51*, 59–70. [CrossRef]
18. Gleason, D.F.; Edmunds, P.J.; Gates, R.D. Ultraviolet Radiation Effects on the Behavior and Recruitment of Larvae from the Reef Coral Porites Astreoides. *Mar. Biol.* **2006**, *148*, 503–512. [CrossRef]
19. Hung, O.S.; Gosselin, L.A.; Thiyagarajan, V.; Wu, R.S.S.; Qian, P.Y. Do Effects of Ultraviolet Radiation on Microbial Films Have Indirect Effects on Larval Attachment of the Barnacle Balanus Amphitrite? *J. Exp. Mar. Bio. Ecol.* **2005**, *323*, 16–26. [CrossRef]
20. Hung, O.S.; Thiyagarajan, V.; Wu, R.S.S.; Qian, P.Y. Effect of Ultraviolet Radiation on Biofilms and Subsequent Larval Settlement of Hydroides Elegans. *Mar. Ecol. Prog. Ser.* **2005**, *304*, 155–166. [CrossRef]
21. Chiang, W.L.; Au, D.W.T.; Yu, P.K.N.; Wu, R.S.S. UV-B Damages Eyes of Barnacle Larvae and Impairs Their Photoresponses and Settlement Success. *Environ. Sci. Technol.* **2003**, *37*, 1089–1092. [CrossRef] [PubMed]
22. Hunsucker, K.Z.; Braga, C.; Gardner, H.; Jongerius, M.; Hietbrink, R.; Salters, B.; Swain, G. Using Ultraviolet Light for Improved Antifouling Performance on Ship Hull Coatings. *Biofouling* **2019**, *35*, 658–668. [CrossRef]
23. Braga, C.; Hunsucker, K.; Gardner, H.; Swain, G. A Novel Design to Investigate the Impacts of UV Exposure on Marine Biofouling. *Appl. Ocean Res.* **2020**, *101*, 102226. [CrossRef]
24. Swain, G.; Herpe, S.; Ralston, E.; Tribou, M. Short-Term Testing of Antifouling Surfaces: The Importance of Colour. *Biofouling* **2006**, *22*, 425–429. [CrossRef] [PubMed]
25. Mason, B.; Beard, M.; Miller, M.W. Coral Larvae Settle at a Higher Frequency on Red Surfaces. *Coral Reefs* **2011**, *30*, 667–676. [CrossRef]
26. Matsumura, K.; Qian, P.-Y. Larval Vision Contributes to Gregarious Settlement in Barnacles: Adult Red Fluorescence as a Possible Visual Signal. *J. Exp. Biol.* **2014**, *217*, 743–750. [CrossRef]
27. Satheesh, S.; Wesley, S.G. Influence of Substratum Colour on the Recruitment of Macrofouling Communities. *J. Mar. Biol. Assoc. U. K.* **2010**, *90*, 941–946. [CrossRef]
28. Taki, Y.; Ogasawara, Y.; Ido, Y.; Yokoyama, N. Color Factors Influencing Larval Settlement of Barnacles, Balanus Amphitrite Subspp. *Nippon Suisan Gakkaishi* **1980**, *46*, 133–138. [CrossRef]
29. Robson, M.A.; Williams, D.; Wolff, K.; Thomason, J.C. The Effect of Surface Colour on the Adhesion Strength of Elminius Modestus Darwin on a Commercial Non-Biocidal Antifouling Coating at Two Locations in the UK. *Biofouling* **2009**, *25*, 215–227. [CrossRef]
30. Zwinkels, J.C.; Noël, M.; Dodd, C.X. Procedures and Standards for Accurate Spectrophotometric Measurements of Specular Reflectance. *Appl. Opt.* **1994**, *33*, 7933–7944. [CrossRef]

31. Citek, K. Anti-Reflective Coatings Reflect Ultraviolet Radiation. *Optometry* **2008**, *79*, 143–148. [CrossRef]
32. ASTM International. *A480/A480M-20a Standard Specification for General Requirements for Flat-Rolled Stainless and Heat-Resisting Steel Plate, Sheet, and Strip*; ASTM International: West Conshohocken, PA, USA, 2020.
33. ASTM International. *6990-05 Standard Practice for Evaluating Biofouling Resistance and Physical Performance of Marine Coating Systems*; Paint-Tests for Formulated Products and Applied Coatings; ASTM International: West Conshohocken, PA, USA, 2005.
34. Kim, D.-K.; Kang, D.-H. Effect of Surface Characteristics on the Bactericidal Efficacy of UVC LEDs. *Food Control.* **2020**, *108*, 106869. [CrossRef]
35. Perepelizin, P.V.; Boltovskoy, D. Effects of 254 Nm UV Irradiation on the Mobility and Survival of Larvae of the Invasive Fouling Mussel Limnoperna Fortunei. *Biofouling* **2014**, *30*, 197–202. [CrossRef]
36. Van Dijk, J.; de Louw, M.D.E.; Kalis, L.P.A.; Morgan, E.R. Ultraviolet Light Increases Mortality of Nematode Larvae and Can Explain Patterns of Larval Availability at Pasture. *Int. J. Parasitol.* **2009**, *39*, 1151–1156. [CrossRef] [PubMed]
37. Tjandraatmadja, G.F.; Burn, L.S.; Jollands, M.J. The effects of ultraviolet radiation on polycarbonate grazing. In *Durability of Building Materials and Components 8*; Lacasseand, M.A., Vanier, D.J., Eds.; National Research Council, Institute for Research in Construction: Ottawa, ON, Canada, 1999; pp. 884–898.
38. Ryan, E.; Turkmen, S.; Benson, S. An Investigation into the Application and Practical Use of (UV) Ultraviolet Light Technology for Marine Antifouling. *Ocean Eng.* **2020**, *216*, 107690. [CrossRef]
39. Braga, C.; Hunsucker, K.; Erdogan, C.; Gardner, H.; Swain, G. The Use of a UVC Lamp Incorporated with an ROV to Prevent Biofouling: A Proof-of-Concept Study. *Mar. Technol. Soc. J.* **2020**, *54*, 76–83. [CrossRef]
40. Korkut, E.; Atlar, M. An Experimental Investigation of the Effect of Foul Release Coating Application on Performance, Noise and Cavitation Characteristics of Marine Propellers. *Ocean Eng.* **2012**, *41*, 1–12. [CrossRef]
41. Sørensen, P.A.; Kiil, S.; Dam-Johansen, K.; Weinell, C.E. Anticorrosive Coatings: A Review. *J. Coat. Technol. Res.* **2009**, *6*, 135–176. [CrossRef]

Journal of
Marine Science and Engineering

MDPI

Review

Substrate Selection of Ascidian Larva: Wettability and Nano-Structures

Euichi Hirose [1,*] and Noburu Sensui [2]

[1] Department of Chemistry, Biology and Marine Science, Faculty of Science, University of the Ryukyus, Nishihara, Okinawa 903-0213, Japan
[2] Department of Human Biology and Anatomy, Graduate School of Medicine, University of the Ryukyus, Nishihara, Okinawa 903-0215, Japan; sensuiso@med.u-ryukyu.ac.jp
* Correspondence: euichi@sci.u-ryukyu.ac.jp

Abstract: Ascidians are marine sessile chordates that comprise one of the major benthic animal groups in marine ecosystems. They sometimes cause biofouling problems on artificial structures underwater, and non-indigenous, invasive ascidian species can potentially and seriously alter native faunal communities. Ascidian larvae are usually tadpole-shaped, negatively phototactic, and adhere on substrates by secreting a glue from their adhesive organs. Although larvae often prefer hydrophobic surfaces, such as a silicone rubber, for settlement, hydrophobic materials are often used to reduce occurrence of fouling organisms on artificial structures. This inconsistency may indicate that an attractive surface for larvae is not always suitable for settlement. Micro-scale structures or roughness may enhance the settlement of ascidian larvae, but settlement is significantly reduced by a nano-scale nipple array (or moth-eye structure), suggesting functional properties of similar structures found on the body surfaces of various invertebrates. The substrate preferences of larvae should be one of the important bases in considering measures against biofouling, and this review also discusses the potential uses of materials to safely reduce the impacts of invasive species.

Keywords: biofouling; larval settlement; substrate preference; water wettability; moth-eye structure; MOSMITE^{TM}; silicone paradox

Citation: Hirose, E.; Sensui, N. Substrate Selection of Ascidian Larva: Wettability and Nano-Structures. *J. Mar. Sci. Eng.* **2021**, *9*, 634. https://doi.org/10.3390/jmse9060634

Academic Editor: Francesca Cima

Received: 19 May 2021
Accepted: 4 June 2021
Published: 7 June 2021

1. Introduction

In nature, sessile organisms are usually distributed unevenly on substrates, suggesting the occurrence of selection. We can consider two selection processes; pre-settlement selection and post-settlement selection. The former is substrate selection for settlement by the mobile phase of settlers, such as larvae. The larvae may carefully select the microhabitat and the substrate material for settlement to have better conditions for survival, using their sensory systems. The present review mainly focuses on this pre-settlement selection. When a larva settles on an unfavorable site, post-settlement selection brings about death or ejection of the settlers from the site. This post-settlement selection is induced by not only physical and chemical factors but also via biological interactions such as inter- and intra-specific competitions.

Substrate selection by larvae has been assessed by both field and laboratory experiments. In field experiments, test substrata are immersed in the field and the amounts of settlers are compared among the substrata under different conditions [1,2]. Results include both pre-settlement selection and post-settlement selection. Laboratory experiments are mainly carried out to examine particular factor(s), such as texture, wettability, or color of substrata, on pre-settlement selection, but as results do not always reflect the natural settlement process, care should be taken in interpreting these results. These experiments can be subdivided into single-choice [3–6], dual-choice [7,8], and multi-choice tests [9–11]. In an example single-choice test, the inner wall of a container was entirely coated with a test substrate and the rate of metamorphosis/settlement of larvae was compared among substrate types. Therefore, larval preferences among the substrates are not directly assessed in this method. In dual- and multi-choice tests, two or more substrates are placed in a container, and the number of settlers or selection indices are compared among the substrates. However, it is difficult to assess the larval preferences for substrates, as larvae often select the inner wall(s) of a container, and thus, the substrate may not be the only assay. We can prevent larval settlement on a container wall by coating with a less preferred material. For instance, ascidian larvae rarely settle on substrates coated with 1.5% agar [12].

The substrate preference of larvae is an important key for controlling larval settlement and for better understanding of evolutionary adaptations in pre-settlement selection. Management of biofouling has been paid great attention, particularly with regards to aquaculture facilities and ship hulls, because of the serious economic losses caused by fouling [13]. Although anti-fouling compounds have been developed and used in coatings to reduce biofouling [14], the massive use of bioactive compounds for settlers may carry a risk of pollution to aquatic ecosystems [15]. The physical properties of a substrate surface, such as structure and wettability, also affect the substrate preferences of settlers. In contrast to chemical approaches, countermeasures based on physical properties may be less effective for settlers but more eco-friendly. To develop such physical approaches against biofouling, it is necessary to elucidate the larval preference(s) of the physical properties of substrates for settlement. This review focuses on wettability and nano-structures of substrate surfaces for the substrate selection of ascidian larvae, while other factors of microhabitat influencing the pre-settlement selection were examined, such as the orientation of the substrate (i.e., vertical or horizontal) [2], nutrient contaminants in water [16], and textures of the substrate [9].

2. Ascidians and Their Larvae

Ascidians are the only sessile group among the extant chordates. They are one of the major benthic animal groups and are suspension feeders in marine ecosystems [17]. Some ascidians are solitary in form throughout their life, and some form colonies following asexual budding. Ascidian species comprise the class Ascidiacea belonging to the subphylum Tunicata (or Urochordata). One of the unique characters of tunicates is biosynthesis of cellulose, and the tunic in ascidians is the cellulosic matrix entirely covering the epidermis and forming an integumentary matrix [18]. Therefore, ascidians always adhere to the substrate via the tunic.

The morphology of ascidians and their larvae were described in detail by Burighel and Cloney [19]. Briefly, ascidian larvae are usually tadpole-shaped and composed of a tail and trunk; the tail has a notochord and muscles that produce swimming force, and the trunk has rudimentary juvenile organs, a sensory vesicle containing a pigmented otolith and/or photolith, and adhesive organs (Figure 1). The larval body is entirely covered by the tunic. Whereas some species produce tail-less larvae through direct development, phylogenetic analyses have shown that this is an apomorphic character [20]. The larva is supposed to sense gravity and light with the otolith and photolith, respectively. By means of laser ablation of the otolith and photolith, Tsuda et al. [21] demonstrated that upward swimming occurs due to negative geotaxis in the initial period of the larval stage, and then, negative phototaxis occurs in the latter period. This upward swimming probably aids in the dispersal of drifting larvae, and negative phototaxis results in larval settlement in shaded sites. Ascidians, particularly juveniles, are vulnerable to ultraviolet (UV) radiation [22], and thus, the light sensitivity of larvae has an important role in selecting the habitat environment in species inhabiting shallow waters. Ascidian larvae usually have three adhesive organs at the apical end of the trunk (arrowheads in Figure 1a), and adhesive compounds containing glycoprotein, sugar residues, and catechol residues are secreted from the collocytes [10]. The adhesive organs also perform a sensory function, and some collocytes are supposed to be evolutionarily derived from neurosecretory cells [23]. Therefore, this organ should play an important part in substrate selection for settlement, sensing surface structures, wettability, and other properties. Following adhesion to the substrate by adhesive secretion, metamorphosing larvae absorb the tail into the trunk (Figure 1b), and extend epidermal ampullae to hold fast to the substrate (arrows in Figure 1c).

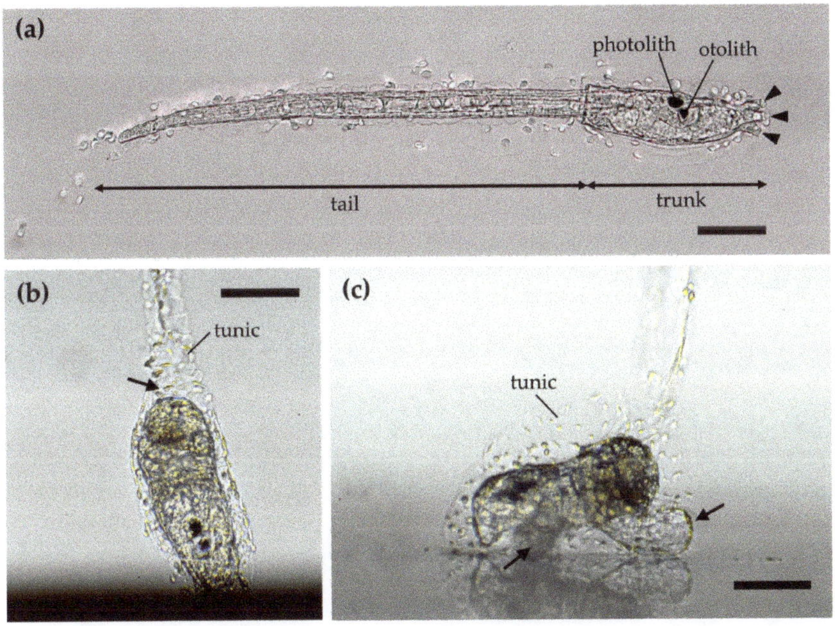

Figure 1. Swimming larva (**a**) and the settlement (**b**,**c**) of *Phallusia philippinensis* (lateral view). (**a**) A swimming larva with a tail. Three adhesive organs are present at the apical end of the trunk (arrowheads). (**b**) A larva adhering to the substrate with the adhesive organs. The tail is absorbed into the trunk (arrow). (**c**) A metamorphosing larva extending epidermal ampullae onto the substrate (arrows). All scale bars indicate 100 μm.

Natural chemical cues for settlement are known in various marine organisms, and many studies have reported on various natural products that affect the settlement and metamorphosis of larvae [24]. In the ascidian *Halocynthia roretzi*, a metamorphosis-inducing substance was isolated from the larvae, the larval-conditioned seawater, and the adult tunic [25], although the working mechanism of this substance is still uncertain. In the colonial ascidian *Botryllus schlosseri*, larvae may recognize their kin and aggregately settle with genetically related individuals [26].

3. Water Wettability

3.1. Measurement of Wettability

Water wettability is generally quantified as the contact angle of a water droplet on the surface. The angle is smaller on more hydrophilic surfaces and larger on more hydrophobic surfaces (Figure 2a). Whereas the contact angle is usually measured in the air, water wettability can be measured in the water as the contact angle of an air bubble (Figure 2b). In this case, the contact angle of a bubble is smaller on more hydrophobic surfaces. To evaluate the surface wettability of substrates, the contact angle of a bubble in seawater is better than the contact angle of a water droplet in the air, as the settlement of marine organisms usually occurs in the seawater. In the same manner, oil wettability (lipophilicity/lipophobicity) can be measured as the contact angle of an oil droplet (Figure 2c). High water wettability means low air wettability, and thus, air bubbles are rarely attached on hydrophilic substrates in water. Accordingly, when we find an object on which many bubbles are attached in shallow water, it likely has a hydrophobic surface.

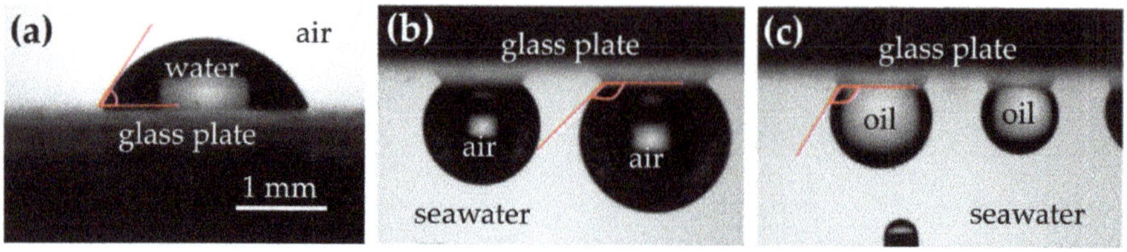

Figure 2. Contact angles (red lines) of a water droplet (**a**), air bubbles (**b**), and oil droplets (**c**) on a glass plate (cover slip).

However, natural objects with many bubbles are rarely encountered in shallow waters, indicating that many natural objects in the water usually have hydrophilic surfaces or a hydrophilic coating, such as a biofilm. The attachment of air bubbles is often a nuisance for marine organisms, because such bubbles cause unfavorable buoyancy and form an obstructive barrier against transport of water-soluble molecules. Figure 3 shows the water-wettability and oil-wettability of some seagrasses and seaweeds inhabiting in the intertidal zone. While a glass (cover slip) without any coatings is a moderately hydrophilic material, the seagrass and seaweeds are also hydrophilic and lipophobic. In some species, mucus exuded from organisms may enhance their hydrophilicity and lipophobicity.

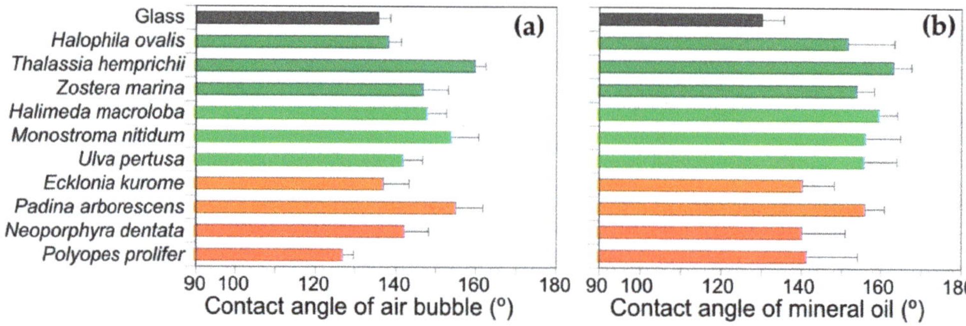

Figure 3. Contact angles of water droplets (**a**) and mineral oil (**b**) on glass (gray), seagrass (green), green alga (yellow green), brown alga (brown), and red alga (red). Fifteen to twenty measurements were carried out for each species in seawater (see Figure 2b,c). Error bars indicate standard deviations.

3.2. Substrate Preference in Wettability

Several studies have reported the substrate preferences of larvae in terms of wettability based on single-choice or multiple-choice tests. The settlement rate of ascidian larvae was shown to be greater on more hydrophobic surfaces than on less hydrophobic surfaces in *Phallusia nigra* (single-choice test) [4] and more larvae attached on more hydrophobic substrates in *Ciona intestinalis* (multi-choice test) [10]. Similarly, larval settlement rates were greater on hydrophobic substrates in the bryozoan *Bugula neritina* [3,4,27], and more zoospores attached on more hydrophobic substrates in the green alga *Ulva linza* [28]. In contrast, settlement rates of pediveligers of the bivalve *Mytilus galloprovicialis* were greater on hydrophilic substrates [6]. Preferences regarding wettability are different in barnacle species; cyprid larvae of *Balanus amphitrite* settled in higher percentages on more hydrophilic surfaces [3,4], although cyprids of *Balanus improvisus* settled in higher percentages on more hydrophobic surfaces [5]. It may be difficult to compare these results simply, because the experimental procedures were different in each study. However, these studies still indicate that larvae or spores of various organisms have settlement preferences in terms of wettability. Comprehensive surveys with a standardized assay are required to clarify the preferences of various species or taxa.

We carried out quadruple-choice tests for substrate selection of the ascidian larvae of *Phallusia philippinensis* using nine commercially available materials as substrates with different water- and oil-wettability [11]. We tested seven combinations of substrates in the quadruple-choice test, and, in each test, Manly's resource selection index was calculated from the number of juveniles that settled on each of four substrates. Multiple comparisons of the selection indices showed significant difference in preference among the substrates. The rank of preference was determined for each of the nine substrates by integrating the test results; the ascidian larvae tended to select more hydrophobic/oleophilic substrates for settlement (Figure 4). We suppose that nine (or more) substrates are more than necessary in order to grasp the substrate preferences regarding wettability and propose a simplified quadruple-choice test using silicone rubber, polyvinyl chloride, glass, and SH2CLHF (3M) as a candidate for a standard assay.

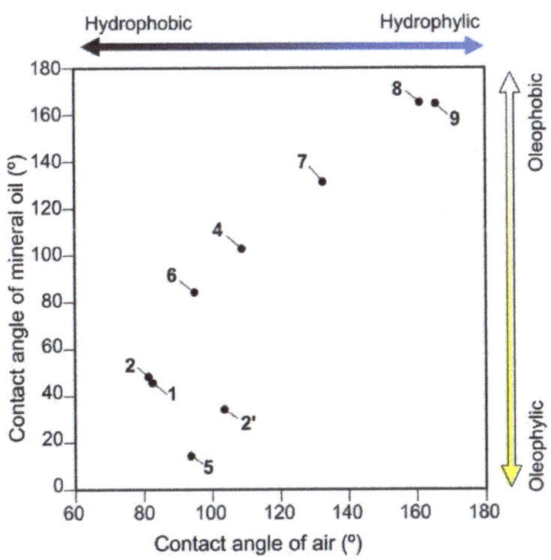

Figure 4. The rank of substrate preference in the settlement of *Phallusia philippinensis* larvae with water- and oil-wettability (modified from [11]). The numbers indicate the rank of preference with the rank of 2 and 2′ being tied. 1, polytetrafluoroethylene; 2, silicone rubber; 2′, polyvinylidene chloride; 4, polychlorotrifluoroethylene; 5, Parafilm (Amcor Ltd.); 6, polyvinyl chloride; 7, glass; 8, 1.5% agar; 9, SH2CLHF (SH: 3M).

Ascidian larvae can detect a small hydrophobic area and settle there. SH2CLHF is a super-hydrophilic material and no larvae settled on the surface. When the SH2CLHF film was slightly cut with a fine surgical knife, a more hydrophobic base film made of polyester was exposed at the cut of less than 50 μm in width. On this film, *P. philippinensis* larvae exclusively attached along the cut line and metamorphosed there (Figure 5). Before settlement, ascidian larvae crawled about on the substrate, with their sensory adhesive organs touching the substrate [23]. It is possible that the larvae sensed the properties of the substrate surface with their adhesive organs and carefully selected the site for settlement.

3.3. *"Silicone Paradox"*

The larvae of ascidians and bryozoans select hydrophobic substrates for settlement, as mentioned above. However, this preference appears inconsistent with the fact that hydrophobic materials, such as silicones, are often used for antifouling on ship hulls and fishing nets. Although silicone rubber ranks second in Figure 4, silicones are also regarded as effective compounds to reduce fouling [29–32]. Zhang et al. [1] reported that superhydrophobic surfaces showed fouling repellency for a few weeks but subsequently lost anti-fouling properties in months. Moreover, the adhesion strength of barnacles on silicone was significantly smaller than that on epoxy resin as a control [33]. In another study, more sporelings of the green alga *Ulva* were detached from hydrophobic surfaces, while diatom *Navicula* cells were more easily detached from hydrophilic surfaces [34]. Thus, some larvae clearly select hydrophobic surfaces that generally show fouling repellency, and this is called the "silicone paradox."

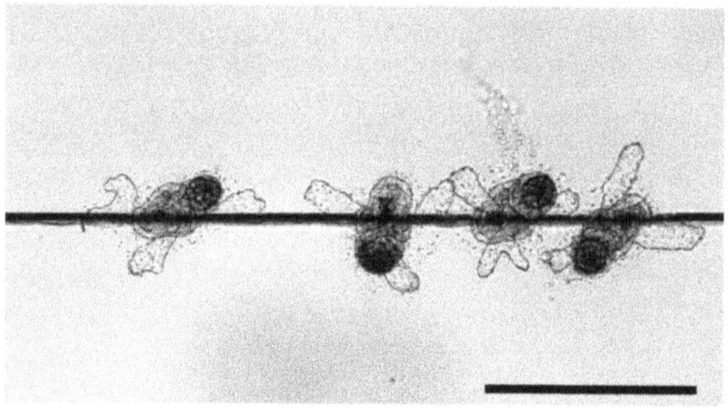

Figure 5. *Phallusia philippinensis* larvae settling exclusively on the cut line of a SH2CLHF film. Scale bar 1 mm.

Considering the low adhesion strength on hydrophobic surfaces, the fouling repellency of silicone probably results from post-settlement selection. On the other hand, the larval preference to hydrophobic surfaces as shown in laboratory experiments is pre-settlement selection of substrate. Therefore, pre-settlement selection favors hydrophobic surfaces, but the settlers have larger risks to be eliminated on more hydrophobic substrates by post-settlement selection. The preference to hydrophobic surfaces appears to be less adaptive for survival, and thus, it is difficult to explain why this preference occurs from an adaptive evolutionary viewpoint.

We think that the preference to hydrophobic surfaces in ascidians does not result from the adaptation to find better sites for settlement, but from a physicochemical mechanism such as an affinity with adhesive compounds secreted from the adhesive organs. In the initial phase of adhesion to a substrate, the larvae touch the substrate with the tips of the adhesive organs. When the larva is detached from the substrate, secreted material remains on the substrate surface in the manner of footprints [10]. If the adhesion strength of the secreted material is higher on hydrophobic surfaces, more larvae will settle on hydrophobic surfaces. After settlement, ascidian juveniles adhere to the substrate via the tunic, but the adhesion strength of the tunic might be lower on hydrophobic surfaces. In an aquatic environment, natural surfaces are usually hydrophilic (e.g., Figure 3), and thus, larvae settle on moderately hydrophilic substrates. For marine organisms, hydrophobic substrates may be almost an unexperienced surface in their evolutionary history, and thus, they have not yet adapted to hydrophobic surfaces even if such a surface is not suitable for post-settlement selection. It will be important to investigate the adhesion mechanism between substrates and the glue secreted from the adhesive organs for proving an answer to this paradox.

4. Surface Structures

4.1. Preference in Surface Structure

Natural substrates often have uneven surfaces, and even artificial, flat substrates may have some roughness at the microscopic level. Several studies have indicated that larval preference depends on the size of the structure. The 400-μm texture on a hydrophobic substrate much enhanced the settlement of *Mytilus* pediveligers [6], and ascidian larvae of *C. intestinalis* and *Botrylloides violaceus* prefer rougher surfaces [9]. In contrast to these micro-scale structures, nano-scale roughness on hydrophobic materials generates a super-hydrophobic surface that show anti-fouling properties against a broad spectrum of settlers including algal cells, algal spores, bryozoans, and barnacles [7]. For a better understanding of effective structures to prevent biofouling, surface structures of marine organisms have recently been focused on as biomimetic models for fouling control [35].

4.2. Nano-Scale Nipple Array

The moth-eye structure, or nano-scale nipple array, was originally discovered on the cornea of the compound eye in nocturnal moths [36]. This structure is an array of nipple-shaped protuberances of about 100 nm or less in height, and is well known to reduce light reflection on the surface, i.e., the moth-eye effect, by forming a gradient of refractivity [36,37]. This lower reflection simultaneously indicates higher transmission of light into the eyes. This is also an advantage in low-light conditions, and comparative studies of nocturnal and diurnal lepidopterans have suggested that greater transmission of UV light from the moth-eye structure may be important for lepidopterans and cause attraction to UV lights [38]. Moreover, some other functions have been demonstrated for this structure. The nipple array of some insects reduces adhesion and water wettability, and this property prevents dust and dew from adhering to the surface [39,40]. A bactericidal property was shown in the nipple array of black silicon, a synthetic material, as well as those on cicada and dragonfly wings [41,42]. Thus, the nipple array can be regarded as a multifunctional structure.

The nipple array has been also found from body surfaces of marine invertebrates belonging to various taxa, such as ascidians [43–45], salps [46–48], endoparasitic copepods [49], echinoderms [50,51], annelids [52], and entoprocts [53]. The histological bases of these nano-structures are different among taxa; the nipple-like protuberances are formed with cellulosic, integumentary matrix (tunic) in tunicates, chitinous exoskeleton in copepods, and microvilli in entoprocts, indicating the functional importance of this structure and its convergent occurrence in the evolutionary history of metazoans. On the other hand, a nipple array does not always occur within some taxa. For instance, nipple arrays were found in the species of some ascidian families (e.g., Polyclinidae and Styelidae) but not in other families (e.g., Didemnidae and Ascidiidae) [45], and, in salps, the nano-structure has only been reported in species that occur at shallow depths in daytime [47]. This suggests that the importance of this functional structure depends on the habitat and lifestyle of the species.

An anti-reflection effect in water certainly exists but is not as great as what was found in the air, as the reflectance of animal body surfaces is naturally small in water due to the small differences of refractive indices between seawater and animal tissue [47,54,55]. Furthermore, synthetic materials mimicking the nipple array have enabled us to evaluate potential functions other than anti-reflection in aquatic conditions; hydrophilic, nano-scale nipple arrays significantly reduce the attachment of air bubbles [56] and suppress cell attachment and phagocytic activity on the structure [57,58]. Thus, the nipple array seems to be a multifunctional structure in water, and may have a significant effect on the fouling of animal body surfaces.

4.3. Settlement Test on Synthetic Nipple Array

MOSMITE™ is a synthetic film for anti-reflection that namely mimics the moth-eye structure [59], with a nano-structure made of acrylic resin coating a base film made of polyethylene terephthalate. To evaluate whether the nipple array affects the substrate preference of ascidian larvae, we performed a dual-choice test with *P. philippinensis* larvae using MOSMITE™ and a flat film made of the same material, i.e., the base film coated with acrylic resin, as control [8]. The ascidian larvae significantly preferred the flat surface to MOSMITE™ for settlement. One of the reasons for this preference is the difference in wettability; MOSMITE™ is more hydrophilic than the flat surface. On the sheets of the same dimension, the surface area of MOSMITE™ is much larger than that of the flat sheet, due to the nano-structures. Under a homogeneous wetting regime, i.e., Wentzel state, a larger surface area enhances the water wettability on a hydrophilic material [60].

Accordingly, MOSMITE™ is more hydrophilic than the flat sheet, even though the surface material is the same in both sheets.

At the beginning of ascidian settlement, the adhesive organs secrete adhesive compounds to attach to the substrate surface [61,62]. On MOSMITE™, the adhesive compound penetrates into the nipple array layer and fills the space among the nipple-like protuberances, whereas the compound simply spread over the flat surface (Figure 6a,b). The adhesion strength on MOSMITE™ is, thus, likely greater than the strength on the flat surface, due to the larger surface area for adhesion. This appears inconsistent with the substrate preference for flat surfaces to MOSMITE™. Following metamorphosis, juveniles extend the epidermal ampullae and hold fast to the substrate via tunic (Figures 1c and 5), and the tunic surface is always overlaid with a cuticular, electron-dense layer. On MOSMITE™, the tunic cuticle adheres to the tip of the nipples only, with the space in the nipple array layer remaining vacant (blue area in Figure 6c), whereas the cuticle covers the flat surface entirely (Figure 6d). Therefore, the settlers on MOSMITE™ would be more easily detached from the substrate than settlers on the flat surface in post-settlement selection. From the viewpoint of evolutionary adaptation, the larvae may have a mechanism to avoid settlement on nano-scale nipple arrays. How can larvae sense nano-structures on substrate surfaces? As mentioned above, natural substrata in an aquatic environment are generally hydrophilic, and, on a hydrophilic material, surfaces with nano-structures are more hydrophilic than flat surfaces. If ascidian larvae select less hydrophilic sites for settlement, they may be able to avoid settlement on nano-structures. This idea can be one of possible explanations for the preference for hydrophobic surfaces causing the silicone paradox.

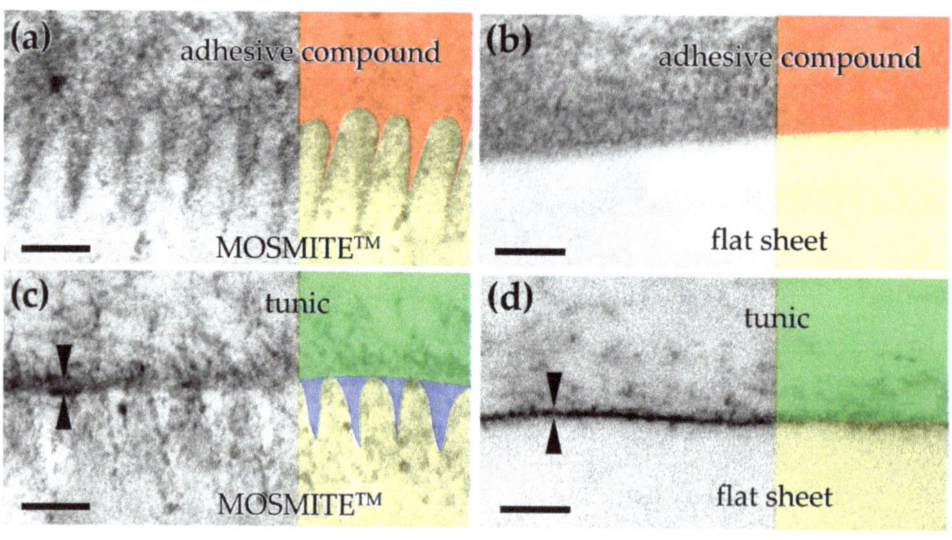

Figure 6. Electron micrographs of the adhesion point between the substrate (yellow) and adhesive compound (red) or tunic (green) in the settlement of *Phallusia philippinensis*. (**a**) Adhesive compound on MOSMITE™. (**b**) Adhesive compound on flat sheet. (**c**) Tunic on MOSMITE™. Vacant space remains between the nipple array layer (blue). (**d**) Tunic on flat sheet. Facing arrowheads indicate tunic cuticle. Scale bars 100 nm.

5. Flypapers May Mitigate Invasive Fouling

Humanity has built, left, or abandoned various artificial structures in the water, including aquaculture facilities and ship hulls as well as ports and seawalls. Artificial structures provide huge areas of a novel microenvironment which marine organisms had never experienced, and some 'winner' species may thrive on these structures. The winner is often a non-indigenous species, causing serious ecological and economic problems, and artificial structures are also a base for the dispersal for non-native, invasive species [63–65]. Many studies have focused on anti-fouling properties, i.e., which substrates are comparatively less fouled and what compounds suppress biofouling. Larval preferences of substrata should also be important to gain a better understanding of the biological processes of settlement, and preference data can be useful in controlling fouling.

Settlers are very often nuisances in aquaculture. They are competitors against the cultured organisms, such as shellfish, for space and food resources, and they often fill up the opening of cages and nets. Moreover, workers have to remove settlers on cultured organisms at harvest. Chemical approaches, i.e., anti-fouling compounds, potentially cause environmental pollution and damage native biota [66–71]. Physical approaches provide less effective but safer countermeasures against fouling of non-indigenous species. Hand elimination of a particular species is the most eco-friendly method but requires a large labor force. We hypothetically propose the use of hydrophobic substrates as a 'flypaper' for settlers. As discussed in this review, the larvae of ascidians and some other invertebrates significantly select more hydrophobic substrates for settlement. If hydrophobic substrates, such as ropes or belts of silicone, are hung nearby aquaculture facilities in the settlement seasons of target species, many larvae can be expected to settle on these hydrophobic substrates. Subsequently, the hydrophobic substrates can be removed from the water and dried/washed to remove settlers. Then, the substrates can be placed in the water again. Of course, this method is effective for the settlers preferring hydrophobic substrates, such as ascidians, and does not radically eliminate the settlers but probably will mitigate the impact of biofouling with a need for less of a labor force than hand elimination. Some settlers would be spontaneously detached from the substrate in post-settlement selection due to weak adhesive strength on hydrophobic surfaces. These detached settlers likely would die on the sea bottom, as they cannot swim after metamorphosis. Moreover, nano-structures on hydrophobic surfaces are expected to enhance the flypaper effect, i.e., attraction and spontaneous detachment of settlers. Theoretically, nano-structures on hydrophobic surfaces reduce water-wettability, and the adhesion strengths are also much reduced due to the smaller adhesion area for settlers, as found in Figure 6c. Even if such a physical approach cannot show enough effect practically, its use combined with chemical approaches may reduce the effective concentrations of antifouling compounds. Currently, this is no more than a plan on paper but it indicates the importance of studies on the mechanisms of substrate selection in fouling organisms.

Author Contributions: E.H. prepared the draft. N.S. provided some figure data. Both authors have read and agreed to the published version of the manuscript.

Funding: This research received no external funding.

Institutional Review Board Statement: This work complies with the ethical guidelines of University of the Ryukyus Animal Experimentation Committee.

Informed Consent Statement: Not applicable.

Data Availability Statement: Not applicable.

Acknowledgments: We are grateful to the special issue editor, Francesca Cima, for providing the opportunity to contribute this review. We thank Mitsubishi Chemical Corporation for generously providing the MOSMITE^{TM} and the flat film. Our thanks also to James Davis Reimer for peer-reviewing and editing the manuscript.

Conflicts of Interest: The authors declare no conflict of interest.

References

1. Zhang, H.; Lamb, R.; Lewis, J. Engineering nanoscale roughness on hydrophobic surface—Preliminary assessment of fouling behaviour. *Sci. Technol. Adv. Mater.* **2005**, *6*, 236–239. [CrossRef]
2. Siddik, A.A.; Al-Sofyani, A.A.; Ba-Akdah, M.A.; Satheesh, S. Invertebrate recruitment on artificial substrates in the Red Sea: Role of substrate type and orientation. *J. Mar. Biol. Assoc. U. K.* **2019**, *99*, 741–750. [CrossRef]

3. Rittschof, D.; Costlow, J.D. Bryozoan and barnacle settlement in relation to initial surface wettability: A comparison of laboratory and field studies. *Sci. Mar.* **1989**, *53*, 411–416.
4. Gerhart, D.J.; Rittschof, D.; Hooper, I.R.; Eisenman, K.; Meyer, A.E.; Baier, R.E.; Young, C. Rapid and inexpensive quantifications of the combined polar components of surface wettability application to biofouling. *Biofouling* **1992**, *5*, 251–259. [CrossRef]
5. Dahlström, M.; Jonsson, H.; Jonsson, P.R.; Elwing, H. Surface wettability as a determinant in the settlement of the barnacle *Balanus Improvisus* (DARWIN). *J. Exp. Mar. Bio. Ecol.* **2004**, *305*, 223–232. [CrossRef]
6. Carl, C.; Poole, A.J.; Sexton, B.A.; Glenn, F.L.; Vucko, M.J.; Williams, M.R.; Whalan, S.; de Nys, R. Enhancing the settlement and attachment strength of pediveligers of *Mytilus galloprovincialis* by changing surface wettability and microtopography. *Biofouling* **2012**, *28*, 175–186. [CrossRef]
7. Scardino, A.J.; Zhang, H.; Cookson, D.J.; Lamb, R.N.; de Nys, R. The role of nano-roughness in antifouling. *Biofouling* **2009**, *25*, 757–767. [CrossRef]
8. Hirose, E.; Sensui, N. Does a nano-scale nipple array (moth-eye structure) suppress the settlement of ascidian larvae? *J. Mar. Biol. Assoc. U. K.* **2019**, *99*, 1393–1397. [CrossRef]
9. Chase, A.L.; Dijkstra, J.A.; Harris, L.G. The influence of substrate material on ascidian larval settlement. *Mar. Pollut. Bull.* **2016**, *106*, 35–42. [CrossRef]
10. Zeng, F.; Wunderer, J.; Salvenmoser, W.; Ederth, T.; Rothbächer, U. Identifying adhesive components in a model tunicate. *Philos. Trans. R. Soc. B Biol. Sci.* **2019**, *374*, 20190197. [CrossRef]
11. Sensui, N.; Hirose, E. Wettability and substrate selection in the larval settlement of the solitary ascidian *Phallusia philippinensis* (Phlebobranchia: Ascidiidae). *Zool. Sci.* **2020**, *37*, 366–370. [CrossRef]
12. Matsunobu, S.; Sasakura, Y. Time course for tail regression during metamorphosis of the ascidian *Ciona intestinalis*. *Dev. Biol.* **2015**, *405*, 71–81. [CrossRef]
13. Bannister, J.; Sievers, M.; Bush, F.; Bloecher, N. Biofouling in marine aquaculture: A review of recent research and developments. *Biofouling* **2019**, *35*, 631–648. [CrossRef] [PubMed]
14. Chambers, L.D.; Stokes, K.R.; Walsh, F.C.; Wood, R.J.K. Modern approaches to marine antifouling coatings. *Surf. Coat. Technol.* **2006**, *201*, 3642–3652. [CrossRef]
15. Turner, A. Marine pollution from antifouling paint particles. *Mar. Pollut. Bull.* **2010**, *60*, 159–171. [CrossRef] [PubMed]
16. Lawes, J.C.; Clark, G.F.; Johnston, E.L. Disentangling settlement responses to nutrient-rich contaminants: Elevated nutrients impact marine invertebrate recruitment via water-borne and substrate-bound cues. *Sci. Total Environ.* **2018**, *645*, 984–992. [CrossRef] [PubMed]
17. Lambert, G. Ecology and natural history of the protochordates. *Can. J. Zool.* **2005**, *83*, 34–50. [CrossRef]
18. Hirose, E. Ascidian tunic cells: Morphology and functional diversity of free cells outside the epidermis. *Invertebr. Biol.* **2009**, *128*, 83–96. [CrossRef]
19. Burighel, P.; Cloney, R.A. Urochordata: Ascidiacea. In *Microscopic Anatomy of Invertebrates Hemichordata, Chaetognatha, and the Invertebrate Chordates*; Harrison, F.W., Ruppert, E., Eds.; Wiley-Liss, Inc.: New York, NY, USA, 1997; Volume 15, pp. 221–347. ISBN 0471561223.
20. Huber, J.L.; da Silva, K.B.; Bates, W.R.; Swalla, B.J. The evolution of anural larvae in molgulid ascidians. *Semin. Cell Dev. Biol.* **2000**, *11*, 419–426. [CrossRef]
21. Tsuda, M.; Sakurai, D.; Goda, M. Direct evidence for the role of pigment cells in the brain of ascidian larvae by laser ablation. *J. Exp. Biol.* **2003**, *206*, 1409–1417. [CrossRef] [PubMed]
22. Bingham, B.L.; Reyns, N.B. Ultraviolet radiation and distribution of the solitary ascidian *Corella inflata* (Huntsman). *Biol. Bull.* **1999**, *196*, 94–104. [CrossRef]
23. Zeng, F.; Wunderer, J.; Salvenmoser, W.; Hess, M.W.; Ladurner, P.; Rothbächer, U. Papillae revisited and the nature of the adhesive secreting collocytes. *Dev. Biol.* **2019**, *448*, 183–198. [CrossRef] [PubMed]
24. Hadfield, M.; Paul, V. Natural chemical cues for settlement and metamorphosis of marine-invertebrate larvae. In *Marine Chemical Ecology*; McClintock, J.B., Baker, B.J., Eds.; CRC Press: Cambridge, UK, 2001; pp. 431–461. ISBN 9781420036602.
25. Tsukamoto, S.; Kato, H.; Hirota, H.; Fusetani, N. Lumichrome. A larval metamorphosis-inducing substance in the ascidian *Halocynthia roretzi*. *Eur. J. Biochem.* **1999**, *264*, 785–789. [CrossRef] [PubMed]
26. Grosberg, R.K.; Quinn, J.F. The genetic control and consequences of kin recognition by the larvae of a colonial marine invertebrate. *Nature* **1986**, *322*, 456–459. [CrossRef]
27. Mihm, J.W.; Banta, W.C.; Loeb, G.I. Effects of adsorbed organic and primary fouling films on bryozoan settlement. *J. Exp. Mar. Biol. Ecol.* **1981**, *54*, 167–179. [CrossRef]
28. Callow, M.E.; Callow, J.A.; Ista, L.K.; Coleman, S.E.; Nolasco, A.C.; Lopez, G.P. Use of self-assembled monolayers of different wettabilities to study surface selection and primary adhesion processes of green algal (*Enteromorpha*) zoospores. *Appl. Environ. Microbiol.* **2000**, *66*, 3249–3254. [CrossRef]
29. Hodson, S.L.; Burke, C.M.; Bissett, A.P. Biofouling of fish-cage netting: The efficacy of a silicone coating and the effect of netting colour. *Aquaculture* **2000**, *184*, 277–290. [CrossRef]
30. Terlizzi, A.; Conte, E.; Zupo, V.; Mazzella, L. Biological succession on silicone fouling-release surfaces: Long-term exposure tests in the Harbour of Ischia, Italy. *Biofouling* **2000**, *15*, 327–342. [CrossRef]

31. Tettelbach, S.T.; Tetrault, K.; Carroll, J. Efficacy of Netminder® silicone release coating for biofouling reduction in bay scallop grow-out and comparative effects on scallop survival, growth and reproduction. *Aquac. Res.* **2014**, *45*, 234–242. [CrossRef]
32. Flemming, H.C. Biofouling in water systems—Cases, causes and countermeasures. *Appl. Microbiol. Biotechnol.* **2002**, *59*, 629–640. [CrossRef]
33. Swain, G.W.; Schultz, M.P. The testing and evaluation of non-toxic antifouling coatings. *Biofouling* **1996**, *10*, 187–197. [CrossRef]
34. Krishnan, S.; Wang, N.; Ober, C.K.; Finlay, J.A.; Callow, M.E.; Callow, J.A.; Hexemer, A.; Sohn, K.E.; Kramer, E.J.; Fischer, D.A. Comparison of the fouling release properties of hydrophobic fluorinated and hydrophilic PEGylated block copolymer surfaces: Attachment strength of the diatom *Navicula* and the green alga *Ulva*. *Biomacromolecules* **2006**, *7*, 1449–1462. [CrossRef] [PubMed]
35. Scardino, A.J.; de Nys, R. Mini review: Biomimetic models and bioinspired surfaces for fouling control. *Biofouling* **2011**, *27*, 73–86. [CrossRef]
36. Bernhard, C.G. Structural and functional adaptation in a visual system. *Endeavour* **1967**, *26*, 79–84.
37. Wilson, S.J.; Hutley, M.C. The optical properties of "moth eye" antireflection surfaces. *Opt. Acta* **1982**, *29*, 993–1009. [CrossRef]
38. Spalding, A.; Shanks, K.; Bennie, J.; Potter, U.; Ffrench-Constant, R. Optical modelling and phylogenetic analysis provide in butterflies and moths. *Insects* **2019**, *10*, 262. [CrossRef] [PubMed]
39. Watson, G.S.; Myhra, S.; Cribb, B.W.; Watson, J.A. Putative functions and functional efficiency of ordered cuticular nanoarrays on insect wings. *Biophys. J.* **2008**, *94*, 3352–3360. [CrossRef]
40. Peisker, H.; Gorb, S.N. Always on the bright side of life: Anti-adhesive properties of insect ommatidia grating. *J. Exp. Biol.* **2010**, *213*, 3457–3462. [CrossRef]
41. Ivanova, E.P.; Hasan, J.; Webb, H.K.; Truong, V.K.; Watson, G.S.; Watson, J.A.; Baulin, V.A.; Pogodin, S.; Wang, J.Y.; Tobin, M.J.; et al. Natural bactericidal surfaces: Mechanical rupture of *Pseudomonas aeruginosa* cells by cicada wings. *Small* **2012**, *8*, 2489–2494. [CrossRef] [PubMed]
42. Ivanova, E.P.; Hasan, J.; Webb, H.K.; Gervinskas, G.; Juodkazis, S.; Truong, V.K.; Wu, A.H.F.; Lamb, R.N.; Baulin, V.A.; Watson, G.S.; et al. Bactericidal activity of black silicon. *Nat. Commun.* **2013**, *4*, 2838. [CrossRef] [PubMed]
43. Hirose, E.; Saito, Y.; Hashimoto, K.; Watanabe, H. Minute protrusions of the cuticle: Fine surface structures of the tunic in ascidians. *J. Morphol.* **1990**, *204*, 67–73. [CrossRef]
44. Hirose, E.; Nishikawa, T.; Saito, Y.; Watanabe, H. Minute protrusions of ascidian tunic cuticle: Some implications for ascidian phylogeny. *Zool. Sci.* **1992**, *9*, 405–412.
45. Hirose, E.; Lambert, G.; Kusakabe, T.; Nishikawa, T. Tunic cuticular protrusions in ascidians (Chordata, Tunicata): A perspective of their character-state distribution. *Zool. Sci.* **1997**, *14*, 683–689. [CrossRef]
46. Hirose, E.; Kimura, S.; Itoh, T.; Nishikawa, J. Tunic morphology and cellulosic components of pyrosomas, doliolids, and salps (Thaliacea, Urochordata). *Biol. Bull.* **1999**, *196*, 113–120. [CrossRef]
47. Hirose, E.; Sakai, D.; Shibata, T.; Nishii, J.; Mayama, H.; Miyauchi, A.; Nishikawa, J. Does the tunic nipple array serve to camouflage diurnal salps? *J. Mar. Biol. Assoc. U. K.* **2015**, *95*, 1025–1031. [CrossRef]
48. Sakai, D.; Kakiuchida, H.; Nishikawa, J.; Hirose, E. Physical properties of the tunic in the pinkish-brown salp *Pegea confoederata* (Tunicata: Thaliacea). *Zool. Lett.* **2018**, *4*, 7. [CrossRef] [PubMed]
49. Hirose, E.; Uyeno, D. Histopathology of a mesoparasitic hatschekiid copepod in hospite: Does *Mihbaicola sakamakii* (Copepoda: Siphonostomatoida: Hatschekiidae) fast within the host fish tissue? *Zool. Sci.* **2014**, *31*, 546–552. [CrossRef] [PubMed]
50. Holland, N.D.; Nealson, K.H. The fine structure of the echinoderm cuticle and the subcuticular bacteria of echinoderms. *Acta Zool.* **1978**, *59*, 169–185. [CrossRef]
51. Flammang, P.; Jangoux, M. Functional morphology of coronal and peristomeal podia in *Sphaerechinus granularis* (Echinodermata, Echinoida) Patrick. *Zoomorphology* **1993**, *113*, 47–60. [CrossRef]
52. Hausen, H. Comparative structure of the epidermis in polychaetes (Annelida). *Hydrobiologia* **2005**, *535–536*, 25–35. [CrossRef]
53. Iseto, T.; Hirose, E. Comparative morphology of the foot structure of four genera of Loxosomatidae (Entoprocta): Implications for foot functions and taxonomy. *J. Morphol.* **2010**, *271*, 1185–1196. [CrossRef] [PubMed]
54. Kakiuchida, H.; Sakai, D.; Nishikawa, J.; Hirose, E. Measurement of refractive indices of tunicates' tunics: Light reflection of the transparent integuments in an ascidian *Rhopalaea* sp. and a salp *Thetys vagina*. *Zool. Lett.* **2017**, *3*, 7. [CrossRef] [PubMed]
55. Sakai, D.; Kakiuchida, H.; Harada, K.; Nishikawa, J.; Hirose, E. Parallel plications may enhance surface function: Physical properties of transparent tunics in colonial ascidians *Clavelina cyclus* and *C. obesa*. *J. Mar. Biol. Assoc. U. K.* **2019**, *99*, 1831–1839. [CrossRef]
56. Hirose, E.; Mayama, H.; Miyauchi, A. Does the aquatic invertebrate nipple array prevent bubble adhesion? An experiment using nanopillar sheets. *Biol. Lett.* **2013**, *9*, 20130552. [CrossRef] [PubMed]
57. Nomura, S.; Kojima, H.; Ohyabu, Y.; Kuwabara, K.; Miyauchi, A.; Uemura, T. Cell culture on nanopillar sheet: Study of HeLa cells on nanopillar sheet. *Jpn. J. Appl. Phys.* **2005**, *44*, 1184–1186. [CrossRef]
58. Ballarin, L.; Franchi, N.; Gasparini, F.; Caicci, F.; Miyauchi, A.; Hirose, E. Suppression of cell-spreading and phagocytic activity on nano-pillared surface: In vitro experiment using hemocytes of the colonial ascidian *Botryllus schlosseri*. *Invertebr. Surviv. J.* **2015**, *12*, 82–88.
59. Moth Eyes High-Performance Film MOSMITETM. Available online: https://www.m-chemical.co.jp/en/products/departments/mcc/hp-films-pl/product/1201267_7568.html (accessed on 11 April 2021).
60. Marmur, A. Wetting on hydrophobic rough surfaces: To be heterogeneous or not to be? *Langmuir* **2003**, *19*, 8343–8348. [CrossRef]

61. Cloney, R.A. Ascidian larvae and the events of metamorphosis. *Am. Zool.* **1982**, *22*, 817–826. [CrossRef]
62. Karaiskou, A.; Swalla, B.J.; Sasakura, Y.; Chambon, J. Metamorphosis in solitary ascidians. *Genesis* **2015**, *47*, 34–47. [CrossRef] [PubMed]
63. Simkanin, C.; Davidson, I.C.; Dower, J.F.; Jamieson, G.; Therriault, T.W. Anthropogenic structures and the infiltration of natural benthos by invasive ascidians. *Mar. Ecol. Prog. Ser.* **2012**, *33*, 499–511. [CrossRef]
64. Airoldi, L.; Turon, X.; Perkol-Finkel, S.; Rius, M.; Keller, R. Corridors for aliens but not for natives: Effects of marine urban sprawl at a regional scale. *Divers. Distrib.* **2015**, *21*, 755–768. [CrossRef]
65. Aldred, N.; Clare, A.S. Mini-review: Impact and dynamics of surface fouling by solitary and compound ascidians. *Biofouling* **2014**, *30*, 259–270. [CrossRef] [PubMed]
66. Cima, F.; Bragadin, M.; Ballarin, L. Toxic effects of new antifouling compounds on tunicate haemocytes I. Sea-nine 211 and chlorothalonil. *Aquat. Toxicol.* **2008**, *86*, 299–312. [CrossRef] [PubMed]
67. Menin, A.; Ballarin, L.; Bragadin, M.; Cima, F. Immunotoxicity in ascidians: Antifouling compounds alternative to organotins-II. The case of Diuron and TCMS pyridine. *J. Environ. Sci. Health Part B* **2008**, *43*, 644–654. [CrossRef] [PubMed]
68. Cima, F.; Ballarin, L. Immunotoxicity in ascidians: Antifouling compounds alternative to organotins: III—The case of copper(I) and Irgarol 1051. *Chemosphere* **2012**, *89*, 19–29. [CrossRef]
69. Cima, F.; Ballarin, L. Immunotoxicity in ascidians: Antifouling compounds alternative to organotins—IV. The case of zinc pyrithione. *Comp. Biochem. Physiol. C Toxicol. Pharmacol.* **2015**, *169*, 16–24. [CrossRef]
70. Amara, I.; Miled, W.; Ben, R.; Ladhari, N. Antifouling processes and toxicity effects of antifouling paints on marine environment. A review. *Environ. Toxicol. Pharmacol.* **2018**, *57*, 115–130. [CrossRef]
71. Cima, F.; Varello, R. Immunotoxicity in ascidians: Antifouling compounds alternative to organotins—V. the case of dichlofluanid. *J. Mar. Sci. Eng.* **2020**, *8*, 396. [CrossRef]

Journal of
Marine Science
and Engineering

MDPI

Article

Effects of Exposure to Trade Antifouling Paints and Biocides on Larval Settlement and Metamorphosis of the Compound Ascidian *Botryllus schlosseri*

Francesca Cima * and **Roberta Varello**

Laboratory of Ascidian Biology, Department of Biology (DiBio), University of Padova, Via U. Bassi 58/B, 35131 Padova, Italy; roberta.varello@phd.unipd.it
* Correspondence: francesca.cima@unipd.it; Tel.: +39-49-827-6198

Abstract: To evaluate the effects of antifouling paints and biocides on larval settlement and metamorphosis, newly hatched swimming larvae of the compound ascidian *Botryllus schlosseri*, a dominant species of soft-fouling in coastal communities, were exposed to (i) substrata coated with seven antifouling paints on the market containing different biocidal mixtures and types of matrices and (ii) sea water containing various concentrations of eight biocidal constituents. All antifouling paints showed high performance, causing 100% mortality and metamorphic inhibition, with ≥75% not-settled dead larvae. All antifouling biocides prevented the settlement of larvae. The most severe larval malformations, i.e., (i) the formation of a bubble encasing the cephalenteron and (ii) the inhibition of tail resorption, were observed after exposure to metal and organometal compounds, including tributyltin (TBT) at 1 μM (325.5 μg L^{-1}), zinc pyrithione (ZnP) at 1 μM (317.7 μg L^{-1}), and CuCl at 0.1 μM (98.99 μg L^{-1}), and to antimicrobials and fungicides, including Sea-Nine 211 at 1 μM (282.2 μg L^{-1}) and Chlorothalonil at 1 μM (265.9 μg L^{-1}). The herbicides seemed to be less active. Irgarol 1051 was not lethal at any of the concentrations tested. Diuron at 250 μM (58.2 mg L^{-1}) and 2,3,5,6-tetrachloro-4-(methylsulphonyl)pyridine (TCMS pyridine) at 50 μM (14.8 mg L^{-1}) completely inhibited larval metamorphosis. These results may have important implications for the practical use of different antifouling components, highlighting the importance of their testing for negative impacts on native benthic species.

Keywords: ascidians; antifouling paints; *Botryllus schlosseri*; booster biocides; EC_{50}; fouling settlement; larval toxicity; metamorphosis; tunicates

Citation: Cima, F.; Varello, R. Effects of Exposure to Trade Antifouling Paints and Biocides on Larval Settlement and Metamorphosis of the Compound Ascidian *Botryllus schlosseri*. *J. Mar. Sci. Eng.* **2022**, *10*, 123. https://doi.org/10.3390/jmse10020123

Academic Editor: Romana Santos

Received: 30 December 2021
Accepted: 15 January 2022
Published: 18 January 2022

1. Introduction

Ascidians are the most common members of the urochordate subphylum. They are the closest relatives to vertebrates, and are the only chordates able to reproduce both sexually and asexually [1]. During the larval stage, known as the "tadpole" stage, they share with vertebrates the same body plan in the tail formed of a dorsal notochord and a tubular nervous system, both flanked by striated muscle. They are sessile occupants of hard substrata in coastal environments, where they often represent the dominant component of soft-fouling. As worldwide filter-feeding organisms living at the water–sediment interface, many solitary and compound species are considered useful bioindicators of various environmental pollutants. The developmental biology of ascidians with particular attention to both the larval stage, which is the dispersal and colonisation stage of the life cycle, and its metamorphosis have been studied extensively [2–5]. After a short dispersal period, during which the ascidian larvae actively swim, moving the tail according to a positive phototropism and a negative geotropism, the larvae select a substratum and attach to it following a photo- and geotropism reversal. They explore and contact the substratum by means of a series of rapid touches with the anterior area of the cephalenteron, which contains special sensory structures, namely "papillae". These structures, protruding from the anterior epidermis and from which an adhesive substance is secreted, act as a control centre for the initiation of metamorphosis and are fundamental for larval settlement. Recently, many inducers of metamorphosis have been identified in papillae as specific gene expressions, growth factors, several neurotransmitters, and transient Ca^{2+} signals [6–8]. The temporary settlement with papillae is rapidly substituted by a stable settlement due to the protrusion of anterior blind-sac vessels, namely "ampullae", which expand onto the substratum and secrete a glue substance from an apical glandular epithelial. At the same time, tail resorption occurs with complete dismantling of the axial complex. Successively, metamorphosis continues inside the cephalenteron with the visceral rotation and the dismantling of other larval structures, such as ocellus and statocyst. Finally, the adult organ primordia complete the development, and the opening of both siphons occurs and the juvenile begins to filter-feed.

Generally, ascidians are pre-disposed to rapid and competitive colonisation of hard substrata. However, geographic invasion and the impact of fouling by ascidians on shipping, aquaculture, pleasure boating, oil and gas installations, and other industries are significant, with numerous species responsible for infesting anthropogenic structures [9]. Although the larval phase is short (24–48 h), the global spread is favoured by translocation via shipping since the hull fouling can be considered the most important vector. The massive ascidian fouling is potentially responsible for physical, economical, and ecological damage since it concerns the coverage of clean surfaces, loss of efficiency of submerged pipelines and harbour/industrial structures, infestation of shellfish and finfish aquacultures, increased fuel consumption during boat navigation, and decreased biodiversity in benthic communities.

For these reasons, beginning from the second half of the 1960s, antifouling compounds have been massively introduced in the formulation of paints to prevent the settlement of the most problematic foulers on submerged structures, such as ship's hulls and propellers, buoys, wharves, and platforms. The most successful antifouling paints at the time contained organotin compounds as biocides, mainly represented by tributyltin (TBT), triphenyltin (TPT), and their derivatives. These compounds proved to be harmful to the benthic marine communities, as they caused severe impacts on the oyster aquaculture and were persistent in the environment in the long term [10–13]. After the total ban on organotin compounds by the International Marine Organisation—Marine Environment Protection Committee (IMO-MEPC) in 1998, and subsequently by the Ordinance No. 782/2003, 14 April 2003, of the European Commission, the paint industry developed substitutive tin-free formulations. The new paint formulations mainly contained biocidal combinations of specific synthesis, e.g., Sea-Nine 211, or pesticides coming from the pharmacology industry (antimicrobials) or agriculture (herbicides, fungicides, insecticides). The aim of these formulations was not only to prevent the settlement of algal propagules and invertebrate larvae of macrofoulers, but also the formation of bacterial and microalgal microfilm, or "biofilm", from which the ecological succession of the hard-substratum community begins.

Because of this choice, a number of substances (Table 1) are at present in various commercial formulations of new-generation antifouling paints. The biocidal compounds play various roles, i.e., as alternatives to organotin compounds or as boosters, with the latter serving to increase the toxic performance of the antifouling paints towards a wider spectrum of fouling organisms. Before the introduction of these compounds in paint formulations, tests of acute and chronic toxicity were performed only on laboratory mammals and freshwater model fish. As a consequence, many new contaminants with potential accumulation and deleterious effects on coastal communities have been introduced worldwide from both direct and indirect pollution sources, which have followed the increase in productivity of agro-industrial, tourism, and commercial shipping sectors.

Table 1. Common biocidal substances used in formulations of antifouling paints in EU countries.

Chemical Name	CAS	Trademark(s)	Other Uses
zinc 2-pyridinethiol-oxide	13463-41-7	Zinc pyrithione, ZnP	Antimicrobial, fungicide in antidandruff shampoo, antiseborrheic, preservative in cosmetics
zinc N-[2(sulfidocarbothioylamino)ethyl]carbamodithioate	9006-42-2	Zineb, Metiram, Amarex, Polyram, Polycarbacin, Parzate, Dithane, Z-78	Fungicide
copper(I) oxide	1317-39-1	Cuprous oxide, Dicopper monoxide, Red copper oxide	Antimicrobial, fungicide, pigment, catalyst
copper(I) thiocyanate	1111-67-7	Cuprous thiocyanate, Copper sulfocyanide, Thiocyanic acid copper (I) salt	Antimicrobial, fungicide, paint additive
4,5-dichloro-2-n-octyl-4-isothiazolin-3-one	64359-81-5	Sea-Nine 211, DCOIT, Kathon 5287, C-9	Fungicide for sealants, PVC, and wood
2,4,5,6-tetrachloroisophthalonitrile	1897-45-6	Chlorothalonil, Bravo Daconil, Faber, Forturf, Nopcocide, Repulse, Termil, Tuffcide	Antimicrobial, fungicide, insecticide, acaricide
α,β-1,2,3,4,7,7-hexachlorobicyclo-[2.2.1]-2-heptene-5,6-bisoxymethylene sulfite	33213-65-9	Endosulfan, Benzoepin, BeositIndan, Sialan, Thiodan, Thiosulfan, Thionex, Thimul	Insecticide, acaricide
2-N-tert-butyl-4-N-cyclopropyl-6-methylsulfanyl-1,3,5-triazine-2,4-diamine	28159-98-0	Irgarol 1051, Cybutryne	Herbicide
3-(3,4-dichlorophenyl)-1,1-dimethylurea	330-54-1	Diuron, Duran, Dynex, Dichlorfenidim, Herbatox, Karmex, Telvar, Vonduron	Herbicide
2,3,5,6-tetrachloro-4-(methylsulphonyl)pyridine	13108-52-6	TCMS pyridine, Davicil, Dowco-282	Fungicide for leather and wood

Recently, toxic effects on aquatic organisms have arisen, with mechanisms of action involving various cell targets [14]. The risk assessment of these emerging contaminants in marine ecosystems is now a priority due to the continuous uncontrolled leaching from antifouling paints and the synergistic interactions, which could affect the primary production and the survival and reproduction of marine fish and invertebrates [15–18].

Organotin compounds and a few alternative biocides are known to provoke embryotoxicity in solitary ascidians such as *Styela plicata* and *Ciona intestinalis* [19,20]. These species are oviparous and both the spawned gametes and embryos might be exposed to biocides in the water column, with important effects on fertilisation and offspring. At present, no study has been performed on compound ascidians, which are ovoviviparous [21–23], and therefore, as a difference from solitary ascidians, only hatched larvae might be exposed. They can be considered a better model than the solitary ascidians for the evaluation of the effects on post-embryonic stages, settlement, and metamorphosis.

Botryllus schlosseri, commonly called the star ascidian, earns its name from the stellate colony formed by a group of asexually reproducing individuals. It is a cosmopolitan compound species, is easy to collect, and breed in aquaria. In peculiar ecosystems with transitional waters such as the Lagoon of Venice, during autumn, more than 90% of the community comprises solitary and colonial ascidians, predominantly botryllids (*Botryllus schlosseri* and *Botrylloides leachii*), forming a stable biocoenosis described as a "*Botryllus* community" [24]. The organism has recently emerged as a simple and important model species for morphogenesis, regeneration, allorecognition, and apoptosis and for studying a variety of biological problems, such as comparative immunobiology, sexual and asexual reproduction, stem cell differentiation, and regeneration [25–27].

In the present study, the compound ascidian *B. schlosseri* was used as an experimental model for the evaluation of the effects of (i) TBT-based and tin-free (alternative to TBT) antifouling paints and (ii) antifouling biocides on larval viability, settlement, and metamorphosis. The biocides considered have been distinguished into three groups: (i) metals and organometals, i.e., CuCl, TBT, and zinc pyrithione (ZnP); (ii) antimicrobials and fungicides, i.e., Sea-Nine 211 and Chlorothalonil; and (iii) herbicides, i.e., Irgarol 1051, Diuron, 2,3,5,6-tetrachloro-4-(methylsulphonyl)pyridine (TCMS pyridine). Experiments were performed in two steps by exposing larvae to substrata coated with seven trade antifouling paints containing mixtures of the above-reported biocides and two different types of matrices (contact-leaching or self-polishing), and to various concentrations of each type of biocide. Different antifouling performances and metamorphic abnormalities have also been considered, and the mechanisms of actions of larval toxicity have been proposed.

2. Materials and Methods

2.1. Larval Collection

B. schlosseri reproductive colonies were collected in spring (April–May) from the southern basin of the Lagoon of Venice and transferred to large aquaria, in which they were reared on glass slides in aerated filtered sea water (FSW) (salinity of 35 ± 1 psu, temperature of 19 ± 0.5 °C, pH 8.1). A semi-static system was used for animal maintenance, with seawater renewal every other day, and colonies were fed with Microbe-Lift®/Phyto-Plus B (Ecological Laboratories, Inc., Cape Coral, FL, USA) and microalgae (*Isochrysis galbana*). Newly hatched swimming larvae were easily identified under a dissection binocular stereomicroscope Wild Heerbrugg with 50× maximum magnification. They were immediately collected with a glass micropipette, counted, and temporarily transferred to a 30 mL glass-evaporating dish.

Five main stages were identified during metamorphosis, from the beginning of settlement to fully metamorphosed oozoids, as listed in Table 2.

Table 2. Main stages of the metamorphosis of *B. schlosseri* larva [1].

<table>
<tr><th colspan="2">Metamorphosis Stages</th></tr>
<tr><td>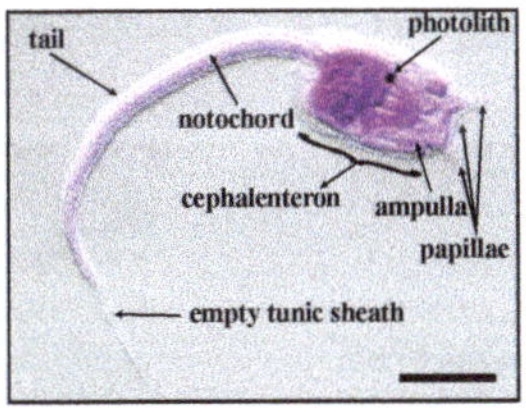
</td><td>Stage 1
– Settled larva with three adhesive papillae
– Initial resorption of tail from its distal region
– No change in cephalenteron organs</td></tr>
<tr><td>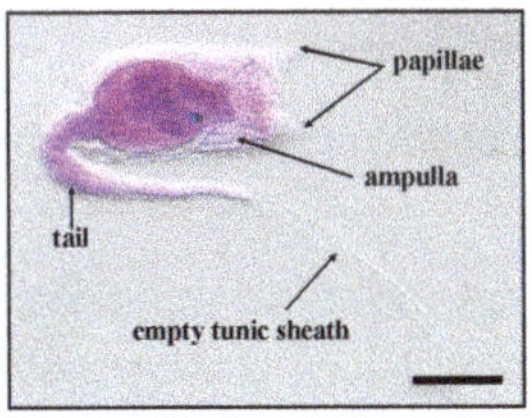
</td><td>Stage 2
– Tail resorption continues
– A long tract of distal empty tunic sheath is recognisable
– Protrusion of eight ampullae contacting the substratum</td></tr>
<tr><td>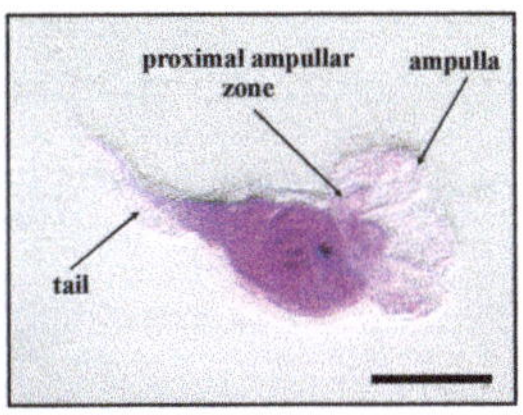
</td><td>Stage 3
– Bell-shaped larva
– Highly protruded ampullae
– Two-thirds of tail resorption
– Initial rotation of the cephalenteron organs</td></tr>
<tr><td>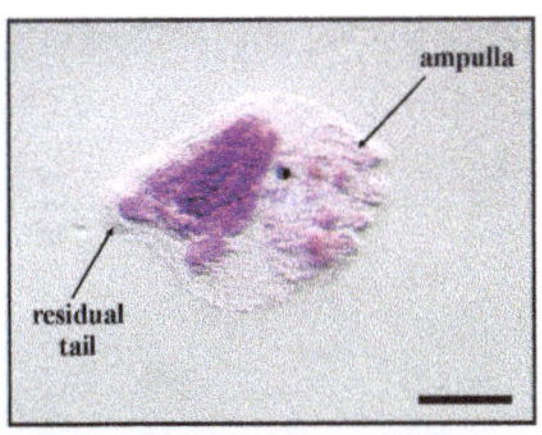
</td><td>Stage 4
– Flattened larva with complete 180° rotation of organs
– Ampullae radiating on the substratum
– Almost complete tail resorption
– Well-developed branchial pharynx
– Enlarged stomach</td></tr>
<tr><td>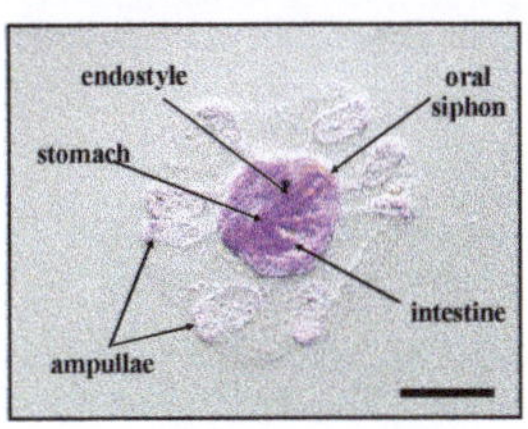
</td><td>Stage 5
– Oozoid
– Complete expansion of ampullae over substratum
– Marginal vessel formation
– Siphon opening and filter-feeding activity
– Presence of a primary bud from asexual reproduction</td></tr>
</table>

[1] In pictures, glutaraldehyde-fixed larvae stained with haematoxylin dye. Bar length: 200 μm.

2.2. Antifouling Paints

In the first series of experiments, seven (A–G) antifouling paints were tested (Table 3) with two replicates for a total exposure of 700 larvae. In each replicate, 50 swimming larvae were put into a glass-crystallising dish filled with 200 mL of FSW and with a 7.5 × 7.5 × 0.15 cm glass plate on the bottom that was previously coated with an antifouling paint. During exposure, the glass containers were covered outside and on the bottom with a black paper simulating a dark substratum, which favoured the settlement of larvae with positive geotropism and negative phototropism (Figure 1). The exposure occurred at

22 °C under artificial light until all larvae (100%) metamorphosed (48 h) in the reference glass-crystallising dishes, the latter without a painted plate on the bottom.

Table 3. Antifouling paints used to coat the glass plates for the assays of larval settlement and metamorphosis.

	Paint	Biocides	Matrix	Use
A	Sigmaplane HB Antifouling	Cu_2O (28%) TBT methacrylate (19%) TBTO (0.5%)	Self-polishing copolymers	Steel and wooden hull of fishing boats and cargo vessels more than 25 m in length (banned since 2003)
B	Marlin Velox TF	Zinc pyrithione (5–10%) Zineb (5–10%) Endosulfan (1–5%)	Contact leaching (hard or insoluble)	Propellers, shafts, and outdrives of fishing boats
C	Veneziani Propeller	CuSCN (7–10%) Zinc pyrithione (7–10%) Diuron (7–10%) Sea Nine 211 (1–3%)	Contact leaching (hard or insoluble)	Propellers, shafts, and outdrives of recreational craft
D	Veneziani Antialga	CuSCN (7–10%) Diuron (7.6%) Sea Nine 211 (2.7%)	Contact leaching (hard or insoluble)	Boattop of high-speed sailboats and powerboats
E	Sikkens Vinyl Antifouling 2000	Cu_2O (41%)	Contact leaching (hard or insoluble)	Steel (not aluminium), wooden, and polyester hull of sailboats and yachts
F	Baseggio Sirena Antivegetativa Universale	Cu_2O (42%) Chlorothalonil (7%) Irgarol 1051 (1.1%)	Self-polishing copolymers	Steel, wooden, and fibreglass hull of fishing boats
G	Veneziani Even Extreme 2 (reactive component)	TCMS pyridine (1–5%) Diuron (1–5%)	Self-polishing based on two-pack Biomatrix technology	Steel, wooden, and fibreglass hull of racing yachts

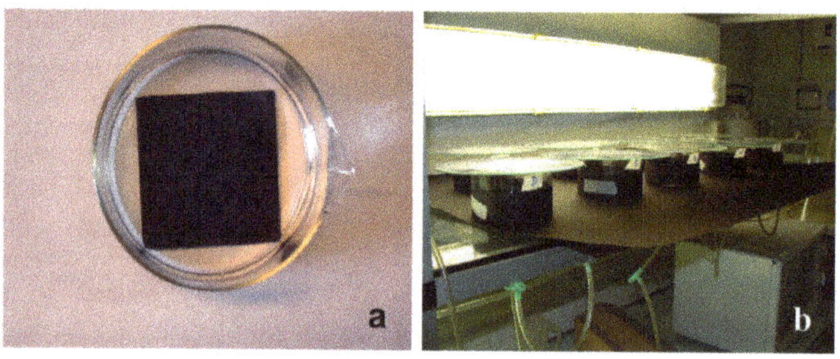

Figure 1. Experimental set-up for the evaluation of the settlement and metamorphosis ability of *B. schlosseri* larvae. (**a**) Glass-crystallising dish with a plate coated with an antifouling paint on the bottom. (**b**) Arrangement of the various glass-crystallising dishes in the thermostatic room at 22 °C.

2.3. *Antifouling Biocides*

In the second series of experiments, eight biocidal compounds were separately tested with three different concentrations and two replicates for a total exposure of 2400 larvae. For every replicate of biocidal concentration, 50 larvae were incubated in a glass-crystallising dish filled with 200 mL of biocidal solution in FSW. Concentration ranges of biocidal solutions were chosen based on previous studies of immunotoxicity in vitro on the haemocytes of this species [28–32]. The biocides considered were tributyltin (TBT) as monochloride (Sigma-Aldrich, Burlington, MA, USA), zinc pyrithione (ZnP, Sigma-Aldrich), copper(I) chloride (CuCl, purified, >99%, Sigma-Aldrich), Sea Nine 211 (Rohm & Haas, Philadelphia,

PA, USA), Chlorothalonil (Fluka), Irgarol 1051 (Riedel-de Haën GmbH, Seelze, Germany), Diuron (Sigma-Aldrich), and TCMS pyridine (Avecia, Manchester, UK).

Stock solutions were prepared at the nominal concentrations of 1 mM in FSW for CuCl; 10 mM in 95% ethanol for TBT, Sea Nine 211, Diuron, and TCMS pyridine; and 10 mM in dimethylsulfoxide (DMSO purum; >99%, Fluka Chemie Gmbh, Buchs, Switzerland) for ZnP, Chlorothalonil, and Irgarol 1051. They were freshly diluted to working concentrations in FSW as follows: 0.1, 1, 10 μM for TBT (corresponding to 32.5, 325.5, 3255 μg L^{-1}); 0.1, 0.5, 1 μM for ZnP (corresponding to 31.7, 158.5, 317.7 μg L^{-1} and 6.5×10^{-7}, 3.2×10^{-6}, 6.5×10^{-6} Zn wt%); 0.01, 0.1, 1 μM for CuCl (corresponding to 9.8, 98.99, 989.9 μg L^{-1} and 1.8×10^{-6}, 1.8×10^{-5}, 1.8×10^{-4} Cu wt%); 0.1, 1, 10 μM for Sea-Nine 211 (corresponding to 28.2, 282.2, 2822 μg L^{-1}); 0.1, 1, 10 μM for Chlorothalonil (corresponding to 26.5, 265.9, 2659 μg L^{-1}); 50, 100, 200 μM for Irgarol 1051 (corresponding to 12.6, 25.3, 50.6 mg L^{-1}); 100, 250, 500 μM for Diuron (corresponding to 23.3, 58.2, 116.5 mg L^{-1}); and 25, 50, 75 μM for TCMS pyridine (corresponding to 7.4, 14.8, 22.2 mg L^{-1}). In controls, larvae were incubated with FSW containing the maximum solvent concentration employed in the experiments with biocides, i.e., 0.02% DMSO or 0.01% 95%-ethanol. These concentrations had no effect on survival or metamorphosis. After all larvae (100%) unexposed to biocides (48 h at 22 °C) reached the oozoid stage, the number of living (motile) and dead (immotile) larvae were recorded, as well as the metamorphosis stage reached (developmental delay) and the presence of abnormalities under a Leica MZ16F stereomicroscope.

2.4. Statistical Analysis

Data are reported as the mean percentage ± standard deviation (SD). The statistical analysis was performed with IBM SPSS Statistics v. 25 software. The probit method was used to calculate the median lethal concentration (LC_{50}) and the median effective concentration (EC_{50}), the latter defined here as the toxicant concentration that reduced normal larvae by 50%, considering both the larval settlement and metamorphosis as endpoints and their 95% confidence intervals. Significant differences ($p < 0.05$) (i) between the control group and test concentrations, (ii) between the control group and test paints, and (iii) among the test concentrations or the paint groups were evaluated with one-way analysis of variance (ANOVA) followed by Fisher's least significant difference (LSD) test and Dunnett's multiple comparison test. In the case of values expressed as percentages, the raw data were analysed after arcsine transformation to achieve normality.

In the experiments of exposure to antifouling paints and biocides, a clustering analysis with average linkage between groups was used to obtain hierarchy dendrograms after a χ^2 test with Yates' p-value correction applied on the contingency table. Regarding the experiments of exposure to antifouling paints, the basic criterion of this test is to verify if the seven paints are significantly associated with different outcomes. Since it leads to rejecting the null hypothesis of independence between paints and outcomes, the clustering analysis can be applied to determine which paints can be considered similar and then grouped into clusters. In the case of the χ^2 test for the evaluation of the different outcomes related to the concentrations of the eight biocidal compounds, only Irgarol 1051 did not give significantly ($p = 0.999$) different outcomes at different concentrations. For all the other biocides, the different concentrations were significantly ($p < 0.001$) associated with different outcomes. For the combinations of biocides and concentrations, clustering analysis groups similar objects in clusters that are most homogeneous within them and most heterogeneous among them.

3. Results and Discussion

3.1. Effects of Antifouling Paints

In the experiments with plates coated with seven antifouling paints, larvae were placed inside glass-crystallising dishes filled with sea water and with a coated plate on the bottom. At the end of exposure, considered when all control larvae in filtered sea water metamorphosed forming completed filter-feeding oozoids (100% survival rate), three effects were evaluated (Figure 2a): (i) complete metamorphosis with the oozoid formation, (ii) settled but dead larvae, and (iii) not-settled but free-floating dead larvae. These effects were also evaluated in three different conditions: (i) individuals floating in the sea water, (ii) individuals settled on the dish's glass bottom surrounding the coated plate, and (iii) individuals settled on the coated paint.

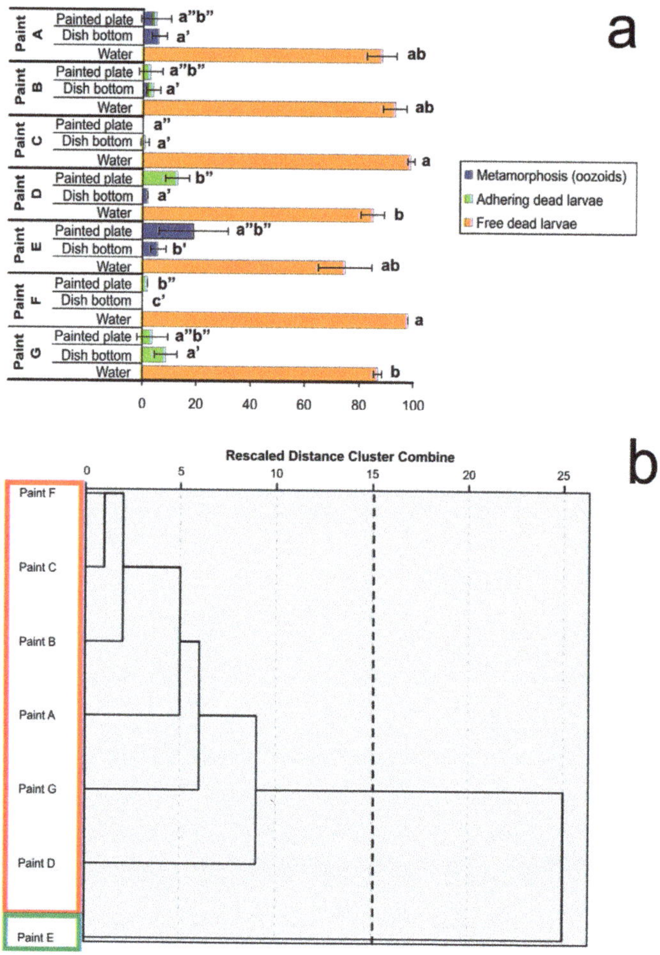

Figure 2. Effects of plates coated with seven antifouling paints on larval settlement and metamorphosis observed after exposure until all control larvae metamorphosed in the reference glass-crystallising dishes forming filter-feeding oozoids. (**a**) Percentages of three effects on metamorphosis, i.e., (i) complete metamorphosis with oozoid formation (blue), (ii) settled but dead larvae (green), and (iii) not-settled but free-floating dead larvae (orange). Larval mortality and abnormality were considered in three conditions: (i) individuals floating in the sea water ("water"), (ii) individuals adhering to the dish glass bottom peripheral to the painted plate ("dish bottom"), and (iii) individuals settled on the antifouling-coated plate ("painted plate"). Significant ($p < 0.05$) differences inside each of the three conditions obtained with Dunnett's test are expressed with different letters. (**b**) Dendrogram obtained from cluster analysis with average linkage between groups. The cutting point at 15 corresponds to the maximum margin showing two clusters bordered by different colours.

All antifouling paints demonstrated a remarkable performance because they caused both high mortality and inhibition of metamorphosis. Paints containing CuSCN + ZnP + Diuron + Sea-Nine 211 (Paint C) and Cu_2O + Chlorothalonil + Irgarol 1051 (Paint F) prevented larval adhesion. Paints containing ZnP + Zineb + Endosulfan (Paint B), CuSCN + Diuron + Sea-Nine 211 (Paint D), and TCMS pyridine + Diuron (Paint G) allowed minor larval adhesion, but prevented metamorphosis. Paints containing TBT (Paint A) or Cu_2O alone (Paint E) occasionally allowed adhesion and metamorphosis, but killed the oozoids. It is noteworthy that the antifouling components generally do not act only through direct contact—the presence of dead and partially metamorphosed larvae, both free-floating and adhering to the glass surrounding the painted plates, suggests a certain degree of leaching of the biocides from the antifouling paints. A large number of free-floating dead larvae were observed in all cases corresponding to $\geq$75% with small numbers of settled larvae, which died without completing metamorphosis (Figure 2a). Nevertheless, the analysis of variance did not permit us to clearly distinguish the different effects of the various antifouling paints.

From the hierarchy dendrogram obtained from the clustering analysis with average linkage between groups (Figure 2b), only two clusters emerged. These clusters represent the mean percentages of five variables, i.e., (i) free-floating dead larvae in sea water, (ii) oozoids on the dish glass bottom, (iii) dead larvae settled on the dish glass bottom, (iv) oozoids on the coated plate, and (v) dead larvae settled on the coated plate.

The first cluster only includes Paint E with 75% of free-floating larvae, 19% of oozoids on the coated plate, and 6% of oozoids on the dish glass bottom. The second cluster is represented by all the other paints with 91.7% of free-floating larvae, 3.8% and 2% of dead larvae settled on the coated plate and on the dish-glass bottom, respectively, and 1.7% and 0.7% of oozoids adhering to the dish glass bottom and to the coated plate, respectively. However, these results are difficult to interpret because the toxic effects are similar. The only relevant consideration is that Paint E, which showed a lower toxic effect than the other paints, is a unique antifouling paint without booster biocides and contains the highest amount of CuCl (41%). This concentration is similar to that of Paint F, which also contained Chlorothalonil and Irgarol 1051 as booster biocides and is based on self-polishing technology rather than on an insoluble matrix. In conclusion, the interactions of a variety of principal and booster biocides on one hand and different matrix technologies on the other hand are complex in commercial paints. The leaching rate of biocides from the paints, which is fundamental for the risk assessment, is often not reported in manufacturers' datasheets. It varies with the mixture of concentrations of biocides in the paint formulation and depends on the interaction with boosters and other additives; environmental abiotic parameters such as temperature, salinity, and pH; and the type of matrix [33,34].

Generally, the most commonly used approach for understanding the action of antifouling compounds is the direct exposure of embryos or larvae of marine invertebrates to various concentrations of a single biocide in sea water with controlled parameters, totally excluding the interaction of matrix, pigments, solvents, and other additives [35–41]. In this way, larval toxicity and the effects on metamorphosis can be clearly observed, elucidating the possible mechanisms of action at both cellular and subcellular levels.

3.2. *Effects of Antifouling Biocides*

At the end of the experiments, all control larvae in filtered sea water metamorphosed, forming complete filter-feeding oozoids (100% survival rate). The order of toxicity of the antifouling biocides assayed with the evaluation of the median lethal and median effective concentrations (LC_{50} and EC_{50}) able to cause mortality and inhibition of larval settlement/metamorphosis, respectively, in *B. schlosseri* (Table 4) is CuCl > ZnP > TBT > Chlorothalonil > Sea Nine 211 > TCMS pyridine > Irgarol 1051 > Diuron.

Table 4. Median lethal concentrations (LC_{50}) of antifouling biocides on *Botryllus schlosseri* larvae and median effective concentrations (EC_{50}) on larval settlement and metamorphosis as endpoints after 48 h exposure at 20–22 °C to antifouling biocides compared with data reported for the solitary ascidian *Ciona intestinalis*.

Biocide	Species	LC_{50}	EC_{50}
TBT	*B. schlosseri*	4.73 µM (1539 µg L^{-1})	0.08 µM (26 µg L^{-1})
	C. intestinalis	ND	0.02 µM (7.1 µg L^{-1}) [42]
ZnP	*B. schlosseri*	0.46 µM (146 µg L^{-1}, 3×10^{-6} Zn wt%)	0.06 µM (19 µg L^{-1}, 4×10^{-6} Zn wt%)
	C. intestinalis	ND	0.11 µM (35 µg L^{-1}, 7×10^{-6} Zn wt%) [43]
Copper(I) chloride	*B. schlosseri*	0.35 µM (34.6 µg L^{-1}, 2×10^{-6} Cu wt%)	0.04 µM (3.96 µg L^{-1}, 2×10^{-6} Cu wt%)
	C. intestinalis	ND	1.61 µM (159 µg L^{-1}, 1×10^{-3} Cu wt%) [44]
Sea-Nine 211	*B. schlosseri*	4.88 µM (1377 µg L^{-1})	0.25 µM (70 µg L^{-1})
	C. intestinalis	ND	0.15 µM (43 µg L^{-1}) [37]
Chlorothalonil	*B. schlosseri*	4.80 µM (1276 µg L^{-1})	0.23 µM (61 µg L^{-1})
	C. intestinalis	ND	0.16 µM (42 µg L^{-1}) [37]
Irgarol 1051	*B. schlosseri*	>200 µM (>50,674 µg L^{-1})	36 µM (9121 µg L^{-1})
	C. intestinalis	ND	>25.60 µM (>6486 µg L^{-1}) [37]
Diuron	*B. schlosseri*	214.96 µM (50,105 µg L^{-1})	64 µM (14,918 µg L^{-1})
	C. intestinalis	ND	ND
TCMS pyridine	*B. schlosseri*	34.99 µM (<10,392 µg L^{-1})	<25 µM (<7375 µg L^{-1})
	C. intestinalis	ND	ND

ND: not determined, i.e., no information reported in the literature. Reference numbers are placed in square brackets.

Both the LC_{50} and EC_{50} found for *B. schlosseri* are higher than the environmental concentrations reported in the literature, with the exception of ZnP and CuCl. The concentration ranges known in the sea water column are <0.02–31.7 µg L^{-1} for ZnP [45], 0.15–26 µg L^{-1} for CuCl [46,47], <0.001–3.3 µg L^{-1} for Sea Nine 211, <0.01–1.4 µg L^{-1} for Chlorothalonil, <0.001–1.7 µg L^{-1} for Irgarol 1051, and <0.001–6.7 µg L^{-1} for Diuron [18]. *B. schlosseri* appears to be more sensitive to both ZnP and CuCl than the solitary ascidian *Ciona intestinalis* [43,44]. On the contrary, *C. intestinalis* is about 4× and 1.5× more sensitive than *B. schlosseri* in terms of the reduction in larval settlement by 50% in the presence of TBT [42] and Sea-Nine 211 or Chlorothalonil [37], respectively, whereas the toxicity concerning Irgarol 1051 is similar [37]. Generally, CuCl and ZnP showed the highest toxicity, causing 50% larval mortality at 0.35 and 0.46 µM (34.6 and 146 µg L^{-1}), respectively, and reducing the larval settlement and metamorphosis by 50% at 0.04 and 0.06 µM (3.96 and 19 µg L^{-1}), respectively. This fact supports the hypothesis that the CuCl and ZnP concentrations usually employed in antifouling paints are too high for their goal and the continuous leaching of these compounds is of great concern considering their impact on coastal ecosystems, particularly the possible long-term negative effects on non-target benthic organisms. Therefore, restrictions and bans regarding the use of copper- and zinc-based antifouling paints are essential for the development of "eco-friendly" antifouling paints [48].

The clustering analysis with average linkage between groups was performed on 25 objects representing the linkages of biocides and test concentrations (Figure 3).

The hierarchy dendrogram obtained shows four clusters with mean percentages of six effects. The effects considered were (i) complete metamorphosis with the oozoid formation (blue, in the legend of Figure 3a), (ii) settled larvae with incomplete metamorphosis or metamorphic delay without abnormalities (green), (iii) incomplete metamorphosis without settlement of dead larvae with abnormalities (orange), (iv) not-settled dead larvae with abnormalities (violet), (v) not-settled dead larvae without recognisable abnormalities (black), and (vi) free-swimming larvae as the result of inhibition or severe delay of metamorphosis (yellow).

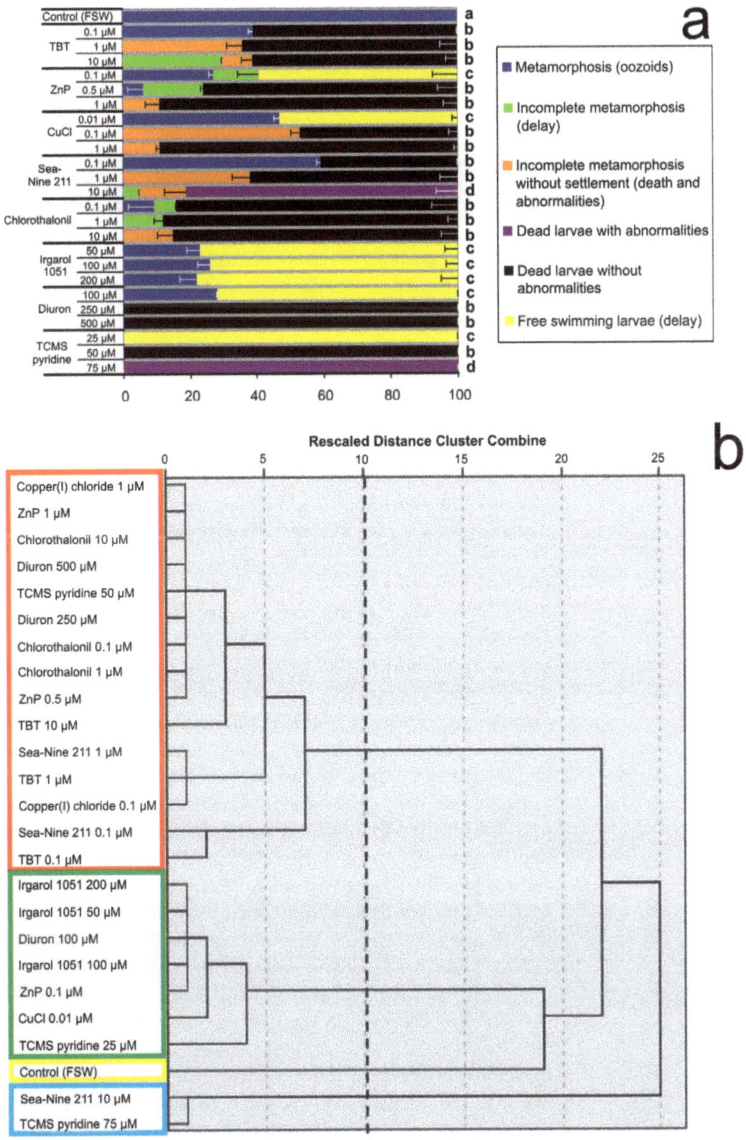

Figure 3. Effects of various concentrations of antifouling biocides in sea water on larval settlement and metamorphosis. (**a**) Percentages of larval mortality and abnormality observed after exposure until all control larvae in FSW completed the metamorphosis forming filter-feeding oozoids. Significant ($p < 0.05$) differences obtained with Dunnett's test are expressed with different letters. (**b**) Dendrogram obtained from cluster analysis with average linkage between groups. The cutting point at 10 corresponds to the maximum margin showing four clusters bordered by different colours.

The first cluster ("a", in Figure 3a; yellow frame, in Figure 3b) is represented by the unique effect observed in the control, i.e., 100% complete metamorphosis and survival. The second cluster ("b", red frame) has 15 objects, represented by all concentrations of TBT and Chlorothalonil, and the middle concentrations of ZnP, CuCl, Sea Nine 211, Diuron, and TCMS pyridine. In this cluster, 76.5% of dead larvae without abnormalities prevail, followed by 11.5% incomplete metamorphosis without settlement, 7.5% complete metamorphosis, and 4.5% incomplete metamorphosis with settlement. The third cluster ("c", green frame) includes seven objects represented by all Irgarol 1051 concentrations, and the lowest concentrations of ZnP, CuCl, Diuron, and TCMS pyridine, with 73.3% free-swimming larvae, 24.7% completed metamorphosis, and 2% incomplete metamorphosis of settled larvae without abnormalities. The fourth cluster ("d", cyan frame) contains two objects, represented by the highest concentrations of Sea Nine 211 and TCMS pyridine, with 90.5% dead larvae with abnormalities, 7% incomplete metamorphosis without settlement, and 2.5% incomplete metamorphosis with settlement.

These results show that although all biocides are able to prevent larval settlement on the substratum and provoke severe metamorphic malformations, the herbicides Diuron and TCMS pyridine at their lowest concentrations and Irgarol 1051 at all concentrations are not lethal.

3.3. *Larval Abnormalities*

The larval abnormalities of the experiments reported in Section 3.2 were analysed under a light microscope (Figure 4). The most severe morphological effects in metamorphic development have been observed after exposure to 1 and 10 μM TBT (Figure 4a), Sea-Nine 211 and Chlorothalonil, 1 μM ZnP, and 0.1 μM CuCl (Figure 4c). In these cases, the initial metamorphosis was evident with the protrusion of ampullae and tail resorption at various degrees, but cephalenteron always appeared to be encased by an anomalous large bubble filled with a colourless fluid with small, scattered cells. Organotin compounds, zinc, copper, and Chlorothalonil are known to cause significant changes in hydromineral fluxes and membrane permeability, mechanisms that maintain osmotic homeostasis [49–52]. The bubble formation in *B. schlosseri* larvae could be the result of the disruption of osmotic control systems due to the direct action on lipid composition and/or ion transport across the plasma membranes. Regarding the tail resorption, TBT, Sea-Nine 211, and Chlorothalonil caused an abnormal extensive detaching of spherical-shaped cells throughout the entire tail length. These compounds are known to directly and indirectly inhibit a series of enzymes. In particular, Ca^{2+}-ATPase is inhibited via biocide interaction with calmodulin [53]. This pump inhibition increases intracellular Ca^{2+} concentrations, which in turn provoke the depolymerisation of cytoskeletal components [54] and induce cell apoptosis [55]. The incomplete tail resorption with the persistence of apoptotic cells dissociated from the axial complex indicates the absence of phagocytosis, which is usually carried out by motile phagocytes able to engulf all dead cells and tissue debris [56], confirming the immunosuppressive activity of these biocides [28,29].

At 0.5 μM ZnP (Figure 4b) and 1 μM CuCl (Figure 4d), the cephalenteron was not encased in a bubble. In the case of ZnP, protrusion of ampullae occurred, but not the tail resorption. The formation of proximal and median bulges of detached cells was recognisable along the tail, probably due to the inhibition of phagocytosis [32]. In the case of CuCl, tail resorption appeared to be nearly completed, but no protrusion of ampullae occurred and the cephalenteron organs underwent a massive regression, recognisable by the appearance of large vacuolated cells instead of the organ primordia.

Many heavy metals are essential to the growth of marine organisms. Low concentrations of copper ions stimulate metamorphosis in ascidians [57], but at high concentrations, metal ions become toxic. Metals can significantly decrease the synthesis of ATP and negatively alter the metabolic activity causing cell death because they act as uncouplers of oxidative phosphorylation or via opening pores in the mitochondrial membrane [58].

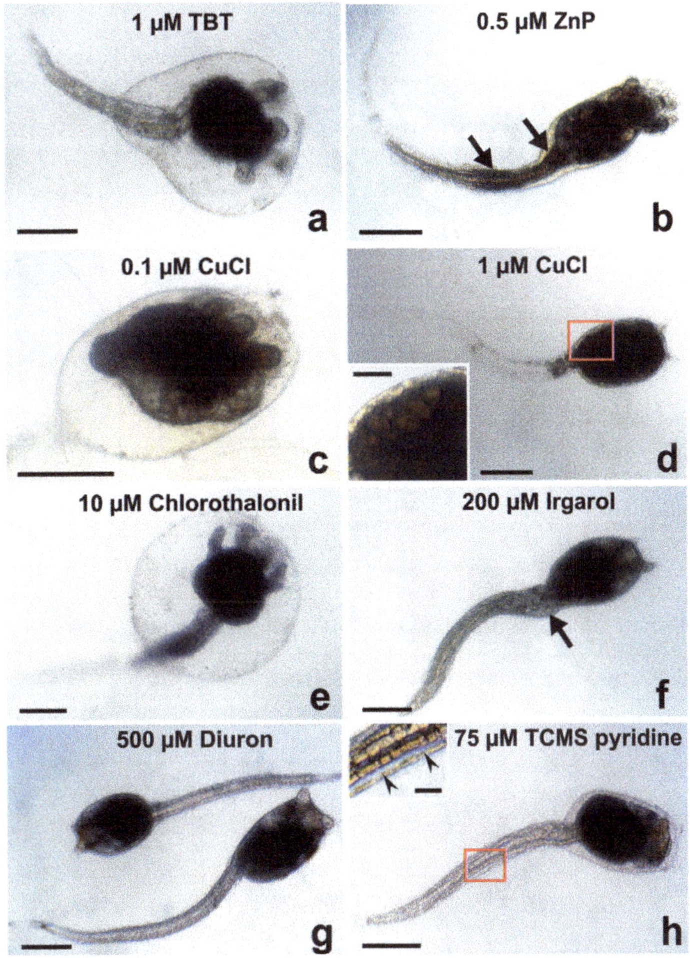

Figure 4. Larval abnormalities in *Botryllus schlosseri* after exposure to antifouling biocides. (**a**) Protruded ampullae inside an anomalous bubble encasing the cephalenteron, and extensive spherical-shaped cells detaching from the tail tissues. (**b**) Protruded ampullae and tail with proximal and median bulges (*arrows*). (**c**) Advanced metamorphosis (protruded ampullae and tail resorption) inhibited by the formation of a bubble encasing the whole body. (**d**) Tail resorption and inhibition of cephalenteron metamorphosis due to extensive organ regression forming vacuolated cells (inset in (**d**)). (**e**) Protruded ampullae inside a bubble encasing the cephalenteron, and normal tail resorption. (**f**) Detaching of cells from the proximal tail zone (*arrow*). (**g**) Absence of metamorphosis. (**h**) Initial protrusion of ampullae, and separation of the lateral tail muscles from the epidermis by large intercellular spaces (*arrowheads* in inset in (**h**)). Bar length: 200 µm in (**a–h**); 30 µm inset in (**d**); 40 µm inset in (**h**).

Regarding the herbicides Irgarol 1051 (Figure 4f), Diuron (Figure 4g), and TCMS pyridine (Figure 4h), the metamorphosis was completely inhibited. The larvae did not swim or settle because they were dead only in the cases of 250 µM Diuron and 50 µM TCMS pyridine. As a minor difference from Diuron, an initial protrusion of ampullar rudiments was observed after exposure to TCMS pyridine. These compounds are known to cause disturbances in the mitochondrial respiratory chains and induce cell apoptosis due to severe oxidative stress [59,60]. After the exposure to 500 μM Diuron (Figure 4g), larvae were without evident malformations but immotile. The absence of signs of tail resorption indicates an alteration of communication with the cephalenteron. It could be triggered through the abnormal activation by xenobiotics of stress-related signal transduction pathways, such as the stress-inducible protein HSP90, which in turn activates the enzyme nitric oxide synthase (NOS). The consequent NO production is responsible for the metamorphosis repression in ascidians, as observed in *Boltenia villosa* larvae, which failed in tail resorption and protrusion of the ampullar rudiments [61]. Cell dissociation occurred in the proximal region of the tail of a small number of larvae after exposure to the highest concentration (200 μM) of Irgarol 1051 (Figure 4f). An extensive separation of the lateral tail muscles from the caudal epidermis by intercellular spaces was recognisable after exposure to 75 μM TCMS pyridine (inset of Figure 4h). These events could be related indirectly to HSP90/NO activation and directly to sudden increases in intracellular Ca^{2+} concentrations. The latter induce both apoptosis through the activation of endonucleases, triggering DNA fragmentation [30,31], and the loss of Ca^{2+}-dependent cell adhesion molecules such as cadherins, causing extensive tissue detachment [62,63].

Finally, it must be emphasised that in all cases of absence of tail resorption, the increase in intracellular calcium ions might cause extensive depolymerisation of the microfilaments of the cells of the caudal epidermis. In such a way, the F-actin responsible for contractions that provide the driving force in tail resorption [2] failed.

4. Conclusions

The experiments carried out on the free-swimming larvae of the compound ascidian *B. schlosseri* provide new evidence for better understanding the effects of trade antifouling compounds on settlement ability, metamorphic development, and viability, arranging them in an order of decreasing toxicity: metals and organometals > antimicrobials and fungicides > herbicides. These results confirm the high toxicity of TBT, ZnP, and copper(I)-based antifouling compounds, but also reveal that Sea-Nine 211 and Chlorothalonil, recently introduced by "green chemistry" in new antifouling paint formulations as TBT alternatives, are effective in causing larval death and abnormalities. Moreover, the herbicides Irgarol 1051, Diuron, and TCMS pyridine, although introduced in paint formulations only to prevent the settlement and growth of algae, showed toxicity towards animals. Since these compounds have been in commerce for decades, and, consequently, widespread in the marine environment, a real risk for the coastal biocoenoses appears to be undeniable. In many cases, their partitioning between water and sediment, modality and half-time of biotic and abiotic degradation, speciation, binding to various ligands, formation of complexes, environmental fate, and bioaccumulation in the trophic chains are not yet well known.

The effects of exposure to antifouling paints are much more difficult to interpret than those of exposure to antifouling biocides directly introduced in the sea water. The latter approach is preferable because it can be better controlled and not affected by the modality of leaching of a biocidal mixture from paints, which could vary with various environmental parameters and also determine a network of synergistic effects with booster biocides, other paint additives, and types of matrices. Paints with a contact-leaching matrix are more resistant to abrasion and rubbing than those with a self-polishing matrix, but they tend to oxidise with time, decreasing the biocidal release. By comparing the biocidal release rates from contact-leaching and self-polishing paint technologies, although the initial value of the latter is low, a constant release soon occurs throughout the lifetime of the paint.

Therefore, according to the Biocides Directive (98/8/EC) [64], more assays of acute and chronic toxicity on various target and non-target marine organisms should be performed before new potential pollutants enter the market, in order to prevent the same errors that already occurred with TBT in both the ecological and economical fields due to the deterioration of environments and habitats.

Author Contributions: Conceptualisation, resources, validation, supervision, F.C.; investigation, methodology, data curation, formal analysis, R.V.; writing—original draft preparation and editing, F.C. and R.V. All authors have read and agreed to the published version of the manuscript.

Funding: This research was supported by grants from the Italian MIUR (DOR 2020) to F.C. and by a funding agreement between P. Zarantonello's "Resimix s.r.l." (Brendola, Vicenza, Italy, http://www.resimix.com/it/) and the University of Padova, Italy, for a Ph.D. fellowship in Biosciences to R.V. (rep. #1488, prot. #186788, 6 May 2019).

Institutional Review Board Statement: The authors followed all applicable international, national, and/or institutional guidelines for the care and use of animals.

Informed Consent Statement: Not applicable.

Data Availability Statement: All data generated or analysed during this study are included in this published article.

Acknowledgments: The authors thank Fabrizio Longo for photos of the stages of *B. schlosseri* metamorphosis of Table 2, and Andrea Sambo, in-charge technician of the staff at the Umberto D'Ancona Hydrobiological Station of Chioggia (Venice, Italy), for assistance in ascidian collection and boat driving.

Conflicts of Interest: The authors declare no conflict of interest. The funders had no role in the design of the study; in the collection, analyses, or interpretation of data; in the writing of the manuscript, or in the decision to publish the results.

References

1. Delsuc, F.; Tsagkogeorga, G.; Lartillot, N.; Philippe, H. Additional molecular support for the new chordate phylogeny. *Genesis* **2008**, *46*, 592–604. [CrossRef] [PubMed]
2. Cloney, R.A. Ascidian larvae and the events of metamorphosis. *Am. Zool.* **1982**, *22*, 817–826. [CrossRef]
3. Karaiskou, A.; Swalla, B.J.; Sasakura, Y.; Chambon, J.-P. Metamorphosis in solitary ascidians. *Genesis* **2014**, *53*, 34–47. [CrossRef] [PubMed]
4. Hotta, K.; Dauga, D.; Manni, L. The ontology of the anatomy and development of the solitary ascidian *Ciona*: The swimming larva and its metamorphosis. *Sci. Rep.* **2020**, *10*, 17916. [CrossRef]
5. Sasakura, Y.; Hozumi, A. Formation of adult organs through metamorphosis in ascidians. *WIREs Dev. Biol.* **2018**, *7*, e304. [CrossRef]
6. Caicci, F.; Zaniolo, G.; Burighel, P.; Degasperi, V.; Gasparini, F.; Manni, L. Differentiation of papillae and rostral sensory neurons in the larva of the ascidian *Botryllus schlosseri* (Tunicata). *J. Comp. Neurol.* **2010**, *518*, 547–566. [CrossRef]
7. Pennati, R.; Rothbächer, U. Bioadhesion in ascidians: A developmental and functional genomics perspective. *Interface Focus* **2015**, *5*, 20140061. [CrossRef]
8. Wakai, M.K.; Nakamura, M.J.; Sawai, S.; Hotta, K.; Oka, K. Two-round Ca^{2+} transient in papillae by mechanical stimulation induces metamorphosis in the ascidian *Ciona intestinalis* type A. *Proc. R. Soc. B* **2021**, *288*, 20203207. [CrossRef]
9. Aldred, N.; Clare, A.S. Mini-review: Impact and dynamics of surface fouling by solitary and compound ascidians. *Biofouling* **2014**, *30*, 259–270. [CrossRef] [PubMed]
10. Bryan, G.W.; Gibbs, P.E.; Burt, G.R.; Hummerstone, L.G. The decline of the gastropod *Nucella lapillus* around southwest England: Evidence for the effects of tributyltin from antifouling paints. *J. Mar. Biol. Assoc. UK* **1986**, *66*, 611–640. [CrossRef]
11. Henderson, A.S.; Salazar, S.M. Flowthrough bioassay studies on the effects of antifouling TBT leachates. In *Organotin: Environmental Fate and Effects*; Champ, M.A., Seligman, P.F., Eds.; Chapman & Hall: London, UK, 1996; pp. 281–303.
12. Hoch, M. Organotin compounds in the environment: An overview. *Appl. Geochem.* **2001**, *16*, 719–743. [CrossRef]
13. Cima, F.; Craig, P.J.; Harrington, C. Organotin compounds in the environment. In *Organometallic Compounds in the Environment*; Craig, P.J., Ed.; Wiley & Sons Ltd.: Chichester, UK, 2003; pp. 101–149.
14. Guardiola, F.A.; Cuesta, A.; Meseguer, J.; Esteban, M.A. Risks of using antifouling biocides in aquaculture. *Int. J. Mol. Sci.* **2012**, *13*, 1541–1560. [CrossRef]
15. Voulvoulis, N.; Scrimshaw, M.D.; Lester, J.N. Alternative antifouling biocides. *Appl. Organomet. Chem.* **1999**, *13*, 135–143. [CrossRef]
16. Evans, S.M.; Birchenough, A.C.; Brancato, M.S. The TBT ban: Out of the frying pan into the fire? *Mar. Pollut. Bull.* **2000**, *40*, 204–211. [CrossRef]
17. Omae, I. Organotin antifouling paints and their alternatives. *Appl. Organomet. Chem.* **2003**, *17*, 81–105. [CrossRef]
18. Konstantinou, I.K.; Albanis, T.A. Worldwide occurrence and effects of antifouling paint booster biocides in the aquatic environment: A review. *Environ. Int.* **2004**, *30*, 235–248. [CrossRef]
19. Cima, F.; Ballarin, L.; Bressa, G.; Martinucci, G.B.; Burighel, P. Toxicity of organotin compounds on embryos of a marine invertebrate (*Styela plicata*; Tunicata). *Ecotoxicol. Environ. Saf.* **1996**, *35*, 174–182. [CrossRef] [PubMed]

20. Zega, G.; Pennati, R.; Candiani, S.; Pestarino, M.; De Bernardi, F. Solitary ascidians embryos (Chordata, Tunicata) as model organisms for testing coastal pollutant toxicity. *Invertebr. Surv. J.* **2009**, *6*, S29–S34.
21. Zaniolo, G.; Burighel, P.; Martinucci, G. Ovulation and placentation in *Botryllus schlosseri* (Ascidiacea): An ultrastructural study. *Can. J. Zool.* **1987**, *65*, 1181–1190. [CrossRef]
22. Manni, L.; Zaniolo, G.; Burighel, P. Ultrastructural study of oogenesis in the compound ascidian *Botryllus schlosseri* (Tunicata). *Acta Zool.* **1994**, *75*, 101–113. [CrossRef]
23. Gasparini, F.; Manni, L.; Cima, F.; Zaniolo, G.; Burighel, P.; Caicci, F.; Franchi, N.; Schiavon, F.; Rigon, F.; Campagna, D.; et al. Sexual and asexual reproduction in the colonial ascidian *Botryllus schlosseri*. *Genesis* **2014**, *53*, 105–120. [CrossRef]
24. Cima, F.; Ballarin, L. A proposed integrated bioindex for the macrofouling biocoenosis of hard substrata in the lagoon of Venice. *Estuar. Coast. Shelf Sci.* **2013**, *130*, 190–201. [CrossRef]
25. Manni, L.; Zaniolo, G.; Cima, F.; Burighel, P.; Ballarin, L. *Botryllus schlosseri*: A model ascidian for the study of asexual reproduction. *Dev. Dyn.* **2007**, *236*, 335–352.
26. Manni, L.; Anselmi, C.; Cima, F.; Gasparini, F.; Voskoboynik, A.; Martini, M.; Peronato, A.; Burighel, P.; Zaniolo, G.; Ballarin, L. Sixty years of experimental studies on the blastogenesis of the colonial tunicate *Botryllus schlosseri*. *Dev. Biol.* **2019**, *448*, 293–308. [CrossRef] [PubMed]
27. Ben-Hamo, O.; Rinkevich, B. *Botryllus schlosseri*—A model colonial species in basic and applied studies. In *Handbook of Marine Model Organisms in Experimental Biology; Established and Emerging*; Boutet, A., Schierwater, B., Eds.; CRC Press, Taylor & Francis Group: Boca Raton, FL, USA, 2021; pp. 385–402.
28. Cima, F.; Ballarin, L.; Bressa, G.; Sabbadin, A. Immunotoxicity of butyltins in tunicates. *Appl. Organomet. Chem.* **1995**, *9*, 567–572. [CrossRef]
29. Cima, F.; Bragadin, M.; Ballarin, L. Toxic effects of new antifouling compounds on tunicate haemocytes—I. Sea-Nine 211™ and chlorothalonil. *Aquat. Toxicol.* **2008**, *86*, 299–312. [CrossRef]
30. Menin, A.; Ballarin, L.; Bragadin, M.; Cima, F. Immunotoxicity in ascidians: Antifouling compounds alternative to organotins—II. The case of diuron and TCMS pyridine. *J. Environ. Sci. Health* **2008**, *43B*, 644–654. [CrossRef] [PubMed]
31. Cima, F.; Ballarin, L. Immunotoxicity in ascidians: Antifouling compounds alternative to organotins: III—The case of copper(I) and Irgarol 1051. *Chemosphere* **2012**, *89*, 19–29. [CrossRef] [PubMed]
32. Cima, F.; Ballarin, L. Immunotoxicity in ascidians: Antifouling compounds alternative to organotins—IV. The case of zinc pyrithione. *Comp. Biochem. Physiol.* **2015**, *169*, 16–24. [CrossRef]
33. Kiil, S.; Dam-Johansen, K.; Weinell, C.E.; Pedersen, M.S.; Codolar, S.A. Estimation of polishing and leaching behaviour of antifouling paints using mathematical modelling: A literature review. *Biofouling* **2003**, *19*, 37–43. [CrossRef]
34. Takahashi, K. Release rate of biocides from antifouling paints. In *Ecotoxicology of Antifouling Biocides*; Arai, T., Harino, H., Ohji, M., Langston, W.J., Eds.; Springer: Tokyo, Japan, 2009; pp. 3–22.
35. Marin, M.G.; Moschino, V.; Cima, F.; Celli, C. Embryotoxicity of butyltin compounds to the sea urchin *Paracentrotus lividus*. *Mar. Environ. Res.* **2000**, *50*, 231–235. [CrossRef]
36. Khandeparker, L.; Desai, D.; Shirayama, Y. Larval development and post-settlement metamorphosis of the barnacle *Balanus albicostatus* Pilsbry and the serpulid polychaete *Pomatoleios kraussii* Baird: Impact of a commonly used antifouling biocide, Irgarol 1051. *Biofouling* **2005**, *21*, 169–180. [CrossRef] [PubMed]
37. Bellas, J. Comparative toxicity of alternative antifouling biocides on embryos and larvae of marine invertebrates. *Sci. Total Environ.* **2006**, *367*, 573–585. [CrossRef] [PubMed]
38. Manzo, S.; Buono, S.; Cremisini, C. Toxic effects of irgarol and diuron on sea urchin *Paracentrotus lividus* early development, fertilization, and offspring quality. *Arch. Environ. Contam. Toxicol.* **2006**, *51*, 61–68. [CrossRef]
39. Mai, H.; Morin, B.; Pardon, P.; Gonzalez, P.; Budzinski, H.; Cachot, J. Environmental concentrations of irgarol, diuron and s-metolachlor induce deleterious effects on gametes and embryos of the Pacific oyster, *Crassostrea gigas*. *Mar. Environ. Res.* **2013**, *89*, 1–8. [CrossRef]
40. Gallo, A.; Tosti, E. Reprotoxicity of the antifoulant chlorothalonil in ascidians: An ecological risk assessment. *PLoS ONE* **2015**, *10*, e0123074. [CrossRef]
41. Dumollard, R.; Gazo, I.; Gomes, I.D.L.; Besnardeau, L.; McDougall, A. Ascidians: An emerging marine model for drug discovery and screening. *Curr. Top. Med. Chem.* **2017**, *17*, 2056–2066. [CrossRef] [PubMed]
42. Antizar-Ladislao, B. Environmental levels, toxicity and human exposure to tributyltin (TBT)-contaminated marine environment. A review. *Environ. Int.* **2008**, *34*, 292–308. [CrossRef]
43. Bellas, J. Toxicity assessment of the antifouling compound zinc pyrithione using early developmental stages of the ascidian *Ciona intestinalis*. *Biofouling* **2005**, *21*, 289–296. [CrossRef]
44. Bellas, J.; Beiras, R.; Vazquez, E. Sublethal effects of trace metals (Cd, Cr, Cu, Hg) on embryogenesis and larval settlement of the ascidian *Ciona intestinalis*. *Arch. Environ. Contam. Toxicol.* **2004**, *46*, 61–66. [CrossRef]
45. Mackie, D.S.; van den Berg, C.M.G.; Readman, J.W. Determination of pyrithione in natural waters by cathodic stripping voltametry. *Anal. Chim. Acta* **2004**, *511*, 47–53. [CrossRef]
46. Ranke, J.; Jastorff, B. Multidimensional risk analysis of antifouling biocides. *Environ. Sci. Pollut. Res.* **2000**, *7*, 105–114. [CrossRef] [PubMed]

47. Kiaune, L.; Singhasemanon, N. Pesticidal copper (I) oxide: Environmental fate and aquatic toxicity. *Rev. Environ. Contam. Toxicol.* **2011**, *213*, 1–26. [PubMed]
48. Lagerström, M.; Ytreberg, E. Quantification of Cu and Zn in antifouling paint films by XRF. *Talanta* **2021**, *223*, 121820. [CrossRef]
49. Hartl, M.G.J.; Hutchinson, S.; Hawkins, L.E. Organotin and osmoregulation: Quantifying the effects of environmental concentrations of sediment-associated TBT and TPhT on the freshwater-adapted European flounder, *Platichthys flesus* (L.). *J. Exp. Mar. Biol. Ecol.* **2001**, *256*, 267–278. [CrossRef]
50. Bregante, M.; Carpaneto, A.; Piazza, V.; Sbrana, F.; Vassalli, M.; Faimali, M.; Gambale, F. Osmoregulated chloride currents in hemocytes from *Mytilus galloprovincialis*. *PLoS ONE* **2016**, *11*, e0167972. [CrossRef] [PubMed]
51. de Polo, A.; Scrimshaw, M.D. Challenges for the development of a biotic ligand model predicting copper toxicity in estuaries and seas. *Environ. Toxicol. Chem.* **2012**, *31*, 230–238. [CrossRef]
52. Haque, M.N.; Eom, H.-J.; Nam, S.-E.; Shin, Y.K.; Rhee, J.-S. Chlorothalonil induces oxidative stress and reduces enzymatic activities of Na^+/K^+-ATPase and acetylcholinesterase in gill tissues of marine bivalves. *PLoS ONE* **2019**, *14*, e0214236. [CrossRef] [PubMed]
53. Cima, F.; Dominici, D.; Mammi, S.; Ballarin, L. Butyltins and calmodulin: Which interaction? *Appl. Organomet. Chem.* **2002**, *16*, 182–186. [CrossRef]
54. Cima, F.; Ballarin, L. Tributyltin induces cytoskeletal alterations in the colonial ascidian *Botryllus schlosseri* phagocytes via interaction with calmodulin. *Aquat. Toxicol.* **2000**, *48*, 419–429. [CrossRef]
55. Cima, F.; Ballarin, L. TBT-induced apoptosis in tunicate haemocytes. *Appl. Organomet. Chem.* **1999**, *13*, 697–703. [CrossRef]
56. Schiaffino, S.; Burighel, P.; Nunzi, M.G. Involution of the caudal musculature during metamorphosis in the ascidian, *Botryllus schlosseri*. *Cell Tissue Res.* **1974**, *153*, 293–305. [CrossRef]
57. Whittaker, J.R. Copper as a factor in the onset of ascidian metamorphosis. *Nature* **1964**, *202*, 1024–1025. [CrossRef] [PubMed]
58. Bragadin, M.; Manente, S.; Marton, D.; Cima, F.; Rigobello, M.P.; Bindoli, A. The interaction of zinc pyrithione with mitochondria from rat liver and a study of the mechanism of inhibition of ATP synthesis. *Appl. Organomet. Chem.* **2003**, *17*, 869–874. [CrossRef]
59. Bragadin, M.; Cima, F.; Ballarin, L.; Manente, S. Irgarol inhibits the synthesis of ATP in mitochondria from rat liver. *Chemosphere* **2006**, *65*, 1898–1903. [CrossRef] [PubMed]
60. Bragadin, M.; Iero, A.; Cima, F.; Ballarin, L.; Manente, S. TCMS inhibits ATP synthesis in mitochondria: A systematic analysis of the inhibitory mechanism. *Toxicol. In Vitro* **2007**, *21*, 1127–1133. [CrossRef] [PubMed]
61. Bishop, C.D.; Bates, W.R.; Brandhorst, B.P. Regulation of metamorphosis in ascidians involves NO/cGMP signaling and HSP90. *J. Exp. Zool.* **2001**, *289*, 374–384. [CrossRef] [PubMed]
62. Ferruzza, S.; Scarino, M.L.; Rotilio, G.; Ciriolo, M.R.; Santaroni, P.; Muda, A.O.; Sambuy, Y. Copper treatment alters the permeability of tight junctions in cultured human intestinal Caco-2 cells. *Am. J. Physiol.* **1999**, *277*, 1138–1148. [CrossRef] [PubMed]
63. Chung-Hsun, L.; I-Hui, C.; Chia-Rong, L.; Chih-Hsien, C.; Ming-Che, T.; Jin-Lian, T.; Hsiu-Fen, L. Inhibition of gap junctional intercellular communication in WB-F344 rat liver epithelial cells by triphenyltin chloride through MAPK and PI3-kinase pathways. *J. Occup. Med. Toxicol.* **2010**, *5*, 17.
64. European Commission. Directive 98/8/EC of the European Parliament and of the Council of 16 February Concerning the Placing of Biocidal Products on the Market. *Off. J. Eur. Comm.* **1998**, *41*, 123.

Journal of
Marine Science and Engineering

MDPI

Review

Biocide vs. Eco-Friendly Antifoulants: Role of the Antioxidative Defence and Settlement in *Mytilus galloprovincialis*

Costantino Parisi [1,2,*], Jessica Sandonnini [1,3], Maria Rosaria Coppola [1,4], Adriano Madonna [1], Fagr Kh. Abdel-Gawad [5], Emidio M. Sivieri [6,†] and Giulia Guerriero [1,*,†]

1 Comparative Endocrinology Laboratories (EClab), Department of Biology, University of Naples Federico II, 80126 Naples, Italy; jsandonnini.eclab@gmail.com (J.S.); coppola.eclab@gmail.com (M.R.C.); adrimad49@gmail.com (A.M.)
2 International Institute of Molecular and Cell Biology in Warsaw, 02-109 Warsaw, Poland
3 Department of Marine Science and Applied Biology, s/n, 03690 Alicante, Spain
4 Department of Experimental Medicine and Center of Excellence for Biomedical Research, University of Genova, 16132 Genova, Italy
5 Centre of Excellence for Advanced Sciences (CEAS), National Research Centre, Water Pollution Research Department, Dokki, Giza 12622, Egypt; fagrabdlgawad@gmail.com
6 Division of Neonatology, The Children's Hospital of Philadelphia, Philadelphia, PA 19104, USA; sivieri@chop.edu
* Correspondence: parisi.eclab@gmail.com (C.P.); giulia.guerriero@unina.it (G.G.)
† These authors contributed equally to this work.

Citation: Parisi, C.; Sandonnini, J.; Coppola, M.R.; Madonna, A.; Abdel-Gawad, F.K.; Sivieri, E.M.; Guerriero, G. Biocide vs. Eco-Friendly Antifoulants: Role of the Antioxidative Defence and Settlement in *Mytilus galloprovincialis*. *J. Mar. Sci. Eng.* **2022**, *10*, 792. https://doi.org/10.3390/jmse10060792

Academic Editor: Francesca Cima

Received: 20 May 2022
Accepted: 6 June 2022
Published: 9 June 2022

Abstract: Antifoulant paints were developed to prevent and reduce biofouling on surfaces immersed in seawater. The widespread use of these substances over the years has led to a significant increase of their presence in the marine environment. These compounds were identified as environmental and human threats. As a result of an international ban, research in the last decade has focused on developing a new generation of benign antifoulant paints. This review outlines the detrimental effects associated with biocide versus eco-friendly antifoulants, highlighting what are effective antifoulants and why there is a need to monitor them. We examine the effects of biocide and eco-friendly antifoulants on the antioxidative defence mechanism and settlement in a higher sessile organism, specifically the Mediterranean mussel, *Mytilus galloprovincialis*. These antifoulants can indirectly assess the potential of these two parameters in order to outline implementation of sustainable antifoulants.

Keywords: biofouling; biocide antifoulant; eco-friendly antifoulant; antioxidative defence; settlement; *Mytilus galloprovincialis*; sustainable development goals; agenda 2030

1. Introduction

Marine biofouling consists of the settlement of microorganisms, plants, and aquatic animals on artificial surfaces immersed in seawater (see for review [1]). It can cause several ecological, human, and economic concerns. In fact, it may introduce invasive species or can promote the settlement and spread of native species with subsequent modifications of biodiversity and/or alteration of the ecological process [2–5]; convey parasites that can affect native species and subsequently humans [6,7]; and induce high ship frictional resistance leading to an increase in fuel consumption and structural damage to ships over time due to corrosion and discoloration [8–10]. The most common antifoulant (AF) paints from the last century contain chemical compounds such as tributyltin oxide and tributyltin fluoride (TBTs), which were the most widely used since the 1950s marking a revolution in AF applications. These compounds are a class of organotin biocides. They are potent fungicides and completely inhibit the growth of most fouling organisms at a very low concentration [11] with oxidative stress phenomena (see for review [12]). Unfortunately, TBT systems adversely affect the environment, inducing endocrine disruption, loss of marine species biodiversity, and pose a human food security threat. The effects of this substance have been known since the 1970s when it was identified as a hormone-like disease agent in an oyster farm in France [13]. Subsequently, the compound's activity has been detected in numerous marine organisms in which it accumulates, eventually posing serious risks to human health through the food chain as demonstrated by an increase in physiological antioxidative defence activity [14–16]. In response to the global ban of TBTs in 2008, the marine coatings industries have developed alternative AF biocides. New alternatives to AF paints were based on copper compounds such as copper oxide and copper thiocyanate with the incorporation of biocide boosters to control Cu-resistant fouling organisms [17,18]. These biocides were intended to be less harmful to the environment than organotin biocides. However, many of the booster biocides are also a threat to the marine environment [19,20]. Some of them can accumulate to high levels, despite claims of rapid degradation, and have a biocidal effect on non-target marine organisms' redox status [21,22]. With the growing concern for environmental protection, these biocides are expected to be phased out. In fact, these issues should be addressed as a priority under proper global governance and executed in line with the Sustainable Development Goals (SDGs) of the United Nations, specifically, to reach goals #3 (health and well-being) and #14 (Life Below Water) [23].

Therefore, the elimination of AF compounds from a wide range of biological organisms (e.g., algae, corals, sponges, and microbes) seems to be the preferred mitigation route, due in part to the high specificity of these compounds for fouling organisms [24,25]. While many commercial fouling release systems are already available on the market, the development of an efficient eco-friendly product entirely based on natural biocides seems still far away. Compounds such as terpenoids, steroids, carotenoids, phenolics, furanones, alkaloids, peptides and lactones all have AF activity [26]. Irritants extracted from land plants such as oleander and bell pepper are also important sources with AF action [1,27]. In nature, marine organisms have developed many biological strategies to interact with microorganisms for protection from pathogens or from being parasitized [28]. In particular, sponges and the associated microbiota produce compounds that interfere with marine biological fouling using quorum sensing (QS) mechanisms [29,30]. The use of marine biological secretions to interfere with bacterial QS can effectively inhibit bacterial biofilm formation, making it difficult for several species of Mytilus including *M. galloprovincialis* to adhere [31,32]. Understanding the molecular mechanism by which these eco-friendly AF compounds act can also help researchers to chemically enhance the functional groups and allow these compounds to achieve high specificity against targets without inducing overexpression of free radicals [4,33]. Their mechanisms include the inhibition of transmembrane transport [10], inhibition of quorum sensing [34] and neurotransmission blocking [35]. They can also inhibit enzymes involved in the production and/or release of adhesive molecules inhibiting cell growth [36]. Studies carried out on the effect of antifoulants are numerous and linked to specific bioindicator organisms. The model organism selected for this review is the Mediterranean mussel, *Mytilus galloprovincialis*, a filter feeder bivalve organism belonging to the phylum Mollusca. It attaches to a variety of surfaces due to adhesive proteins produced and stockpiled in its foot and then secreted into the byssal groove such as Mgfp-1, 2 and 3 (*Mytilus galloprovincialis* foot protein 1, 2 and 3). Collagen, cellulose, chitin, and mineral deposits also play an important role in the adhesive process. In addition, the enzymatic oxidation of specific polyphenolic proteins such as dihydroxyphenylalanine (DOPA) provides the distinctive resistance to moisture in mussel underwater adhesion. DOPA forms complexes with metal ions, oxides (Fe^{3+}, Mn^{3+}) and semimetals such as silicon, giving the mussel the ability to adhere to rocks and smooth surfaces [37]. This organism represents an important tool for biomonitoring environmental pollution; it is considered a sentinel organism, serving as a bio-indicator to evaluate chemical pollutants in marine environments. In fact, this species is capable of bio-accumulating relevant quantities of different xenobiotics (see for review: [38–40]). *M. galloprovincialis* has been used in several toxicological investigations due to its versatility, which makes it possible to work both on retrieved species and laboratory-grown individuals [41–45]. As part of mussels and filter feeders, they are exposed to a wide range of natural stressors that result in biochemical and physiological changes [46]. Mussels are also known to exhibit signs of oxidative stress by overexpression of free radicals when exposed to toxic compounds, which are harmful to cells because they cause disruption of cell membrane fluidity by peroxidation, as well as DNA damage in the form of degradation, base deletions and mutations [46–48]. Indeed, antioxidant enzymes which counteract oxidative stress are used on this biofouling model as contaminant/stress-related biomarkers [49–52]. Thus, the primary objective of the present review is to summarise and collect information available in the literature on the effect of eco-friendly and biocide AFs and their role in the antioxidative defence and settlement in *M. galloprovincialis* as a biofouling model organism.

2. Biocide Antifoulants and Their Effect on *Mytilus galloprovincialis*

Biocides in antifoulants are active substances designed to destroy, deter, control or prevent the action of any harmful organism which is considered a detrimental and unwanted biological presence for humans, animals and the environment [1,53,54]. Antifouling efficacy is affected by many factors, including physical, chemical, the type of biocide present in the coating, as well as other factors. These in turn determine the polishing behavior of the coating and its ability to control the biocide [1,53–56]. It has been shown in the literature how these substances can cause antioxidant enzymatic alterations, through cellular reactive oxygen species (ROS) in *M. galloprovincialis*, compromising larval metamorphosis [57] and causing an inhibition in reaching a well-developed veliger larval stage [58] and/or inhibiting the settlement process [59].

Copper (Cu) and copper-based derivatives have long been used as biocidal agents and have found use in AF paints to prevent colonisation of ship hulls by fouling organisms [60]. It is a naturally occurring trace metal that is essential for the proper functioning of biological systems, but it is toxic when present in excessive concentrations, as demonstrated by numerous ecotoxicological studies reporting disruption of many biological functions in marine organisms [20,61–63].

Cima and Varello [20] assessed the potential disruptive effects of copper-based (Cu_2O) AF paints on the biodiversity of coastal macro-fouling communities, including *M. galloprovincialis* (Table 1). Wooden and steel panels were coated with four copper(I)-based AFs; paint A, a mix of TBT compounds (Cu_2O, TBT methacrylate and antifouling agent tributyltin (TBTO)); paint B, copper(I) (Cu_2O); paint C, copper(I) as cuprous thiosyante (CuSCN) mixed with dichlofluanid; paint D, Cu_2O mixed with s-triazine herbicide (Irgarol 1051)/organochlorine fungicide (chlorothalonil). In particular, paint B was the only one based on contact-leaching technology containing only copper (I) oxide without boosters. Besides having lower AF efficiency on the macro-fouling communities, paint B had a significant inhibitory effect on *M. galloprovincialis* on the wooden panels. Paint D, had the higher biocidal power and also prevented development of mollusks on the wood panels. Thus, they found that the co-presence of organic booster biocides and the type of polymeric matrix can significantly increase the performance of the AF. Their results showed how copper-based paints have different effects on the settlement and growth of key macro-fouling species.

Table 1. Summarized results and data references for biocide antifoulants occurring in *Mytilus galloprovincialis*.

Biocide Antifoulant	Exposure Time	Toxicity Index/Concentration	Sample	Effect on Antioxidative Defence and Settlement	Reference
Paint B	10 months	Cu_2O (41 wt.%) Cu leaching rate: 25.53 μg cm^{-2}/d	Total body	Settlements inhibition	Cima and Varello [20]
Paint D	10 months	Cu_2O (42 wt.%) Chlorothalonil (7 wt.%)	Total body	Settlements inhibition	
A-3 A-4 A-5	45 days 45 days 45 days	20 wt.% Cu_2O: 17.9 μg/L 30 wt.% Cu_2O: 21.2 μg/L 40 wt.% Cu_2O: 22.4–42.2 μg/L	Byssus	Byssus production inhibition as %Cu_2O increase	Kojima et al. [59]
Cu^{2+}	15 days	10 μg/L	Digestive gland/Gills	SOD↑, CAT↑, GPX↑	Gomes et al. [64,65]
CuO NPs	15 days	10 μg/L	Digestive gland/Gills	SOD↑, CAT↑, GPX↑	
Cu	4 days 4 days	5 μg/L 15 μg/L	Gills	GST↑	Perić and Burić [66]
Cu + Chp	4 days 4 days	Cu 5 μg/L, Chp 0.05 μg/L Cu 15 μg/L, Chp 0.05 μg/L	Gills	GST↑	
B[a]P	7 days	10 μg/L	Gills	CAT↑, tGPx↑,GST↑	Maria and Bebianno [67]
Cu	7 days	5, 10, 25 μg/L	Gills Digestive gland	SOD↓ GPx↓, GR↓, CAT↓, tGPx↓, GSH↓, SOD↓	
B[a]P + Cu	7 days	B[a]P 10 μg/L, Cu 5, 10, 25 μg/L	Digestive gland Gills	GR↓, CAT↓, tGPx↓, SOD↓, GST↑ SOD↑, CAT↑, GR↑	
ZnPT	96 h	20 and 40 μg/L	Digestive gland/Gills	SOD↑, GSH↓	Katalay et al. [68]
Mexel 432®	14 days	0.5, 1, 2 mg/L	Digestive gland/Gills	GST↑, CAT↓	López-Galindo et al. [69]
NaClO	14 days	0.1, 0.2, 0.5 mg/L	Digestive gland Gills	GST↑, CAT↓ GST↑, CAT↑	
NaClO + Mexel 432®	14 days	NaClO: 0.1, 0.2, 0.5 mg/L, Mexel 432®: 0.5, 1, 2 mg/L	Gills	GST↑, CAT↑	
ZnO NPs	14 days	100 μg/L	Digestive gland/Gills	GST↑	Bouzidi et al. [70]
TiO_2 NPs	14 days	100 μg/L	Digestive gland	SOD↑, CAT↑	
Diuron	14 days	50 and 100 μg/L	Digestive gland/Gills	SOD↑, CAT↑, GST↑	
ZnONP + Diuron	14 days	100 μg/L	Digestive gland/Gills	SOD↑, CAT↑, GST↑	
Antifoulants containing: Cu_2O, aromatic hydrocarbons, zinc oxide, rosin, zineb	12 months	n.d.	Gonads	Settlements inhibition, GST↑, *gpx4*↑	Madonna et al. [4]

Abbreviations: Cu^{2+}, cupric ion or copper ions; CuO NPs Copper oxide nanoparticles; Cu, copper; Chp, Chlorpyrifos; B[a]P, Benzo(a)pyrene; ZnPT, Zinc Pyrithione; NaClO, Sodium hypochlorite; ZnO NPs, Zinc Oxide Nanoparticle; TiO_2 NPs, Titanium dioxide nanoparticles;Cu_2O, Copper(I) oxide; wt.%, weight percent; n.d., not determined, (i.e., no information reported in manufacturer's datasheet or literature); SOD, superoxide dismutase; ↑ increase; ↓ decrease activity; CAT, catalase; GPX, glutathione peroxidase; GST, glutathione S-transferase; tGPX, total glutathione peroxidase; GR, glutathione reductase; GSH, glutathione; *gpx4*, glutathione peroxidase 4.

Moreover, six different AFs containing 0, 5, 10, 20, 30 and 40 wt.% of Cu_2O were tested by Kojima and colleagues [59] on the formation of filaments produced by the bissal apparatus in *M. galloprovincialis* (Table 1). This apparatus is composed of several secretory glands: the collagen gland produces the bissal collagen core, the accessory gland is responsible for the protein cortex, the phenolic glands produce the protein adhesive, and another glandular system is known to produce a sulfur-rich mucopolysaccharide [71]. Each paint formulation was coated on one surface for 45 days. A behavioural test was then conducted using mussels *M. galloprovincialis* that were glued to the coated surface of each aged test plate. The number of byssus threads produced by each mussel generally decreased as the Cu_2O content of the coating increased. In fact, a positive correlation was observed between the Cu_2O content in the paints and the inhibition of byssus threads production showing significant AF efficacy at greater than 20 wt.% of Cu_2O. Thus, the authors showed that AF containing a minimum threshold of 20 wt.% of Cu_2O have a repellent effect on the mussels settlement.

Copper oxide nanoparticles (CuO NP) are often used for their antimicrobial properties in AF paints. Moreover, given the widespread use of copper nanoparticles in various industrial and commercial applications, they will inevitably end up in the aquatic environment.

In this regard, Gomes and colleagues [64,65] investigated the effects of copper nanoparticles in the digestive gland and gills of *M. galloprovincialis* and assessed oxidative stress and level of metal bioaccumulation (Table 1). Mussels of *M. galloprovincialis* were exposed to Cu as nanoparticles (CuO NPs) and Cu^{2+} for 15 days. In the digestive gland of mussels exposed to CuO NPs, they showed an increase in superoxide dismutase (SOD), Catalase (CAT), Glutathione peroxidase (GPX) activity and lipid peroxide (LPO) levels after 7 days of exposure while metallothionein (MT) levels increased linearly with time. In the gills exposed to CuO NPs, SOD activity increased linearly in the first 7 days, CAT was only induced after 3 days while GPX was induced after one week. LPO and MT increased linearly with time of exposure, while inhibition of acetylcholinesterase (AChE) was observed only on the last exposure day (at the end of the experiment, day 15). Regarding Cu^{2+} exposure in the digestive gland, the mussels exhibited an increase in SOD activity after the first three days and remained unchanged throughout the exposure period while CAT activity increased with exposure time. Significant increases in GPX activity and LPO levels were detected in the digestive gland from day 7 of exposure while MT was observed only at the end of the exposure period (day 15). In the gills, SOD activity was observed during the whole experiment, while CAT and GPX activity only increased after 3 days of exposure. MT levels in the gills also increased in the first week of exposure while LPO increased linearly throughout the entire exposure period. Inhibition of AChE was observed only on the last exposure day. Results of antioxidant enzyme activities showed responses in the digestive gland and gills of exposed mussels demonstrating a susceptibility to CuO NP toxicity and Cu^{2+}.

Further studies were carried out on compounds in combination with copper. Perić and Burić [66] studied the combinatory effect of Cu and the organophosphorus pesticide chlorpyrifos (Chp) in *M. galloprovincialis* after short-term exposure to sublethal concentrations (Table 1). Chp is a common organophosphorus pesticide, a controversial compound due to its wide consumption and environmental implications [72]. Previous studies have reported that Chp is capable of mediating cytotoxicity through various mechanisms [73] such as oxidative damage further inducing systemic diseases such as inflammation and immune response [74]. Thus, the mussels were exposed for 4 days to Cu, and Chp both separately as well as to their binary mixtures at different concentrations. AChE, GST activities, MT content and LPO levels were used as biomarkers of the oxidative response using digestive gland and gill tissue. Exposure to Cu or Chp alone did not induce changes in AChE activity, whereas Cu alone caused a significant increase in GST activity and LPO level. Exposure to lower and higher concentrations of Chp alone resulted in an increase in the MT content and in the LPO level, respectively. The mixture of a lower concentration of Chp when combined with Cu significantly increased GST activity, while a higher concentration of

Cu in combination with Chp, resulted in a significant decrease of LPO. Higher concentrations of either Cu or Chp alone both decreased AChE activity. Thus, the authors in this study showed that low and environmentally relevant concentrations of Cu and Chp can alter the biological response in *M. galloprovincialis* showing that the activity of GST was the most sensitive biomarker that revealed oxidative damage imposed by both the single compounds and their mixtures. Subsequent studies on combined toxicological effects using copper together with benzo[a]pyrene (B[a]P) were performed. This latter, is a carcinogenic polynuclear aromatic hydrocarbon belonging to a class of compounds produced by any incomplete combustion of organic material and are therefore present all over the world due to anthropogenic activity [75].

Maria and Bebianno [67] studied the effects of single and binary mixtures of B[a]P and copper (Cu) on the antioxidant system and the lipid peroxidative response in the gills and digestive system of *M. galloprovincialis* (Table 1). They observed that different antioxidant sets are altered in gills and digestive system in relation to the dose-response to mixtures used. In gills, the B[a]P exposure induces an increase in CAT, total Glutathione Peroxidase (tGPx), GST, while Cu exposure induces an increase in MR and LPO and a decrease in SOD. In the digestive system, the B[a]P exposure induces a decrease of SOD, glutathione reductase (GR) and MT, while Cu exposure induces a decrease of GPx, GR, CAT, tGPx, glutathione (GSH) and SOD. In addition, the authors also tested the combined effect of these two compounds on the same tissues. In fact, in gills, the effect of B[a]P + Cu compounds significantly increased SOD, CAT and GR, while there were no significant changes in other antioxidants. However, increasing the Cu concentration altered LPO tGPx and GST showing a mixture-dependent increase. In the digestive gland, the enzymatic activity of GR, CAT and tGPx decreases significantly and the activity of LPO increases, favouring the inhibition of SOD, while GST activity increases. Thus, the authors showed that mussels *M. galloprovincialis* exposed to a combination of B[a]P and Cu were characterised by the strongest antioxidant response. They observed a better pro-oxidant power, while Cu concentration is higher with a B[a]P/Cu mixture suggesting a lowering of the mussel's immune defence to the copper contaminant.

Zinc pyrithione, 2-mercapto-pyridine-1-oxide complex (ZnPT), is one of the most widely used alternative biocides in AF paints, toxic to a wide range of marine organisms, including bivalves. It was introduced to the market as a replacement for organotin biocides (TBT) by Arch Chemicals (Norwalk, CT, USA) in 1991, and is currently considered the best candidate to replace TBT in AF paints [76]. There is little information on the toxicity of ZnPT in bivalves. The effect of ZnPT is also known to be toxic at the embryonic development stages in other organism belonging to the same genus such us the *Mytilus edulis* or other organisms such as the sea urchin (*Paracentrotus lividus*). In fact, Bellas and colleagues [77] demonstrated through the results of the embryo-larval toxicity test with *M. edulis* that ZnPT significantly influenced normal embryonic development with dose-dependent effect.

Katalay and colleagues [68] tested the effect after acute exposure to ZnPT using hepatopancreas and gill of *M. galloprovincialis* (Table 1). They observed an increase in SOD enzyme activity in both tissues treated by ZnPT with a significant GSH inhibition in gills. Furthermore, the authors observed loss of tubular epithelial integrity of digestive tubules, ciliate fractures, fusions, erosions and loss of epithelial cells in gill filaments. Histological changes in the tissues increased with increasing ZnPT concentration. ZnPT also caused a dose-dependent increase in cells undergoing apoptosis, particularly in the hepatopancreas and gill tissues. Thus, the authors observed that this specific type of biocide activates the antioxidative response in *M. galloprovincialis*. More precisely, an increase in SOD and a decrease in GSH was found in the gills, which are the first tissues in contact with the external environment, and a decrease in SOD in the hepatopancreatic tissues, which is the site of metabolic activity, demonstrating how this biocide can disturb the organs responsible for filtration and storage/bioaccumulation. In line with these results, Marcheselli and colleagues [78] have previously demonstrated how zinc pyrithione and its main secondary products, Zn and ionized pyrithione (PT−), rapidly accumulated in the tissues of exposed mussels, identifying the gills and digestive gland as important targets in the biological pathway of contaminants.

López and colleagues [69] tested two AFs: sodium hypochlorite (NaClO) and an alkylamine surfactant (Mexel 432®) that is used as a biocide in the cooling systems of power plants to reduce biofouling (Table 1). Cooling water from power plants is usually not treated to remove the biocide before discharge into the sea, which can have negative consequences for nearby aquatic life [79]. Sodium hypochlorite (NaClO) is a derivative of chlorine and one of the most widely used disinfecting agents, due to its great ability to inhibit the proliferation of microorganisms in general. Using the common mussel *M. galloprovincialis* as the study organism, the authors assessed enzyme activity and responses to oxidative stress. Mussels were exposed to different concentrations of NaClO and alkylamine surfactant. The digestive gland and gills of each mussel were dissected after 1, 3, 7 and 14 days of exposure. Both AFs caused a response of the enzymes CST, CAT AChE and LPO. NaClO initiated the oxidation process in the gills, increasing LPO levels, GST and CAT activity and decreasing AChE activity. In the digestive gland, GST activity also increased, while CAT activity was decreased but there were no other signs of damage (such as increased LPO levels). No clear trends in antioxidant enzymes (GR and GPX activity) were observed in the digestive gland, despite an initial inhibition of GR and an increase in GPX. Mussels exposed to Mexel 432 experienced increased GST activity in the digestive gland as well as in the gill, while inhibiting CAT; LPO levels were slightly elevated in the digestive gland. Thus, the authors showed how the overall exposure to NaClO and Mexel 432 reflected in a toxic response mainly in the gills of *M. galloprovincialis*, causing a strong impact on oxidative response.

The herbicide Diuron is widely used for weed control in many types of crops. It reaches water bodies through various pathways and can negatively threaten non-target organisms. Ons Bacha and colleagues [80] studied this type of biocide to assess the antioxidant activity on the mollusc *M. galloprovincialis* during seven days of exposure (Table 1). Biomarkers of oxidative stress were assessed in the gills and digestive gland of the mollusc by observing changes in the activities of enzymes such as CAT, GST, AChE and MDA. The results obtained showed that Diuron altered biomarkers of oxidative stress in both the gills and digestive gland.

Combinatory effects using Diuron with nanoparticle NPs (ZnO NPs and TiO2 NPs) at sub-lethal doses were assessed on the marine mussel *M. galloprovincialis*. Bouzidi and colleagues [70] tested these two known antifouling biocides on digestive glands and gills (Table 1). SOD, CAT and GST were used as toxicological biomarkers. Upon Diuron exposure, SOD, CAT, GST, and LPO activity increased both in the digestive gland and gills directly with the compound concentration. At higher exposure, TiO_2 and ZnO NPs increased SOD, CAT, GST and LPO activity in the digestive gland while in the gills only GST and LPO activity increased. The combined effects of Diuron with the two NPs in different redox markers showed concentration-dependent effects. The mixture of higher concentration of Diuron with ZnO NPs (ZnONP2 + Di2) increased significantly the activities of SOD, CAT and GST and level of LPO in the digestive gland and gills. Thus, the authors showed how the antioxidant defense mechanisms can be complex in response to herbicide and nanoparticle compounds used in AFs.

3. Eco-Friendly Antifoulants and Their Effect on *Mytilus galloprovincialis*

Over the years the application of antifoulant paints has been criticised due to the damage they cause to the environment [81]. As people become more environmentally aware and regulations become more stringent, eco-friendly antifoulant systems have become the focus of current research [82]. The development of alternative AF solutions and emerging non-toxic strategies are generating a new class of protective coatings (Table 2), which follow two main approaches:

I. Applying non-biocidal strategy which is the most environmentally friendly and well represented by dirt-release coatings.

II. Applying biocidal strategy based on the controlled release of harmless bioactive agents and on chemical immobilisation strategy avoiding the release of biocidal agents.

Non-stick coatings are a non-toxic alternative due to many factors including low surface free energy, low glass transition temperature, low micro-rugosity, and a high mobility of polymer chains. Therefore, fouling species cannot safely attach to the surface and can be easily removed hydro-dynamically [84].

In this context, Madonna and colleagues [4] assessed the biological effects of AFs containing biocide-free paint and metal biocide using *M. galloprovincialis* gonads (Table 2). They painted submerged panels with eco-friendly AFs based on silicone and hydrogel which are fouling release coatings. In parallel, another set of panels were painted with well-known AF biocides such as copper oxide, aromatic hydrocarbons, zinc oxide, rosin and zineb. They observed an alteration in antioxidant defence in relation to the type of AF used. The GST enzymatic activity and the *gpx4* transcriptional activity were higher in the *M. galloprovincialis* gonads treated with biocidal AFs than in eco-friendly treated organisms. The results obtained indicate that the levels of *gpx4* and GST in *M. galloprovincialis* increased with exposure to paint containing the biocides copper and zinc proving that these two elements are promoters of oxidative stress while eco-friendly AFs did not activate the antioxidant response. In fact, GST enzyme activity and *gpx4* transcriptional activity are involved in detoxification response and apoptosis regulation respectively, so their alteration in biocide-treated *Mytilus* may indicate an acceleration of these processes. As antioxidant increases are mainly observed in the gonads, this study showed how the use of AFs containing biocides such as copper or zinc affects the reproductive state of the organism. In addition, the authors also assessed the AF effectiveness on the settlement of the benthic community. They identified, in total, 20 species belonging to 8 different phyla: Arthropoda, Mollusca, Porifera, Anellidae, Bryozoa, Cnidaria, Cordati, and Chlorophyta. They found lower diversity and richness values in panels treated with AF paint containing metal biocide, confirming their effectiveness as well as the level of toxicity that these types of compounds exert on the fouling community. This is in line with findings by Miller and colleagues [85] stating that coatings containing a mixture of these two biocides significantly reduce the settlement reported as abundance and biodiversity. The work carried out by Madonna and colleagues [4] allowed them to define the biodiversity of the submerged benthic community as related to the biofouling observed in their sampling zone and also allowed an understanding of how organisms such as *M. galloprovincialis* respond to the stresses induced by two types of antifoulants, i.e. eco-friendly and biocide AFs. They showed that an eco-friendly AF is ecologically more sustainable and does not cause oxidative stress as caused by toxic agents.

Natural products are suggested as an alternative to toxic biocides in AF paints for biofouling control [86]. Recent advances in AF research on natural products obtained from both marine and terrestrial sources have highlighted their biofouling inhibitory activities on marine microorganisms [87].

Bioactive compounds are synthesised by organisms in small and complex mixtures making their extraction and purification processes very laborious [88,89]. In addition, in order to derive bioactive compounds from marine organisms, a large number of organisms would need to be harvested from the sea. This may be of concern from a biodiversity conservation point of view [90].

Table 2. Summarized results and data references for eco-friendly antifoulants occurring in *Mytilus galloprovincialis*.

Eco-Friendly Antifoulant	Exposure Time	Toxicity Index/Concentration	Sample	Effect on Antioxidative Defence and Settlement	Reference
Silicone and hydrogel-based	12 months	nd.	Gonads	Settlement inhibition	Madonna et al. [4]
Napyradiomycin A1	15 h	EC_{50} = 0.655 [0.300; 0.906] µg/mL	Total body	Settlement inhibition	Pereira et al. [83]
18-Hydroxynapyradiomycin A1	15 h	EC_{50} = 1.999 [1.581; 2.547] µg/mL LC_{50} > 12 µg/mL	Total body	Settlement inhibition	
Napyradiomycins SF2415B3	15 h	EC_{50} = 1.092 [0.225; 2.933] µg/mL LC_{50} > 12 µg/mL	Total body	Settlement inhibition	
Napyradiomycin A2	15 h	EC50 = 6.339 [5.602; 7.181] µg/mL LC_{50} > 12 µg/mL	Total body	Settlement inhibition	
16-Oxonapyradiomycin A2	15 h	EC_{50} = 4.331 [2.911; 7.091] µg/mL LC_{50} > 12 µg/mL	Total body	Settlement inhibition	
4-Dehydro-4a-dechloro-napyradiomycin A2	15 h	EC_{50} = 4.331 [2.911; 7.091] µg/mL LC_{50}>12 µg/mL	Total body	Settlement inhibition	
4-Dehydro-4a-dechloronapyradiomycin SF2415B3	15 h	EC_{50} = 1.092 [0.225; 2.933] µg/mL LC_{50} = >12 µg/mL	Total body	Settlement inhibition	
Napyradiomycin B3	15 h	EC_{50} = 0.727 [0.065; 1.406] µg/mL LC_{50} > 12 µg/mL	Total body	Settlement inhibition	
Napyradiomycin a80915A	15 h	EC_{50} = 0.947 [0.586; 1.473] µg/mL LC_{50} > 12 µg/mL	Total body	Settlement inhibition	
Napyradiomycin a80915C	15 h	EC_{50} = 0.102 [0.072; 0.140] µg/mL LC_{50} > 12 µg/mL	Total body	Settlement inhibition	
4-Dehydro-4a dechloronapyradiomycin B3	15 h	EC_{50} = 0.727 [0.065; 1.406] µg/mL LC_{50} > 12 µg/mL	Total body	Settlement inhibition	
4-Dehydro-4a-dechloronapyradiomycin A80915A	15 h	EC_{50} = 0.947 [0.586; 1.473] µg/mL LC_{50} > 12 µg/mL	Total body	Settlement inhibition	
Rutin 2″,2‴,3′,3″,3‴,4′,4″,4‴,7-nonasulfate	Concentration- response analysis	EC_{50} = 22.59 µM; 36.84 µg/mL	Total body	Settlement inhibition	Almeida et al. [33]
3,6-bis (β-D-glucopyranosyl)xanthone persulfate	Concentration- response analysis	EC_{50} = 23.19 µM; 31.74 µg/mL	Total body	Settlement inhibition	
gallic acid persulfate	concentration- response analysis	EC_{50} = 17.65 µM; 8.4 µg/mL	Total body	Settlement inhibition	

Abbreviations: n.d., not determined, (i.e., no information reported in manufacturer's datasheet or literature); EC_{50}, Half maximal effective concentration, median [interquartile range]; LC_{50}, Lethal Concentration at 50% of the tested population.

Among the various natural products used as AFs are napiradiomycins derivatives. Napiradiomycins are an interesting group of natural halogenated secondary metabolites produced mainly by bacteria of the Streptomycetaceae family. Almost 47 different napiradiomycins have currently been described [91–94]. They belong to a class of hybrid isoprenoids and/or meroterpenoids previously known for their antimicrobial and antitumor activities [92–94]. In fact, in silico–toxicity predictions of napiradiomycins suggested similar toxicity to marketed drugs and AF biocides with a low bioaccumulation factor and no mutagenicity [83]. These compounds had already been shown to be effective at concentrations below 25 μg/mL [95].

Lacret and colleagues [96] showed that a novel napiradiomycin isolated from the genus *Streptomyces*, exhibited potent antimicrobial activity against microorganisms. Pereira and colleagues [83] attempted to identify new compounds with inhibitory activity against biofouling (Table 2). They isolated and tested 12 compounds from napiradiomycins. The authors assessed inhibitory capacity of napiradiomycins isolated from *Streptomyces aculeolatus* against micro- and macro-fouling species. They isolated five molecules from PTM-029 *S. aculeolatus* strains having a methyl group in the core structure at position 7 and seven molecules from PTM-420 *S. aculeolatus* strains containing hydrogen atom at position 7 having a hydrogen atom in that position of their structure. In order to assess the AF activity, the authors tested these group compounds on different organisms, including higher sessile organisms. Larval settlement tests conducted on *M. galloprovincialis* showed that both napyradomycins groups were effective in preventing their settlement with values ranging from 0.10 to 6.34 μg/mL. They found that the most promising AF affecting larvae were those with EC50 below 1 μg/mL (EC_{50}, half maximal effective concentration). In the same study, the authors also investigated the same napiradiomycin groups on lower organisms and tested their inhibitory activity on bacterial propagation and bacterial biofilm formation confirming their activity as AFs. Thus, Pereira and colleagues have shown napiradiomycins as potential AF agents. Given their potent antibacterial, anti-biofilm, and anti-sediment activities, they revealed the ability to prevent the settlement of higher organisms.

Nature has implemented several metabolic strategies, such as sulfation, to prevent toxic action during physiological and pathological processes. It has been observed that some sulphate secondary metabolites from marine organisms such as flavonoids, coumarins, cinnamic acids and sulphate sterols, have an AF effect while also being safe for the environment [97–102]. Since it is difficult and unsustainable to obtain quantities for commercial use directly from nature, the synthesis of non-natural small sulphate molecules could be an effective alternative for new and relevant non-toxic AF agents. These polysaccharide compounds are highly hydrophilic and thus able to form hydrogels that incorporate water. As shown by many studies, coatings with amphiphilic properties have a high potential for their use as inert surface coatings. This can be achieved by chemically modifying polysaccharide structures with hydrophobic molecules [85].

Almeida and colleagues [33] synthesised sulphate polyphenols and tested them for AF potential as assessed by anti-settlement activity in *M. galloprovincialis* plantigrade post-larvae (Table 2). From this class of polyphenols they isolated the following three compounds: rutin persulfate 3,6-bis(β-D-glucopyranosyl), xanthone persulfate and gallic acid persulfate. These compounds exhibited therapeutic ratios (LC 50/EC 50) greater than 15, which is consistent with the standard requirement for the level of efficacy of natural AF agents as established by the U.S. Navy Program [86]. They first showed that none of these isolated compounds were lethal on *M. galloprovincialis* at any of the concentrations tested compared to a biocide AF (commercial AF agent ECONEA®).

The authors found that gallic acid persulfate resulted in the most effective anti-settling activity on the *M. galloprovincialis* thus presenting the highest potential as an effective AF agent. When comparing the structures of the above three tested compounds, the presence of the benzoic acid scaffold in the gallic acid persulfate compared to the other two compounds led to the hypothesis that it is related to a more interesting AF activity, as already observed [103] on other organisms such as *Escherichia coli*. In addition, this compound has a similar structure to a naturally occurring AF, zosteric acid, which is a metabolite of the marine herb *Zostera marina*. Zosteric acid has been found to have antibacterial [104–107] and antifungal [108,109] activity and prevents attachment of higher order organisms [110] at non-toxic concentrations. Its AF activity has been attributed primarily to the sulfate–ester group [106]. Recently, it has been shown that the anti-biofilm activity of zosteric acid against *E. coli* is related to the cinnamic acid scaffold [103,111].

In addition, these natural compounds were tested to quantify oxidative damage as reflected by acetylcholinesterase and tyrosinase which are involved in the settlement mechanism of *M. galloprovincialis* [112,113]. Adhesive plaques of DOPA-containing mussels are also produced by tyrosinase [114]. No effect was observed in the activity of these enzymes indicating that at least these two enzymatic processes seemed to remain unchanged after exposure to these AFs. Therefore, the authors found that the same compounds effective on mussels were less effective on biofouling bacteria. This led to the conclusion that the anti-settling action of sulphate compounds appears to be organism specific, more related to the metabolic pathways of *M. galloprovincialis* than to the biofouling succession cascade that begins with biofilm colonisation [115], while not altering the antioxidant response of the mussels. In addition, this study advanced the importance of the synthesis of small, non-natural sulphated molecules to generate a new class of non-toxic AF agents.

In nature, there are found not only compounds derived from bacteria with AF potential, but also AF compounds derived from algae (among other organisms). Benthic marine organisms are subject to intense competition for space [116]. It is assumed that benthic microalgae produce natural products showing AF activity. A natural product derived from benthic microalgae was recently discovered: the cyclic imin protamine. It was randomly identified during extraction of pinnatoxins from *Vulcanodinium rugosum* dinoflagellate [117]. Cyclic imines are a class of polycyclic ethers with a spirocyclic imine ring and are found only in microalgae [118]. They are generally fast-acting bioactives which inhibit neuromuscular transmission via blocking of nicotinic acetylcholine receptors and exhibit pro-apoptotic activity in cells [119]. Apoptosis plays a key role in the metamorphic processes of many biofouling organisms [120]. Given the high biocidal potential of portimin, its defensive activity against other benthic organisms may also be plausible [121].

In this regard, Brooke and colleagues [121] evaluated the AF activity of protamine against *M. galloprovincialis* (Table 2). Furthermore, a distinct AF action was seen by three related compounds: gymnodimine-A, 13-desmethyl spirolide C and pinnatoxin-F.

Larval development and metamorphosis were selected as endpoints of the bioassay because apoptosis is an integral part of these processes in marine invertebrates [122–125].

In fact, using *M. galloprovincialis* they found that the effect, as observed visually, of portamin on embryo/larvae morphology was concentration dependent. At a concentration lower than 0.15 ng/mL of protamine the organisms were intact but with incomplete embryogenesis. At higher concentrations (0.3 ng/mL) the organisms were disaggregated/disintegrated. Furthermore, none of the other three protamine related compounds were effective against *M. galloprovincialis*. Brooke and colleagues also tested inhibitory concentrations (EC50) of protamine against other marine macrofouling organisms, such as *Ciona savignyi*, *Spirobranchus cariniferus* and *Amphibalanus improvisus*, showing its potential as an AF. Thus, the authors found an impressive high potency of protamine against the settlement of *M. galloprovincialis* embryos/larvae which were absent at the end of the bioassay assuming that this compound can induce a massive apoptosis event to disintegrate and/or disaggregate *M. galloprovincialis* embryos.

4. Conclusions

Antifoulant paints were developed to prevent and reduce biofouling of surfaces in seawater. They were used as a surface treatment since ancient times. By the second half of the 20th century, TBT-based compounds were detected to be harmful to the ecosystem. Thus, as a substitute for highly toxic compounds, less damaging copper-based compounds were developed. Although less toxic, they are biocides designed to eliminate and reduce organisms' settlement on surfaces. The widespread use of these substances over the years has led to a significant increase of their presence in the marine environment.

Due to an increasing number of toxicological studies, these compounds were identified as environmental and human threats. Thus, they were subject to regulation as in the case of organotin compounds which were completely banned by the International Marine Organization - Committee for the Protection of the Marine Environment (IMO-MEPC) in 1998, and subsequently by the Ordinance no. 782/2003, 14 April 2003, of the European Commission. Subsequently, new, naturally occurring compounds with much lower toxicity have been developed to replace the use of these toxic agents. They are designed to have the same efficiency as biocides without having a significant impact on the environment.

In this data collection, we have highlighted that biocide paints can cause antioxidant enzymatic alterations through reactive oxygen cell species and they can interfere with the formation of byssus in *M. galloprovincialis* leading to anti-settlement activity. We also discussed a dose-response relationship, i.e. different antioxidative responses caused by increasing biocide exposure in different tissues. In fact, antioxidative defence activation was found in gills and in the digestive gland, demonstrating how these biocides can disrupt organs responsible for filtration, storage/bioaccumulation and sustainability of reproductive processes/resources. In addition, some natural products derived from bacteria and algae have proved to be a valid ecological alternative to the canonical antifoulants on the market. They showed an inhibitory capacity against micro- and macro-fouling, with antibacterial, anti-biofilm and anti-settlement activity. In some cases, they exhibited similar toxicity to marketed drugs and antifouling biocides with a low bioaccumulation factor and no mutagenicity. Furthermore, through the analysis of biomarkers implicated in the antioxidative response in *Mytilus galloprovincialis*, it was found that they did not induce any oxidative damage. Taken together, the data obtained by these investigations emphasizes the discrepancy between biocide and eco-friendly antifoulants while also acknowledging that they both achieve the primary goal of inhibiting settlement. Moreover, using antioxidative markers we are able to detect the degree of environmental impact of antifoulants and their potential damage to biodiversity and human food security. Accordingly, to this end, transcriptomic antioxidative biomarker investigations as well as enzymatic investigations may provide useful indicators as to which future toxicological studies to focus on in order to evaluate environmentally friendly compounds used as antifoulants.

Since antifoulant paints have global economic and environmental impacts, the implementation of effective eco-friendly strategies is needed. In fact, the term antifoulant is no longer synonymous with biocidal. Bio-inspired chemical strategies are the most promising for developing a new generation of antifoulants.

Collaborations among governments, scientists, biologists, chemists, engineers and many others from various fields have motivated advancements in eco-friendly strategies for sustainable development to improve health and well-being (SDG 3), climate action (SDG 13), and Life Below Water (SDG 14).

Author Contributions: Conceptualization, C.P., J.S., M.R.C. and G.G.; validation, C.P., J.S., M.R.C., A.M., F.K.A.-G., E.M.S. and G.G.; Writing—Original Draft preparation, C.P., J.S., M.R.C. and G.G.; writing—review and editing, C.P., G.G. and E.M.S.; Supervision F.K.A.-G. and G.G.; provided financial support, G.G. All authors have read and agreed to the published version of the manuscript.

Funding: This research received no external funding.

Acknowledgments: This work was performed within the framework of the Memorandum of Understanding between the National Research Centre of Giza, Suez Canal University (Egypt) and Federico II University (Italy).

Conflicts of Interest: The authors declare no conflict of interest.

References

1. Jin, H.; Tian, L.; Bing, W.; Zhao, J.; Ren, L. Bioinspired Marine Antifouling Coatings: Status, Prospects, and Future. *Prog. Mater. Sci.* **2022**, *124*, 100889. [CrossRef]
2. Gentilucci, M.; Parisi, C.; Coppola, M.R.; Majdoubi, F.-Z.; Madonna, A.; Guerriero, G. Influence of Mediterranean Sea Temperature Increase on Gaeta Gulf (Tyrrhenian Sea) Biodiversity. *Proc. Zool. Soc.* **2021**, *74*, 91–103. [CrossRef]
3. Sandonnini, J.; Del Pilar Ruso, Y.; Cortés Melendreras, E.; Barberá, C.; Hendriks, I.E.; Kersting, D.K.; Giménez Casalduero, F. The Emergent Fouling Population after Severe Eutrophication in the Mar Menor Coastal Lagoon. *Reg. Stud. Mar. Sci.* **2021**, *44*, 101720. [CrossRef]
4. Madonna, A.; Balzano, A.; Rabbito, D.; Hasnaoui, M.; Moustafa, A.A.; Guezgouz, N.; Vittorioso, A.; Majdoubi, F.-Z.; Olanrewaju, O.S.; Guerriero, G. Biological Effects Assessment of Antibiofouling EDCs: Gaeta Harbor (South Italy) Benthic Communities' Analysis by Biodiversity Indices and Quantitative Gpx4 Expression. *Proc. Zool. Soc.* **2021**, *74*, 591–604. [CrossRef] [PubMed]
5. Parisi, C.; De Marco, G.; Labar, S.; Hasnaoui, M.; Grieco, G.; Caserta, L.; Inglese, S.; Vangone, R.; Madonna, A.; Alwany, M.; et al. Biodiversity Studies for Sustainable Lagoon: Thermophilic and Tropical Fish Species vs. Endemic Commercial Species at Mellah Lagoon (Mediterranean, Algeria). *Water* **2022**, *14*, 635. [CrossRef]
6. Conn, D.B. Aquatic Invasive Species and Emerging Infectious Disease Threats: A One Health Perspective. *AI* **2014**, *9*, 383–390. [CrossRef]
7. Sadan, N.E.; Akash, P.S.; Kumar, P.G.S. Biofouling Impacts and Toxicity of Antifouling Agents on Marine Environment: A Qualitative Study. *Sustain. Agric. Food Environ. Res.* **2021**, *10*, 2492. [CrossRef]
8. Lejars, M.; Margaillan, A.; Bressy, C. Fouling Release Coatings: A Nontoxic Alternative to Biocidal Antifouling Coatings. *Chem. Rev.* **2012**, *112*, 4347–4390. [CrossRef]
9. Richard, K.; Hunsucker, K.; Gardner, H.; Hickman, K.; Swain, G. The Application of UVC Used in Synergy with Surface Material to Prevent Marine Biofouling. *JMSE* **2021**, *9*, 662. [CrossRef]
10. Cima, F.; Varello, R. Effects of Exposure to Trade Antifouling Paints and Biocides on Larval Settlement and Metamorphosis of the Compound Ascidian Botryllus Schlosseri. *JMSE* **2022**, *10*, 123. [CrossRef]
11. Omae, I. Organotin Antifouling Paints and Their Alternatives. *Appl. Organometal. Chem.* **2003**, *17*, 81–105. [CrossRef]
12. de Campos, B.G.; Figueiredo, J.; Perina, F.; de Souza Abessa, D.M.; Loureiro, S.; Martins, R. Occurrence, Effects and Environmental Risk of Antifouling Biocides (EU PT21): Are Marine Ecosystems Threatened? *Crit. Rev. Environ. Sci. Technol.* **2021**, 1–32. [CrossRef]
13. Swain, G. Biofouling Control: A Critical Component of Drag Reduction. In Proceedings of the International Symposium on Sea Water Drag Reduction, Newport, RI, USA, 22–23 July 1998; pp. 155–161.
14. Guerriero, G.; Bassem, S.M.; Abdel-Gawad, F.K. Biological Responses of White Sea Bream (Diplodus Sargus, Linnaeus 1758) and Sardine (Sardine Pilchardus, Walbaum 1792) Exposed to Heavy Metal Contaminated Water. *Emir. J. Food Agric.* **2018**, *30*, 688–694. [CrossRef]
15. Wu, Y.; Wang, C.; Wang, Y.; Zhao, Y.; Chen, Y.; Zuo, Z. Antioxidant Responses to Benzo[a]Pyrene, Tributyltin and Their Mixture in the Spleen of Sebasticus Marmoratus. *J. Environ. Sci.* **2007**, *19*, 1129–1135. [CrossRef]
16. Li, Z.-H.; Li, P.; Shi, Z.-C. Chronic Exposure to Tributyltin Induces Brain Functional Damage in Juvenile Common Carp (Cyprinus Carpio). *PLoS ONE* **2015**, *10*, e0123091. [CrossRef]
17. Mozafari, M.; Seyedpour, S.F.; Salestan, S.K.; Rahimpour, A.; Shamsabadi, A.A.; Firouzjaei, M.D.; Esfahani, M.R.; Tiraferri, A.; Mohsenian, H.; Sangermano, M.; et al. Facile Cu-BTC Surface Modification of Thin Chitosan Film Coated Polyethersulfone Membranes with Improved Antifouling Properties for Sustainable Removal of Manganese. *J. Membr. Sci.* **2019**, *588*, 117200. [CrossRef]
18. Tian, J.; Xu, K.; Hu, J.; Zhang, S.; Cao, G.; Shao, G. Durable Self-Polishing Antifouling Cu-Ti Coating by a Micron-Scale Cu/Ti Laminated Microstructure Design. *J. Mater. Sci. Technol.* **2021**, *79*, 62–74. [CrossRef]
19. Soares, K.L.; Barbosa, S.C.; Primel, E.G.; Fillmann, G.; Diaz-Cruz, M.S. Analytical Methods for Antifouling Booster Biocides Determination in Environmental Matrices: A Review. *Trends Environ. Anal. Chem.* **2021**, *29*, e00108. [CrossRef]
20. Cima, F.; Varello, R. Potential Disruptive Effects of Copper-Based Antifouling Paints on the Biodiversity of Coastal Macrofouling Communities. *Environ. Sci. Pollut. Res.* **2022**. [CrossRef]
21. Thomas, K.V.; Brooks, S. The Environmental Fate and Effects of Antifouling Paint Biocides. *Biofouling* **2010**, *26*, 73–88. [CrossRef]
22. Paz-Villarraga, C.A.; Castro, Í.B.; Fillmann, G. Biocides in Antifouling Paint Formulations Currently Registered for Use. *Environ. Sci. Pollut. Res.* **2022**, *29*, 30090–30101. [CrossRef] [PubMed]
23. Kim, H.-J. Strategic Actions for Sustainable Vessel Hull Coatings in Line with the UN SDGs. *JAMET* **2021**, *45*, 231–242. [CrossRef]
24. Qian, P.-Y.; Li, Z.; Xu, Y.; Li, Y.; Fusetani, N. Mini-Review: Marine Natural Products and Their Synthetic Analogs as Antifouling Compounds: 2009–2014. *Biofouling* **2015**, *31*, 101–122. [CrossRef] [PubMed]
25. Wang, K.-L.; Dou, Z.-R.; Gong, G.-F.; Li, H.-F.; Jiang, B.; Xu, Y. Anti-Larval and Anti-Algal Natural Products from Marine Microorganisms as Sources of Anti-Biofilm Agents. *Mar. Drugs* **2022**, *20*, 90. [CrossRef]
26. Almeida, J.R.; Vasconcelos, V. Natural Antifouling Compounds: Effectiveness in Preventing Invertebrate Settlement and Adhesion. *Biotechnol. Adv.* **2015**, *33*, 343–357. [CrossRef]

27. Gu, Y.; Yu, L.; Mou, J.; Wu, D.; Xu, M.; Zhou, P.; Ren, Y. Research Strategies to Develop Environmentally Friendly Marine Antifouling Coatings. *Mar. Drugs* **2020**, *18*, 371. [CrossRef]
28. Puglisi, M.P.; Sneed, J.M.; Sharp, K.H.; Ritson-Williams, R.; Paul, V.J. Marine Chemical Ecology in Benthic Environments. *Nat. Prod. Rep.* **2014**, *31*, 1510–1553. [CrossRef]
29. Borges, A.; Simões, M. Quorum Sensing Inhibition by Marine Bacteria. *Mar. Drugs* **2019**, *17*, 427. [CrossRef]
30. Saurav, K.; Borbone, N.; Burgsdorf, I.; Teta, R.; Caso, A.; Bar-Shalom, R.; Esposito, G.; Britstein, M.; Steindler, L.; Costantino, V. Identification of Quorum Sensing Activators and Inhibitors in The Marine Sponge Sarcotragus Spinosulus. *Mar. Drugs* **2020**, *18*, 127. [CrossRef]
31. Toupoint, N.; Mohit, V.; Linossier, I.; Bourgougnon, N.; Myrand, B.; Olivier, F.; Lovejoy, C.; Tremblay, R. Effect of Biofilm Age on Settlement of *Mytilus edulis*. *Biofouling* **2012**, *28*, 985–1001. [CrossRef]
32. Cannuel, R.; Beninger, P.G.; McCombie, H.; Boudry, P. Gill Development and Its Functional and Evolutionary Implications in the Blue Mussel *Mytilus edulis* (Bivalvia: Mytilidae). *Biol. Bull.* **2009**, *217*, 173–188. [CrossRef] [PubMed]
33. Almeida, J.R.; Correia-da-Silva, M.; Sousa, E.; Antunes, J.; Pinto, M.; Vasconcelos, V.; Cunha, I. Antifouling Potential of Nature-Inspired Sulfated Compounds. *Sci. Rep.* **2017**, *7*, 42424. [CrossRef] [PubMed]
34. Dobretsov, S.; Teplitski, M.; Bayer, M.; Gunasekera, S.; Proksch, P.; Paul, V.J. Inhibition of Marine Biofouling by Bacterial Quorum Sensing Inhibitors. *Biofouling* **2011**, *27*, 893–905. [CrossRef] [PubMed]
35. Lind, U.; Alm Rosenblad, M.; Hasselberg Frank, L.; Falkbring, S.; Brive, L.; Laurila, J.M.; Pohjanoksa, K.; Vuorenpää, A.; Kukkonen, J.P.; Gunnarsson, L.; et al. Octopamine Receptors from the Barnacle Balanus Improvisus Are Activated by the α_2 -Adrenoceptor Agonist Medetomidine. *Mol. Pharmacol.* **2010**, *78*, 237–248. [CrossRef]
36. Bayer, M.; Hellio, C.; Maréchal, J.-P.; Frank, W.; Lin, W.; Weber, H.; Proksch, P. Antifouling Bastadin Congeners Target Mussel Phenoloxidase and Complex Copper(II) Ions. *Mar. Biotechnol.* **2011**, *13*, 1148–1158. [CrossRef]
37. Silverman, H.G.; Roberto, F.F. Understanding Marine Mussel Adhesion. *Mar. Biotechnol.* **2007**, *9*, 661–681. [CrossRef]
38. Fasulo, S.; Guerriero, G.; Cappello, S.; Colasanti, M.; Schettino, T.; Leonzio, C.; Mancini, G.; Gornati, R. The "SYSTEMS BIOLOGY" in the Study of Xenobiotic Effects on Marine Organisms for Evaluation of the Environmental Health Status: Biotechnological Applications for Potential Recovery Strategies. *Rev. Environ. Sci. Bio/Technol.* **2015**, *14*, 339–345. [CrossRef]
39. Piscopo, M.; Trifuoggi, M.; Notariale, R.; Labar, S.; Troisi, J.; Giarra, A.; Rabbito, D.; Puoti, R.; Brundo, M.V.; Basile, A.; et al. Protamine-like Proteins Analyses as Emerging Biotechnique for Cadmium Impact Assessment on Male Mollusk *Mytilus galloprovincialis* (Lamarck 1819). *Acta Biochim. Pol.* **2018**, *65*, 259–267. [CrossRef]
40. Curpan, A.-S.; Impellitteri, F.; Plavan, G.; Ciobica, A.; Faggio, C. Review: Mytilus galloprovincialis: An Essential, Low-Cost Model Organism for the Impact of Xenobiotics on Oxidative Stress and Public Health. *Comp. Biochem. Physiol. Part C Toxicol. Pharmacol.* **2022**, *256*, 109302. [CrossRef]
41. Vassalli, Q.A.; Caccavale, F.; Avagnano, S.; Murolo, A.; Guerriero, G.; Fucci, L.; Ausió, J.; Piscopo, M. New Insights into Protamine-Like Component Organization in *Mytilus galloprovincialis'* Sperm Chromatin. *DNA Cell Biol.* **2015**, *34*, 162–169. [CrossRef]
42. Piscopo, M.; Notariale, R.; Rabbito, D.; Ausió, J.; Olanrewaju, O.S.; Guerriero, G. Mytilus galloprovincialis (Lamarck, 1819) Spermatozoa: Hsp70 Expression and Protamine-like Protein Property Studies. *Environ. Sci. Pollut. Res.* **2018**, *25*, 12957–12966. [CrossRef] [PubMed]
43. Cappello, T.; Giannetto, A.; Parrino, V.; Maisano, M.; Oliva, S.; De Marco, G.; Guerriero, G.; Mauceri, A.; Fasulo, S. Baseline Levels of Metabolites in Different Tissues of Mussel *Mytilus galloprovincialis* (Bivalvia: Mytilidae). *Comp. Biochem. Physiol. Part D Genom. Proteom.* **2018**, *26*, 32–39. [CrossRef] [PubMed]
44. Giannico, O.V.; Baldacci, S.; Desiante, F.; Basile, F.C.; Franco, E.; Fragnelli, G.R.; Diletti, G.; Conversano, M. PCDD/Fs and PCBs in *Mytilus galloprovincialis* from a Contaminated Area in Italy: The Role of Mussel Size, Temperature and Meteorological Factors. *Food Addit. Contam. Part A* **2022**, 1–13. [CrossRef] [PubMed]
45. Scalici, M.; Traversetti, L.; Spani, F.; Malafoglia, V.; Colamartino, M.; Persichini, T.; Cappello, S.; Mancini, G.; Guerriero, G.; Colasanti, M. Shell Fluctuating Asymmetry in the Sea-Dwelling Benthic Bivalve Mytilus galloprovincialis (Lamarck, 1819) as Morphological Markers to Detect Environmental Chemical Contamination. *Ecotoxicology* **2017**, *26*, 396–404. [CrossRef]
46. Parisi, C.; Guerriero, G. Antioxidative Defense and Fertility Rate in the Assessment of Reprotoxicity Risk Posed by Global Warming. *Antioxidants* **2019**, *8*, 622. [CrossRef]
47. Guerriero, G.; Trocchia, S.; Abdel-Gawad, F.K.; Ciarcia, G. Roles of Reactive Oxygen Species in the Spermatogenesis Regulation. *Front. Endocrinol.* **2014**, *5*, 56. [CrossRef]
48. Afsa, S.; De Marco, G.; Giannetto, A.; Parrino, V.; Cappello, T.; ben Mansour, H.; Maisano, M. Histological Endpoints and Oxidative Stress Transcriptional Responses in the Mediterranean Mussel Mytilus galloprovincialis Exposed to Realistic Doses of Salicylic Acid. *Environ. Toxicol. Pharmacol.* **2022**, *92*, 103855. [CrossRef]
49. Guerriero, G.; Trocchia, S.; Ciccodicola, A.; Ciarcia, G. The Antioxidant Phospholipid Hydroperoxide Glutathione Peroxidase (GPx4/PHGPx) in the Frog Hypothalamus of Pelophylax Bergeri as Tool for Its Bioconservation Assessment. *Exp. Clin. Endocrinol. Diabetes* **2013**, *121*, P5. [CrossRef]
50. D'Errico, G.; Vitiello, G.; De Tommaso, G.; Abdel-Gawad, F.K.; Brundo, M.V.; Ferrante, M.; De Maio, A.; Trocchia, S.; Bianchi, A.R.; Ciarcia, G.; et al. Electron Spin Resonance (ESR) for the Study of Reactive Oxygen Species (ROS) on the Isolated Frog Skin (Pelophylax Bergeri): A Non-Invasive Method for Environmental Monitoring. *Environ. Res.* **2018**, *165*, 11–18. [CrossRef]

51. Guerriero, G.; Parisi, C.; Abdel-Gawad, F.K.; Hentati, O.; D'Errico, G. Seasonal and Pharmaceutical-Induced Changes in Selenoprotein Glutathione Peroxidase 4 Activity in the Reproductive Dynamics of the Soil Biosentinel Podarcis Sicula (Chordata: Reptilia). *Mol. Reprod. Dev.* **2019**, *86*, 1378–1387. [CrossRef]
52. Guerriero, G.; D'Errico, G. Effect of Oxidative Stress on Reproduction and Development. *Antioxidants* **2022**, *11*, 312. [CrossRef] [PubMed]
53. Meseguer Yebra, D.; Kiil, S.; Weinell, C.E.; Dam-Johansen, K. Presence and Effects of Marine Microbial Biofilms on Biocide-Based Antifouling Paints. *Biofouling* **2006**, *22*, 33–41. [CrossRef] [PubMed]
54. Yebra, D.M.; Kiil, S.; Weinell, C.E.; Dam-Johansen, K. Dissolution Rate Measurements of Sea Water Soluble Pigments for Antifouling Paints: ZnO. *Prog. Org. Coat.* **2006**, *56*, 327–337. [CrossRef]
55. Finnie, A.A.; Williams, D.N. Paint and Coatings Technology for the Control of Marine Fouling. In *Biofouling*; Drr, S., Thomason, J.C., Eds.; Wiley-Blackwell: Oxford, UK, 2009; pp. 185–206. ISBN 978-1-4443-1546-2.
56. Bressy, C.; Hellio, C.; Maréchal, J.P.; Tanguy, B.; Margaillan, A. Bioassays and Field Immersion Tests: A Comparison of the Antifouling Activity of Copper-Free Poly(Methacrylic)-Based Coatings Containing Tertiary Amines and Ammonium Salt Groups. *Biofouling* **2010**, *26*, 769–777. [CrossRef]
57. Pagano, M.; Stara, A.; Aliko, V.; Faggio, C. Impact of Neonicotinoids to Aquatic Invertebrates—In Vitro Studies on *Mytilus galloprovincialis*: A Review. *J. Mar. Sci. Eng.* **2020**, *8*, 801. [CrossRef]
58. dos Santos, J.V.N.; Martins, R.; Fontes, M.K.; de Campos, B.G.; Silva, M.B.M.D.P.E.; Maia, F.; Abessa, D.M.D.S.; Perina, F.C. Can Encapsulation of the Biocide DCOIT Affect the Anti-Fouling Efficacy and Toxicity on Tropical Bivalves? *Appl. Sci.* **2020**, *10*, 8579. [CrossRef]
59. Kojima, R.; Kobayashi, S.; Satuito, C.G.P.; Katsuyama, I.; Ando, H.; Seki, Y.; Senda, T. A Method for Evaluating the Efficacy of Antifouling Paints Using *Mytilus galloprovincialis* in the Laboratory in a Flow-Through System. *PLoS ONE* **2016**, *11*, e0168172. [CrossRef]
60. Turner, A. Marine Pollution from Antifouling Paint Particles. *Mar. Pollut. Bull.* **2010**, *60*, 159–171. [CrossRef]
61. Cotou, E.; Henry, M.; Zeri, C.; Rigos, G.; Torreblanca, A.; Catsiki, V.-A. Short-Term Exposure of the European Sea Bass Dicentrarchus Labrax to Copper-Based Antifouling Treated Nets: Copper Bioavailability and Biomarkers Responses. *Chemosphere* **2012**, *89*, 1091–1097. [CrossRef]
62. Filimonova, V.; Gonçalves, F.; Marques, J.C.; De Troch, M.; Gonçalves, A.M.M. Biochemical and Toxicological Effects of Organic (Herbicide Primextra®Gold TZ) and Inorganic (Copper) Compounds on Zooplankton and Phytoplankton Species. *Aquat. Toxicol.* **2016**, *177*, 33–43. [CrossRef]
63. Abdel-Gawad, F.K.; Khalil, W.K.B.; Bassem, S.M.; Kumar, V.; Parisi, C.; Inglese, S.; Temraz, T.A.; Nassar, H.F.; Guerriero, G. The Duckweed, Lemna Minor Modulates Heavy Metal-Induced Oxidative Stress in the Nile Tilapia, Oreochromis Niloticus. *Water* **2020**, *12*, 2983. [CrossRef]
64. Gomes, T.; Pinheiro, J.P.; Cancio, I.; Pereira, C.G.; Cardoso, C.; Bebianno, M.J. Effects of Copper Nanoparticles Exposure in the Mussel Mytilus galloprovincialis. *Environ. Sci. Technol.* **2011**, *45*, 9356–9362. [CrossRef] [PubMed]
65. Gomes, T.; Pereira, C.G.; Cardoso, C.; Pinheiro, J.P.; Cancio, I.; Bebianno, M.J. Accumulation and Toxicity of Copper Oxide Nanoparticles in the Digestive Gland of Mytilus galloprovincialis. *Aquat. Toxicol.* **2012**, *118–119*, 72–79. [CrossRef] [PubMed]
66. Perić, L.; Burić, P. The Effect of Copper and Chlorpyrifos Co-Exposure on Biomarkers in the Marine Mussel Mytilus galloprovincialis. *Chemosphere* **2019**, *225*, 126–134. [CrossRef] [PubMed]
67. Maria, V.L.; Bebianno, M.J. Antioxidant and Lipid Peroxidation Responses in *Mytilus galloprovincialis* Exposed to Mixtures of Benzo(a)Pyrene and Copper. *Comp. Biochem. Physiol. Part C Toxicol. Pharmacol.* **2011**, *154*, 56–63. [CrossRef]
68. Katalay, S.; Guner, A.; Dagdeviren, M.; Yigitturk, G.; Yavasoglu, A.; Gunal, A.C.; Karabay Yavasoglu, N.U.; Oltulu, F. Oxidative Stress-Induced Apoptotic Changes after Acute Exposure to Antifouling Agent Zinc Pyrithione (ZnPT) in *Mytilus galloprovincialis* Lamark (Mediterranean Mussels) Tissues. *Chem. Ecol.* **2022**, *38*, 356–373. [CrossRef]
69. López-Galindo, C.; Vargas-Chacoff, L.; Nebot, E.; Casanueva, J.F.; Rubio, D.; Mancera, J.M.; Solé, M. Sublethal Responses of the Common Mussel (*Mytilus galloprovincialis*) Exposed to Sodium Hypochlorite and Mexel®432 Used as Antifoulants. *Ecotoxicol. Environ. Saf.* **2010**, *73*, 825–834. [CrossRef]
70. Bouzidi, I.; Sellami, B.; Mezni, A.; Hedfi, A.; Almalki, M.; Pacioglu, O.; Boufahja, F.; Mougin, K.; Beyrem, H. Nanoparticles Influence the Herbicide Diuron Mediated Toxicity on Marine Mussel *Mytilus galloprovincialis*: Single and Mixture Exposure Study. *Mater. Res. Express* **2021**, *8*, 085005. [CrossRef]
71. Coyne, K.J.; Qin, X.-X.; Waite, J.H. Extensible Collagen in Mussel Byssus: A Natural Block Copolymer. *Science* **1997**, *277*, 1830–1832. [CrossRef]
72. Huang, X.; Cui, H.; Duan, W. Ecotoxicity of Chlorpyrifos to Aquatic Organisms: A Review. *Ecotoxicol. Environ. Saf.* **2020**, *200*, 110731. [CrossRef]
73. Naime, A.A.; Lopes, M.W.; Colle, D.; Dafré, A.L.; Suñol, C.; da Rocha, J.B.T.; Aschner, M.; Leal, R.B.; Farina, M. Glutathione in Chlorpyrifos-and Chlorpyrifos-Oxon-Induced Toxicity: A Comparative Study Focused on Non-Cholinergic Toxicity in HT22 Cells. *Neurotox. Res.* **2020**, *38*, 603–610. [CrossRef] [PubMed]
74. Reuter, S.; Gupta, S.C.; Chaturvedi, M.M.; Aggarwal, B.B. Oxidative Stress, Inflammation, and Cancer: How Are They Linked? *Free. Radic. Biol. Med.* **2010**, *49*, 1603–1616. [CrossRef] [PubMed]

75. Gómez-Mendikute, A.; Cajaraville, M.P. Comparative Effects of Cadmium, Copper, Paraquat and Benzo[a]Pyrene on the Actin Cytoskeleton and Production of Reactive Oxygen Species (ROS) in Mussel Haemocytes. *Toxicol. Vitr.* **2003**, *17*, 539–546. [CrossRef]
76. Doose, C.A.; Ranke, J.; Stock, F.; Bottin-Weber, U.; Jastorff, B. Structure–Activity Relationships of Pyrithiones – IPC-81 Toxicity Tests with the Antifouling Biocide Zinc Pyrithione and Structural Analogs. *Green Chem.* **2004**, *6*, 259–266. [CrossRef]
77. Bellas, J. Toxicity Assessment of the Antifouling Compound Zinc Pyrithione Using Early Developmental Stages of the Ascidian *Ciona intestinalis*. *Biofouling* **2005**, *21*, 289–296. [CrossRef] [PubMed]
78. Marcheselli, M.; Rustichelli, C.; Mauri, M. Novel Antifouling Agent Zinc Pyrithione: Determination, Acute Toxicity, and Bioaccumulation in Marine Mussels (*Mytilus galloprovincialis*). *Environ. Toxicol. Chem.* **2010**, *29*, 2583–2592. [CrossRef] [PubMed]
79. Pan, S.-Y.; Snyder, S.W.; Packman, A.I.; Lin, Y.J.; Chiang, P.-C. Cooling Water Use in Thermoelectric Power Generation and Its Associated Challenges for Addressing Water-Energy Nexus. *Water-Energy Nexus* **2018**, *1*, 26–41. [CrossRef]
80. Bacha, O.; Khazri, A.; Mezni, A.; Mezni, A.; Touaylia, S. Protective Effect of the *Spirulina platensis* against Toxicity Induced by Diuron Exposure in *Mytilus galloprovincialis*. *Int. J. Phytoremediat.* **2021**, *24*, 778–786. [CrossRef]
81. McVay, I.R.; Maher, W.A.; Krikowa, F.; Ubrhien, R. Metal Concentrations in Waters, Sediments and Biota of the Far South-East Coast of New South Wales, Australia, with an Emphasis on Sn, Cu and Zn Used as Marine Antifoulant Agents. *Environ. Geochem. Health* **2019**, *41*, 1351–1367. [CrossRef]
82. Pan, J.; Xie, Q.; Chiang, H.; Peng, Q.; Qian, P.-Y.; Ma, C.; Zhang, G. "From the Nature for the Nature": An Eco-Friendly Antifouling Coating Consisting of Poly(Lactic Acid)-Based Polyurethane and Natural Antifoulant. *ACS Sustain. Chem. Eng.* **2020**, *8*, 1671–1678. [CrossRef]
83. Pereira, F.; Almeida, J.R.; Paulino, M.; Grilo, I.R.; Macedo, H.; Cunha, I.; Sobral, R.G.; Vasconcelos, V.; Gaudêncio, S.P. Antifouling Napyradiomycins from Marine-Derived Actinomycetes Streptomyces Aculeolatus. *Mar. Drugs* **2020**, *18*, 63. [CrossRef] [PubMed]
84. Nendza, M. Hazard Assessment of Silicone Oils (Polydimethylsiloxanes, PDMS) Used in Antifouling-/Foul-Release-Products in the Marine Environment. *Mar. Pollut. Bull.* **2007**, *54*, 1190–1196. [CrossRef] [PubMed]
85. Miller, R.J.; Adeleye, A.S.; Page, H.M.; Kui, L.; Lenihan, H.S.; Keller, A.A. Nano and Traditional Copper and Zinc Antifouling Coatings: Metal Release and Impact on Marine Sessile Invertebrate Communities. *J. Nanopart. Res.* **2020**, *22*, 129. [CrossRef]
86. Qian, P.-Y.; Xu, Y.; Fusetani, N. Natural Products as Antifouling Compounds: Recent Progress and Future Perspectives. *Biofouling* **2009**, *26*, 223–234. [CrossRef]
87. Dobretsov, S.; Abed, R.M.M.; Teplitski, M. Mini-Review: Inhibition of Biofouling by Marine Microorganisms. *Biofouling* **2013**, *29*, 423–441. [CrossRef]
88. Joana Gil-Chávez, G.; Villa, J.A.; Fernando Ayala-Zavala, J.; Basilio Heredia, J.; Sepulveda, D.; Yahia, E.M.; González-Aguilar, G.A. Technologies for Extraction and Production of Bioactive Compounds to Be Used as Nutraceuticals and Food Ingredients: An Overview: Production of Nutraceutical Compounds: A Review. *Compr. Rev. Food Sci. Food Saf.* **2013**, *12*, 5–23. [CrossRef]
89. Lam, K.S. Discovery of Novel Metabolites from Marine Actinomycetes. *Curr. Opin. Microbiol.* **2006**, *9*, 245–251. [CrossRef]
90. Yang, L.H.; Miao, L.; Lee, O.O.; Li, X.; Xiong, H.; Pang, K.-L.; Vrijmoed, L.; Qian, P.-Y. Effect of Culture Conditions on Antifouling Compound Production of a Sponge-Associated Fungus. *Appl. Microbiol. Biotechnol.* **2007**, *74*, 1221–1231. [CrossRef]
91. Soria-Mercado, I.E.; Prieto-Davo, A.; Jensen, P.R.; Fenical, W. Antibiotic Terpenoid Chloro-Dihydroquinones from a New Marine Actinomycete. *J. Nat. Prod.* **2005**, *68*, 904–910. [CrossRef]
92. Cheng, Y.-B.; Jensen, P.R.; Fenical, W. Cytotoxic and Antimicrobial Napyradiomycins from Two Marine-Derived *Streptomyces* Strains: Napyradiomycins from Two Marine-Derived *Streptomyces* Strains. *Eur. J. Org. Chem.* **2013**, *2013*, 3751–3757. [CrossRef]
93. Wu, Z.; Li, S.; Li, J.; Chen, Y.; Saurav, K.; Zhang, Q.; Zhang, H.; Zhang, W.; Zhang, W.; Zhang, S.; et al. Antibacterial and Cytotoxic New Napyradiomycins from the Marine-Derived Streptomyces Sp. SCSIO 10428. *Mar. Drugs* **2013**, *11*, 2113–2125. [CrossRef] [PubMed]
94. Farnaes, L.; Coufal, N.G.; Kauffman, C.A.; Rheingold, A.L.; DiPasquale, A.G.; Jensen, P.R.; Fenical, W. Napyradiomycin Derivatives, Produced by a Marine-Derived Actinomycete, Illustrate Cytotoxicity by Induction of Apoptosis. *J. Nat. Prod.* **2014**, *77*, 15–21. [CrossRef] [PubMed]
95. Che, Q.; Zhu, T.; Qi, X.; Mándi, A.; Kurtán, T.; Mo, X.; Li, J.; Gu, Q.; Li, D. Hybrid Isoprenoids from a Reeds Rhizosphere Soil Derived Actinomycete *Streptomyces* Sp. CHQ-64. *Org. Lett.* **2012**, *14*, 3438–3441. [CrossRef] [PubMed]
96. Lacret, R.; Pérez-Victoria, I.; Oves-Costales, D.; Cruz, M.; Domingo, E.; Martín, J.; Díaz, C.; Vicente, F.; Genilloud, O.; Reyes, F. MDN-0170, a New Napyradiomycin from Streptomyces Sp. Strain CA-271078. *Mar. Drugs* **2016**, *14*, 188. [CrossRef]
97. Ashton, G.V.; Davidson, I.C.; Geller, J.; Ruiz, G.M. Disentangling the Biogeography of Ship Biofouling: Barnacles in the Northeast Pacific: Biogeography of Ship Biofouling Barnacles. *Glob. Ecol. Biogeogr.* **2016**, *25*, 739–750. [CrossRef]
98. Callow, J.A.; Callow, M.E. Trends in the Development of Environmentally Friendly Fouling-Resistant Marine Coatings. *Nat. Commun.* **2011**, *2*, 244. [CrossRef]
99. Pettengill, J.B.; Wendt, D.E.; Schug, M.D.; Hadfield, M.G. Biofouling Likely Serves as a Major Mode of Dispersal for the Polychaete Tubeworm *Hydroides Elegans* as Inferred from Microsatellite Loci. *Biofouling* **2007**, *23*, 161–169. [CrossRef]
100. Piola, R.F.; Johnston, E.L. The Potential for Translocation of Marine Species via Small-Scale Disruptions to Antifouling Surfaces. *Biofouling* **2008**, *24*, 145–155. [CrossRef]
101. Ware, C.; Berge, J.; Sundet, J.H.; Kirkpatrick, J.B.; Coutts, A.D.M.; Jelmert, A.; Olsen, S.M.; Floerl, O.; Wisz, M.S.; Alsos, I.G. Climate Change, Non-indigenous Species and Shipping: Assessing the Risk of Species Introduction to a High- A Rctic Archipelago. *Divers. Distrib.* **2014**, *20*, 10–19. [CrossRef]

102. Yamaguchi, T.; Prabowo, R.E.; Ohshiro, Y.; Shimono, T.; Jones, D.; Kawai, H.; Otani, M.; Oshino, A.; Inagawa, S.; Akaya, T.; et al. The Introduction to Japan of the Titan Barnacle, *Megabalanus coccopoma* (Darwin, 1854) (Cirripedia: Balanomorpha) and the Role of Shipping in Its Translocation. *Biofouling* **2009**, *25*, 325–333. [CrossRef]
103. Cattò, C.; Dell'Orto, S.; Villa, F.; Villa, S.; Gelain, A.; Vitali, A.; Marzano, V.; Baroni, S.; Forlani, F.; Cappitelli, F. Unravelling the Structural and Molecular Basis Responsible for the Anti-Biofilm Activity of Zosteric Acid. *PLoS ONE* **2015**, *10*, e0131519. [CrossRef] [PubMed]
104. Newby, B.Z.; Cutright, T.; Barrios, C.A.; Xu, Q. Zosteric Acid—An Effective Antifoulant for Reducing Fresh Water Bacterial Attachment on Coatings. *J. Coat. Technol. Res.* **2006**, *3*, 69–76. [CrossRef]
105. Polo, A.; Foladori, P.; Ponti, B.; Bettinetti, R.; Gambino, M.; Villa, F.; Cappitelli, F. Evaluation of Zosteric Acid for Mitigating Biofilm Formation of Pseudomonas Putida Isolated from a Membrane Bioreactor System. *Int. J. Mol. Sci.* **2014**, *15*, 9497–9518. [CrossRef] [PubMed]
106. Todd, J.S.; Zimmerman, R.C.; Crews, P.; Alberte, R.S. The Antifouling Activity of Natural and Synthetic Phenol Acid Sulphate Esters. *Phytochemistry* **1993**, *34*, 401–404. [CrossRef]
107. Xu, Q.; Barrios, C.A.; Cutright, T.; Zhang Newby, B. Evaluation of Toxicity of Capsaicin and Zosteric Acid and Their Potential Application as Antifoulants. *Environ. Toxicol.* **2005**, *20*, 467–474. [CrossRef]
108. Villa, F.; Remelli, W.; Forlani, F.; Vitali, A.; Cappitelli, F. Altered Expression Level of Escherichia Coli Proteins in Response to Treatment with the Antifouling Agent Zosteric Acid Sodium Salt: Escherichia Coli Response to Zosteric Acid Sodium Salt. *Environ. Microbiol.* **2012**, *14*, 1753–1761. [CrossRef]
109. Villa, F.; Albanese, D.; Giussani, B.; Stewart, P.S.; Daffonchio, D.; Cappitelli, F. Hindering Biofilm Formation with Zosteric Acid. *Biofouling* **2010**, *26*, 739–752. [CrossRef]
110. Ram, J.L.; Purohit, S.; Newby, B.Z.; Cutright, T.J. Evaluation of the Natural Product Antifoulant, Zosteric Acid, for Preventing the Attachment of Quagga Mussels – a Preliminary Study. *Nat. Prod. Res.* **2012**, *26*, 580–584. [CrossRef]
111. Kurth, C.; Cavas, L.; Pohnert, G. Sulfation Mediates Activity of Zosteric Acid against Biofilm Formation. *Biofouling* **2015**, *31*, 253–263. [CrossRef]
112. Faimali, M.; Falugi, C.; Gallus, L.; Piazza, V.; Tagliafierro, G. Involvement of Acetyl Choline in Settlement of Balanus Amphitrite. *Biofouling* **2003**, *19*, 213–220. [CrossRef]
113. Mansueto, V.; Cangialosi, M.V.; Arukwe, A. Acetylcholinesterase Activity in Juvenile *Ciona intestinalis* (Ascidiacea, Urochordata) after Exposure to Tributyltin. *Caryologia* **2012**, *65*, 18–26. [CrossRef]
114. Hellio, C.; Bourgougnon, N.; Gal, Y.L. Phenoloxidase (E.C. 1.14.18.1) from the Byssus Gland of *Mytilus edulis:* Purification, Partial Characterization and Application for Screening Products with Potential Antifouling Activities. *Biofouling* **2000**, *16*, 235–244. [CrossRef]
115. Hadfield, M.G.; Paul, V.J. Natural Chemical Cues for Settlement and Metamorphosis of Marine Invertebrate Larvae. *Mar. Chem. Ecol.* **2001**, *13*, 431–461.
116. Trepos, R.; Cervin, G.; Pile, C.; Pavia, H.; Hellio, C.; Svenson, J. Evaluation of Cationic Micropeptides Derived from the Innate Immune System as Inhibitors of Marine Biofouling. *Biofouling* **2015**, *31*, 393–403. [CrossRef] [PubMed]
117. Selwood, A.I.; Wilkins, A.L.; Munday, R.; Shi, F.; Rhodes, L.L.; Holland, P.T. Portimine: A Bioactive Metabolite from the Benthic Dinoflagellate Vulcanodinium Rugosum. *Tetrahedron Lett.* **2013**, *54*, 4705–4707. [CrossRef]
118. Otero, A.; Chapela, M.-J.; Atanassova, M.; Vieites, J.M.; Cabado, A.G. Cyclic Imines: Chemistry and Mechanism of Action: A Review. *Chem. Res. Toxicol.* **2011**, *24*, 1817–1829. [CrossRef]
119. Cuddihy, S.L.; Drake, S.; Harwood, D.T.; Selwood, A.I.; McNabb, P.S.; Hampton, M.B. The Marine Cytotoxin Portimine Is a Potent and Selective Inducer of Apoptosis. *Apoptosis* **2016**, *21*, 1447–1452. [CrossRef]
120. Sasso, S.; Pohnert, G.; Lohr, M.; Mittag, M.; Hertweck, C. Microalgae in the Postgenomic Era: A Blooming Reservoir for New Natural Products. *FEMS Microbiol. Rev.* **2012**, *36*, 761–785. [CrossRef]
121. Brooke, D.G.; Cervin, G.; Champeau, O.; Harwood, D.T.; Pavia, H.; Selwood, A.I.; Svenson, J.; Tremblay, L.A.; Cahill, P.L. Antifouling Activity of Portimine, Select Semisynthetic Analogues, and Other Microalga-Derived Spirocyclic Imines. *Biofouling* **2018**, *34*, 950–961. [CrossRef]
122. Perry, K.; Lynn, J. Detecting Physiological and Pesticide-Induced Apoptosis in Early Developmental Stages of Invasive Bivalves. *Hydrobiologia* **2009**, *628*, 153–164. [CrossRef]
123. Shikuma, N.J.; Antoshechkin, I.; Medeiros, J.M.; Pilhofer, M.; Newman, D.K. Stepwise Metamorphosis of the Tubeworm *Hydroides Elegans* Is Mediated by a Bacterial Inducer and MAPK Signaling. *Proc. Natl. Acad. Sci. USA* **2016**, *113*, 10097–10102. [CrossRef] [PubMed]
124. Chambon, J.-P.; Soule, J.; Pomies, P.; Fort, P.; Sahuquet, A.; Alexandre, D.; Mangeat, P.-H.; Baghdiguian, S. Tail Regression in *Ciona intestinalis* (Prochordate) Involves a Caspase-Dependent Apoptosis Event Associated with ERK Activation. *Development* **2002**, *129*, 3105–3114. [CrossRef] [PubMed]
125. Chen, Z.-F.; Matsumura, K.; Wang, H.; Arellano, S.M.; Yan, X.; Alam, I.; Archer, J.A.C.; Bajic, V.B.; Qian, P.-Y. Toward an Understanding of the Molecular Mechanisms of Barnacle Larval Settlement: A Comparative Transcriptomic Approach. *PLoS ONE* **2011**, *6*, e22913. [CrossRef] [PubMed]

Journal of
Marine Science and Engineering

MDPI

Article

Experimental Assessment of a Conducting Polymer (PEDOT) and Microbial Biofilms as Deterrents and Facilitators of Macro-Biofouling: Larval Settlement of the Barnacle *Notobalanus flosculus* (Darwin, 1854) from Central Chile

Simone Baldanzi [1,2,3,*], Ignacio T. Vargas [2,4], Francisco Armijo [2,5], Miriam Fernández [2,6] and Sergio A. Navarrete [2,6,7]

1 Facultad de Ciencia del Mar y de Recursos Naturales, Universidad de Valparaíso, Av. Borgoño 16344, Viña del Mar 2520000, Chile
2 Marine Energy Research and Innovation Center (MERIC), Las Condes 7550000, Chile; itvargas@ing.puc.cl (I.T.V.); jarmijom@uc.cl (F.A.); mfernandez@bio.puc.cl (M.F.); snavarrete@bio.puc.cl (S.A.N.)
3 Centro de Observación Marino para Estudios de Riesgos del Ambiente Costero (COSTA-R), Universidad de Valparaíso, Valparaíso 2520000, Chile
4 Departamento de Ingeniería Hidráulica y Ambiental, Pontificia Universidad Católica de Chile, Santiago 7820436, Chile
5 Facultad de Química y de Farmacia, Pontificia Universidad Católica de Chile, Av. Vicuña Mackenna 4860, Macul 7820436, Chile
6 Estación Costera de Investigaciones Marinas (ECIM), Pontificia Universidad Católica de Chile, Osvaldo Marin 1672, Las Cruces 2690000, Chile
7 Center for Applied Ecology and Sustainability (CAPES), Pontificia Universidad Católica de Chile, Av. Bernardo O'Higgins 340, Santiago 2690000, Chile
* Correspondence: simone.baldanzi@uv.cl

Citation: Baldanzi, S.; Vargas, I.T.; Armijo, F.; Fernández, M.; Navarrete, S.A. Experimental Assessment of a Conducting Polymer (PEDOT) and Microbial Biofilms as Deterrents and Facilitators of Macro-Biofouling: Larval Settlement of the Barnacle *Notobalanus flosculus* (Darwin, 1854) from Central Chile. *J. Mar. Sci. Eng.* **2021**, *9*, 82. https://doi.org/10.3390/jmse9010082

Received: 7 October 2020
Accepted: 6 January 2021
Published: 14 January 2021

Abstract: Maritime enterprises have long sought solutions to reduce the negative consequences of the settlement and growth of marine biofouling (micro- and macro-organisms) on virtually all surfaces and materials deployed at sea. The development of biofouling control strategies requires solutions that are cost-effective and environmentally friendly. Polymer-based coatings, such as the poly (3,4-ethylenedioxythiophene) (PEDOT) and its potential applications, have blossomed over the last decade thanks to their low cost, nontoxicity, and high versatility. Here, using multiple-choice larval settlement experiments, we assessed the efficacy of PEDOT against the balanoid barnacle *Notobalanus flosculus* one of the most common biofouling species in Southeastern Pacific shores, and compared results against a commercially available antifouling (AF) coating, and biofilms at different stages of succession (1, 2, 4 and 8 weeks). We show that larval settlement on PEDOT-coated surfaces was similar to the settlement on AF-coated surfaces, while larvae settled abundantly on roughened acrylic and on early-to-intermediate stages of biofilm (one to four weeks old). These results are promising and suggest that PEDOT is a good candidate for fouling-resistant coating for specific applications at sea. Further studies to improve our understanding of the mechanisms of barnacle larval deterrence, as well as exposure to field conditions, are encouraged.

Keywords: biofouling; environmental protection; coastal waters; larval settlement-biofilm interactions; *Notobalanus flosculus*; Chile

1. Introduction

Marine biofouling refers to the settlement and growth of organisms on artificial structures deployed in the ocean and is an age-old problem that has been a target for control since the beginning of human maritime enterprises [1]. In the marine environment biofouling includes the biofilms of microorganisms that rapidly settle on virtually all materials as well as large bodied invertebrates and macroalgae and can attain large biomass per unit area [2]. Biofilms create diverse, extensive challenges for industries dedicated to the development of new technologies, such as marine renewable energy (MRE, [3,4]) and great economic losses to the navigation and aquaculture sectors [5,6]. The transformation and rapid expansion of the aquaculture and MRE sectors [7–9] must cope with the challenges imposed by marine biofouling [10].

Several factors act simultaneously to define biofouling risks: (a) the characteristics of local biofouling species [11], (b) the properties of the material, such as surface topography (e.g., roughness), wettability and colour [12], (c) the type of application of the material (small rigid sensor, large moving energy converter) (d) the local environmental conditions, such as temperature, local productivity, hydrodynamics and oxygen conditions [13] and (e) the biotic interactions among species during different phases of the ecological succession, such as the initial biofilm–larval interactions [4,14,15]. It is, therefore, fundamental to understand the complex interactions, often of inter-kingdom origins [16] arising among different organisms involved in the biofouling process and between those organisms and abiotic forces [14]. In fact, the existence of a wide range of fouling species demands a variety of antifouling coating strategies, which can be summarized as follows [17]: (i) fouling-resistant coatings that prevent adhesion of biofilm and/or algae, (ii) fouling-release coatings, which allow an initial weak foulant-surface adhesion, followed by an easy removal by the application of mechanical forces, and (iii) fouling-degrading coatings, which degrade adsorbed organic material and/or kill biofilm by the action of bactericides. While the application of fouling-degrading coatings, (e.g., copper and zinc-based anti-fouling coatings) and recurrent maintenance has been the standard approach in the industry, this is either impractical for many applications (e.g., sensors, soft flexible materials, etc.) or questioned for environmental consequences (e.g., the inevitable release of chemical pollutants to the environment), making this approach highly questionable and subject to increasing restrictions [4,18]. Thus, the race is on to find alternative, environmentally friendly coatings and strategies to reduce the "biofouling problem", improve sustainability of the industries and promote the development of new marine technologies and instrumentation [19].

Fouling-resistant and fouling-release strategies using polymer-based coatings [17] have blossomed over the last decade, thanks to advances in medical science [20,21] and nanotechnologies [22]. These coatings are low cost, nontoxic, biocompatible, highly versatile, and their functionalities and architectures can be easily modified, allowing interfacial adjustments of the antifouling properties [17]. Among polymer coatings, the conducting polymer poly (3,4-ethylenedioxythiophene) (PEDOT) and its composites have been used as antifouling and anticlotting coatings in medical applications [20,23,24] and biocorrosion in marine environments [25–27]. An important factor associated to its application are the experimental conditions necessary to obtain the PEDOT on different surfaces, which are related to its adherence, conductivity, and stability. Usually three electrochemical methods are used to obtain PEDOT, which are cyclic voltammetry (application of a linearly variable voltage), chronopotentiometry (constant current), and chronoamperometry (constant potential) [28]. In addition, the effects of other experimental variables have been observed (e.g., solvent, starting unit, monomer and supporting electrolyte kind and concentration, temperature effect) on the mechanism and process of electropolymerization [29,30]. It is therefore imperative to specifically study each system to optimize the electropolymerization conditions according to the intended use for the conducting polymer electrodeposited on the working electrode. Nonetheless, PEDOT seems to be particularly promising to prevent marine fouling because it prevents adsorption of non-specific proteins associated with a biofilm [31,32]. Recently, laboratory experiments showed that PEDOT can delay bacterial formations by about 35 days in submerged PEDOT-coated coupons compared to uncoated ones [27]. To our knowledge, there is no evidence of the efficiency of conducting polymers in general and PEDOT in particular, on preventing direct colonization by macrofouling, the most harmful components of biofouling for many applications. Here we provide an experimental assessment to start filling this important information gap for the development of environmentally friendly antifouling strategies in the maritime industry. Moreover, little information is available on strategies that prevent macrofouling colonization of submerged materials in general, in regions characterised by high productivity, such as the Humboldt Upwelling Ecosystem, in the Southern Pacific [33].

Several studies have shown that the settlement and fast growth of fouling species in the Humboldt Upwelling Ecosystem can reach very high biomass accumulation rates [34–36], among the highest reported in the world. Understanding biofouling dynamics and mechanisms to prevent it in this productive region is a major concern for the success of the MRE and aquaculture industries [36]. The ecological succession at wave exposed sites was characterised by initial settlement of microbial biofilms that were rapidly colonized by fast growing invertebrate larvae, leading to deterministic final stages of succession dominated by barnacles and tunicates of large biomass [36]. Further experiments found that diverse materials, deployed above and below the thermocline, were colonized indistinctively by late successional species (barnacles and tunicates of large biomass) and that standard, copper-based antifouling paint was an effective deterrent after seven months of exposure [36]. Here we advance our understanding on biofouling dynamics by testing the antifouling efficiency of PEDOT with settling larvae of barnacle *Notobalanus flosculus*, an intertidal/shallow subtidal barnacle [37] found in abundance in biofouling communities [36,38]. Using multiple choice experiments, we tested PEDOT efficiency as a fouling-resistant coating against barnacle settlement using both a positive control substratum (roughened acrylic plates) and a negative control (self-polishing Cu_2O-based antifouling paint). Furthermore, since larval-biofilm interactions are complex and some barnacle species actively select substrate with different biofilm composition [14,15] (or even without bacterial deposition, Roberts et al. 1991), we evaluated settlement preferences for different stages of biofilm (i.e., biofilms with different age of deposition).

2. Materials and Methods

2.1. Study Species and Sampling Area

We chose the common balanoid barnacle *N. flosculus* [39] as test model for our experiments because the species is one of the most frequent barnacle present in the biofouling of wave-exposed habitats of central Chile [36]. The species is found along the entire coast of central-northern Chile, and it can attain high abundances from the low intertidal rocky shore to shallow subtidal environments [37]. Although it can be displaced in late stages of ecological succession by the larger barnacle *Austromegabalanus psittacus* and the tunicate *Pyura chilensis* [36], larval culture and availability make it an ideal laboratory test model species for south-eastern Pacific waters. Moreover, understanding the interaction between anti-fouling materials and barnacles can shed light on other important barnacle fouling species. Specimens of *N. flosculus* were collected at Las Cruces, Central Chile (S33.5. W71.6) from the rocky intertidal shores during Summer 2019. This site was chosen for three reasons: (i) availability and proximity to the facilities of the larval laboratory of ECIM (Pontificia Universidad Católica de Chile), (ii) existence of biofouling pilot experiments on *N. flosculus* at the same location, and (iii) the existence of Wave Energy Converter projects in the same area (https://lascrucesem.cl).

2.2. Settlement Substrates

In the laboratory experiments described below, we tested five different settlement substrates, including biofilms. First, since many field studies have used roughened acrylic plates to monitor barnacle settlement [34–36,40], we used this material as a positive reference for larval settlement to contrast settlement on PEDOT. We used an AF paint to assess whether it inhibits barnacle settlement. Most commercially available antifouling paints use copper (cuprous oxide, Cu_2O) as biocide. To prevent excessive Cu_2O leaching within the experimental aquaria, we opted for a coating with Self-Polishing Copolymer technology (SPC), which has a controlled and stable hydrolysis release over time (SeaVoyage CDP100, Sherwin Williams; https://www.sherwin.cl/industrial/marino/). The paint was applied on roughened acrylic plates following manufacturer instructions. Although the product labeling does not specify the presence of copper, the coating does have copper as active compound and the manufacturer ensures that encapsulation of Cu_2O guarantees long-term retention in the matrix with negligible leaching (below detection; see manufacturer instructions attached as supplementary material). Considering the short duration of our experiments and water movement, leaching could not be detected. In any case, AF paint plates were deployed in the numbers in all aquaria, thus preventing any potentially confounding effects. Both roughened acrylic plates and AF paint on which larvae are known to settle and not settle, respectively, can be considered as the positive and negative references for our broad hypothesis (Ho: Larval settlement is similar among the different materials). As organic coating, we used PEDOT, deposited on stainless steel (SS) plates (see details below). Finally, two different biofilm stages were used, (i) early and (ii) late biofilm stages (see details below), grown on roughened acrylic plates.

2.3. Preparation of PEDOT Coated Stainless Steel

The electrochemical synthesis of PEDOT was conducted in a conventional three-electrode cell following the procedures described by Aguirre and colleagues [27], using a cylindrical mesh of AISI 316L SS (surface area 42 cm^2) as counter electrode, Ag/AgCl (KCl saturated) as reference electrode, and the AISI 304 SS coupons as working electrodes, exposing a surface area of $\approx$ 20 cm^2. The conducting polymer was electrodeposited from an organic solution containing 0.1 mol L^{-1} monomers of 3,4-ethylenedioxythiophene (EDOT) and 0.10 mol L^{-1} $LiClO_4$ in acetonitrile (CH_3CN). To prevent the oxidation of the monomers prior to electropolymerization, the solution was de-aerated by purging with nitrogen gas. Deposition of PEDOT on the working electrode was achieved by cyclic voltammetry, applying 10 potentiodynamic cycles at a scan rate of 0.05 $V{\cdot}s^{-1}$ in a potential window between -0.7 to 1.3 V. These experiments were carried out using an OrigaLys potentiostat (OrigaFlex OGF500).

2.4. Surface Roughness and Wettability of PEDOT, Acrylic and AF Coating

We measured surface roughness (SR) and wettability (W) for the PEDOT, acrylic and AF plates. The SR was measured using a BioLogic Optical Surface Profiler (OSP470) along two perpendicular transects within a 4 $\times$ 2 cm area on a randomly chosen plate of each substrate. All measurements were then expressed as RMS (mean square root) and given in μm^2. Surface wettability was measured by determining the water contact angle (WCA) following ASTM D7334-08 (ASTM International, 2013), using water-type II reagent (distilled) after ASTM D1193-06 (ASTM International, 2011). The W value was used to define hydrophobic and hydrophilic properties of the substrate from critical surface tension theory, assigning hydrophilic property to WCA < 45°, hydrophobic property to WCA > 90°, and intermediate property to 45° < WCA < 90°. Finally, the quality of polymer adhesion was tested by means of a tape-and-peel test, conducted according to ASTM D3359-09ε2 standard (method A, ASTM International).

2.5. Biofilm Deposition and Characterization

The biofilms were grown on 40 finely roughened acrylic plates using emery paper (grit 240) and exposed in aquaria (50 $\times$ 30 $\times$ 17 cm) to the running sea water system of ECIM, which pumps seawater from ca. 1.5 m depth and filtered at 1 mm. Prior to exposure to running seawater, plates were left in 95% ethanol overnight in the dark. Acrylic plates were then left undisturbed in the aquaria under natural sunlight, photoperiod, and seawater temperature, allowing biofilm formation. Different biofilm ages were used in the two trials described below, early, and late, since the exact age of the biofilm varied between experiments as mandated by the availability of competent barnacle larvae (see below).

At the end of the biofilm depositions and before deployment in experimental arenas, each plate was placed in a clean Petri dish and photographed with a digital camera (Nikon D80 set at f/2, 1/60 s exposure, ISO 400, 50 mm focal length, no flash, and recording in RAW format). Lighting was controlled using overhead lights. The photo images (Figure 1A) were then analysed using the software Image-J2 [41] to calculate the % of biofilm cover per plate using the standard technique described by Otsu [42]. Briefly, this technique establishes a threshold to delineate foreground (biofilm) from background (clean acrylic surface) by reducing a grayscale image (8-bit image) to a binary image and assuming pixel values form a bimodal histogram. Threshold value were chosen algorithmically using the Otsu's method of thresholding, which is one of the most referenced algorithmic techniques. Once photographed, biofilms were kept in aquaria with still FSSW in the dark avoiding successive biomass growth for a maximum of three days.

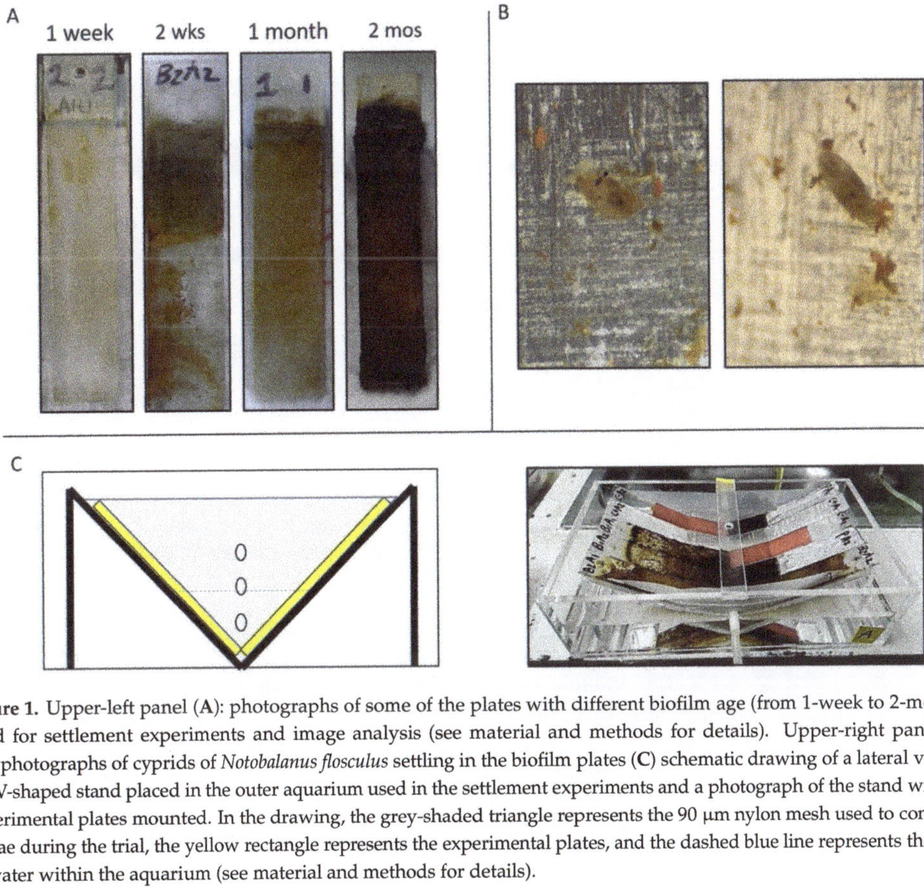

Figure 1. Upper-left panel (**A**): photographs of some of the plates with different biofilm age (from 1-week to 2-months) used for settlement experiments and image analysis (see material and methods for details). Upper-right panel (**B**): two photographs of cyprids of *Notobalanus flosculus* settling in the biofilm plates (**C**) schematic drawing of a lateral view of the V-shaped stand placed in the outer aquarium used in the settlement experiments and a photograph of the stand with the experimental plates mounted. In the drawing, the grey-shaded triangle represents the 90 μm nylon mesh used to constrain larvae during the trial, the yellow rectangle represents the experimental plates, and the dashed blue line represents the level of water within the aquarium (see material and methods for details).

2.6. Barnacle Collection, Spawning and Larval Rearing

Animals were collected manually by removing approximately 10 small boulders colonized by adult *N. flosculus* from the low intertidal zone. In the laboratory rocks were inspected under a stereoscope and other species were carefully removed with tweezers and scalpels. External shells of *N. flosculus* and naked rocks were gently brushed cleaned using a diluted solution (1:50) of ethanol to reduce the microbial load (i.e., biofilm) and then rinsed carefully using pre-filtered (1 μm), UV sterilized seawater (FSSW). Rocks were cut into units of 10 cm long max with 20 to 30 individual barnacles and randomly assigned to five different aquaria (20 L each) connected to a flow-through system provided with serial polyethylene filtering—(from 20 to 1 μm) and UV light sterilizing apparatus (FSSW). The water leaving each aquarium was collected in a 15 × 15 × 20 cm (W × L × H) plastic boxes equipped with 100 μm mesh for larval retention. The entire system (aquaria with adult individuals and larval retention filters) was placed inside a temperature-controlled chamber (hereafter incubation chamber) which allowed full control of light intensity, photoperiod (14:10 h dark:light) and temperature (15 °C, simulating the temperature at the study site when animals were collected) to stimulate spawning (see below). Barnacles were fed ad libitum with a daily dose of the microalgae *Skeletonema* sp. of approximately 2×10^6 cells mL^{-1} and recently hatched nauplius of *Artemia salina*. Spawning in the laboratory facilities was achieved using a light-stress technique. After a week of acclimation to laboratory conditions in the incubation chamber, animals were kept completely in the dark and starved for 48 h. After this period, light was applied at full intensity (6 led bulbs of 2520 lumens each) for the following 24 h, which led to the release of larvae in delayed batches over a period of about a week. Hatched larvae were collected daily and placed in pre-autoclaved 10 × 10 × 2.5 cm glass containers half-filled with 150 mL of FSSW. The number of total larvae per container depended on the number of larvae released per batch, with a maximum concentration of 100 larvae per container. Larvae were kept separated by day of collection to rear similar cohorts within each container. Following Jonsson and collaborators [43] we identified 6 nauplius stages over the larval developmental time starting from nauplius 2–3 until cypris stage which is competent to settle (Figure 1B). Larvae were fed with a microalgal mix (*Skeletonema* sp. and *Isochrysis* sp.) given at a concentration of 1 to 2×10^6 cells mL^{-1}. The FSSW of the glass containers was changed daily.

2.7. Settlement Choice Experiments

2.7.1. Experimental Set-Up

Settlement choice experiments consisted in offering larvae the five selected materials and biofilms conditions. Experiments were performed using a flow-through, acrylic aquarium modified from [44]. Briefly, aquarium consisted of an inner V-shaped stand in which experimental plates could be placed in a 120° angle facing each other (Figure 1C). To the sides of each stands a 90 μm nylon mesh, which was glued allowing FSSW to flow through while retaining larvae. The design of the experimental chambers ensured that the experimental plates were the only substrate available to settle (settlement on nylon mesh was zero; see also [43]). The stand was placed inside an aquarium fed with FSSW from a header tank. Water temperature during the experiments was kept constant at 15 °C by placing the header tank and the experimental aquarium within an incubation chamber (same used above for the nauplius rearing).

Each of the five substrates (treatment) was offered on two 10 × 2 cm plates that covered the entire available surface the V-shape stands. A total of 10 plates (five substrates) were therefore available in each aquarium. Using a table of random numbers, where each number represented a plate, we placed one plate of each substrate on the left and the other on the right-hand side of the V-shape stand, in this manner preventing potential bias due to unforeseen gradients. Before analyses, the number of settled larvae on the two plates was added within each replicate aquarium.

2.7.2. Laboratory Trials

As larvae developed and turned into the cyprid stage, they were placed inside separate glass containers (same as above) filled with 150 mL of FSSW and left in the dark at 6 °C for a maximum of five days [44] until a total 360 individual cyprids were obtained (60 from a first batch and 300 from a second batch).

Two separated settlement experimental trials were performed, using the two separate batches of cyprids, so the only difference between the two trials was the age of the biofilms, which cannot be controlled with precision. In the first trial (named experiment 1), we had 2-month old biofilm and 2-week old biofilm, besides the PEDOT, AF coating and acrylic plates. A total of 60 cyprids were used in experiment 1. In the second experiment (named experiment 2), we had 1-month old biofilm and 1-week old biofilm, besides the other three materials. A total of 300 cyprids were used in the experiment 2. In both experiments, cyprids were carefully placed in three (experiment 1) and five (experiment 2) replicated aquaria with the settlement plates and left undisturbed with flowing FFSW seawater, in the dark at 15 °C, for one week. After this incubation time, aquaria were inspected under a stereoscope for the evidence/confirmation of larval settlement. Those larvae found attached to each substrate were gently stimulated using a micropipette to ensure that settlement had occurred. After this preliminary inspection, and if the majority (>80%) of larvae had settled, plates were individually transferred to a large Petri dish filled with FSSW and settlers were carefully counted, and their general condition assessed. The number of larvae settled was registered for each plate and aquarium. Larvae that did not settle within the experimental time were recorded but not considered in analyses. Assessment of statistical significance of larval settlement among plates was conducted through separate Pearson chi-square tests for each experiment. The test compared observed numbers of larvae settled on the different substrates against the expectation of equal frequencies on all substrates. Larvae which had settled within each replicated plate on the aquaria were averaged before the analysis.

3. Results

3.1. Physical Properties of Materials and Biofilm Coverage

The cyclic voltammetry profiles recorded during EDOT electro-polymerization at the AISI 304 SS surface in ACN+LiClO4 supporting electrolyte are shown in Figure 2, as well as a proposed electrochemical polymerization mechanism. In the first cycle (red line) at 1.0 V onset the oxidation of the monomer and a nucleation current loop can be observed. This is the expected result for a 2D nucleation followed by 3D growth [27]. Later, in the consecutive cycles, it was possible to observe the redox processes of the polymer. There is a current increase with the number of scans until a stable cyclic voltammogram (blue line) is reached, which can be correlated with the doping–undoping processes of the polymer. This behavior indicates the gradual formation of a conductive film on the AISI 304 SS coupon and has been previously reported for PEDOT and other conducting polymers [29]. In addition, under the same conditions for obtaining the conducting polymer, a thickness of ~2 μm has been reported [27].

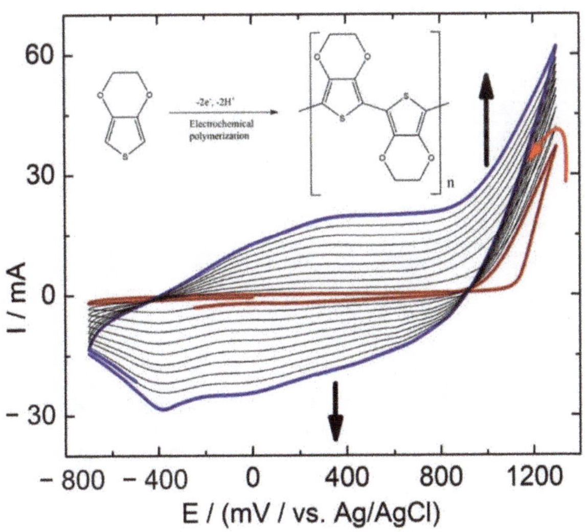

Figure 2. Cyclic voltammograms recorded during 10 cycles of EDOT electropolymerization in the potential range from −0.7 to 1.3 V on 304 SS at 0.05 $V{\cdot}s^{-1}$ from a 0.1 $mol{\cdot}L^{-1}$ EDOT + 0.1 $mol{\cdot}L^{-1}$ LiClO4 solution, in CH3CN. Inset: Proposed electrochemical polymerization mechanism of EDOT.

The physical properties (SR and W) of the experimental substrates are summarized in Table 1. Surface roughness was highest for the acrylic plate (0.332 μm^2) and lowest for the PEDOT (0.118 μm^2) with intermediate roughness observed for AF painting (0.243 μm^2). Wettability was lowest for PEDOT, which was categorized as the most hydrophobic substrate, while acrylic and AF painting had similar intermediate hydrophobicity values (see Table 1).

Table 1. Results of the analysis of surface roughness and wettability of the two materials (Acrylic, PEDOT) and antifouling (AF) paint. WCA = Water Contact Angle; High = WCA > 90°; Intermediate = 45 ° < WCA < 90°.

Plate	Surface Roughness (μm^2)	Wettability (WCA)	Hydrophobicity
Acrylic	0.332	81.9	Intermediate
PEDOT	0.118	103.4	Hydrophobic
AF	0.243	81.13	Intermediate

The mean cover (%) of biofilm, measured with image analysis, increased rapidly with biofilm age, from slightly over 10% in 1-week-old biofilm, to less than 50% in 2- to 4-week-old biofilms, to 100% cover in 2-month-old biofilm (Figure 3A).

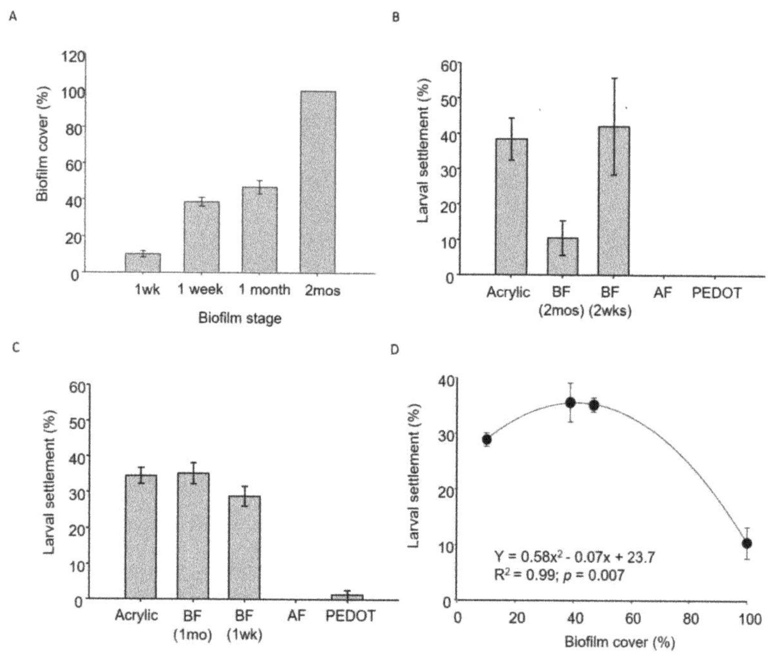

Figure 3. (**A**) Biofilm deposition cover (%) on acrylic plates calculated from the image analysis of the four different stages of biofilm, (**B**) proportion of barnacle settlement in the 5 experimental plates for the experiment 1, (**C**) proportion of barnacle settlement in the 5 experimental plates for the experiment 2, (**D**) results from the polynomial quadratic OLS regression between larval settlement (mean ± SE) and biofilm cover. BF = biofilm; AF = antifouling paint. Experimental time: 1 week.

3.2. Settlement Experiments

In the experiment 1, from a total of 60 larvae (20 larvae per aquarium), 52 (87%) settled within the one-week experimental time. Of these, 42% settled on the 2-week-old biofilm, 38% on the acrylic plates, and 10% settled on the 2-month-old biofilm. No larvae settled on PEDOT or AF-painted plates (Figure 3B).

In the experiment 2, from a total of 300 larvae (60 per aquarium), 215 (72%) settled within the experimental time. Of these, 35.2% settled on the 1-month old biofilm, 34.5% on acrylic plates, 28.9% on 1-week-old biofilm while 1.4% settled on the PEDOT (Figure 3C), corresponding to three larvae observed in three plates. No larvae settled on the AF paint. Since no larvae settled on AF paint on any of the experiments and only three larvae settled on PEDOT (pooling results from experiment 1 and 2), these two substrates were not included in the statistical comparisons and must be considered significantly different to those in which positive settlement was observed [45]. In both experiments, the frequency of settlement in the three substrates compared in the analyses (acrylic plate, early and late biofilms) was significantly different from the equal settlement expectation. In experiment 1, larval settlement on 2-month-old biofilm was lower than expected ($\chi^2_{(2, 52)}$ = 11.2311: $p = 0.0034$) and lower than on acrylic or 2-week-old biofilm (Figure 3). In experiment 2, settlement across treatment was more even and was not significantly different than expected ($\chi^2_{(2, 212)}$ = 1.623; $p = 0.443$).

Since we observed differences in settlement among different stages of biofilm, which were probably related to the differences in biofilm age and total cover (Figure 3A), we examined whether total larvae settled on biofilm (%) was associated to biofilm cover (Figure 3D). We used percentage of larvae settled as the response variable because the total number of larvae available to settle was different between the two experiments. The relationship was significant and non-linear (significant quadratic polynomial fit), with increased larval settlement at intermediate ages of the biofilm (between 2 weeks and 1 month) than when biofilm was only 1 week old or 2 months old.

4. Discussion

Our experiments showed that (i) the PEDOT is a good candidate for a fouling-resistant coating, as it showed extremely low to nil barnacle settlement, similar to the commercially available AF painting, (ii) *N. floscolus* is a good model species for biofouling experiments and to investigate larval settlement-biofilm interactions, as well as for implementation of environmentally friendly AF strategies, (iii) the cypris of *N. floscolus* prefer biofilms of intermediate age (between 2 weeks and 1 month), suggesting that larvae of this species perceive differences and actively choose among microbial communities of different age and composition.

A central issue of this study is the use of an AF paint as negative reference that contain copper (CU_2O), although it has a delayed releasing technology (see supplementary material). Copper could have affected larval settlement in the plates within each aquaria, especially those closer to the AF treatment. We recognize this as a limitation to the discussion of our results, although we found that randomizing the plates within each aquarium should have homogenized this effect. In fact, even assuming that copper was circulating in the assays, we have still showed differential settlement on the different plates, with low-to-zero settlement on PEDOT and higher settlement on acrylic and biofilms.

An important result that extends previous results, was the high efficiency of the conducting polymer coating PEDOT to inhibit *N. flosculus* larval settlement. A potential mechanism for this deterrent activity may be associated with the previously documented prevention of microbial biofilms by a PEDOT coating [27]. The previous authors showed that PEDOT can delay biofilm formation of at least 35 days, suggesting therefore that PEDOT should be a candidate for further experimentation, including field testing. The hydrophobic characteristic of PEDOT possibly also played an important role in settlement inhibition. Indeed, hydrophobicity of the material has been shown to inhibit larval settlement in *Balanus amphitrite* ([12,46], but see also [47] for the opposite pattern) and other important fouling organisms such as mussels [48] as well as spores of marine macroalgae [49]. The efficiency of PEDOT as antifouling coating against other taxa, which accounts for a large fraction of the final biofouling biomass in submerged structures, like the giant barnacle *Austromegabalanus psittacus* and the tunicate *Pyura chilensis* [36], should also be a priority if PEDOT is to be developed as a marine, environmentally friendly application.

The conducting polymers are interdisciplinary materials with cohesive aggregation of various areas, viz., electrochemistry, microbiology, environmental engineering, material sciences, biochemistry, and many other related areas, and are therefore a prime subject for research at the interface with marine biology.

We showed that the surface roughness among PEDOT coated surfaces, AF and roughened acrylic largely differ (see Table 1), with PEDOT showing lower roughness than both positive and negative references. Given that rougher surfaces can generate higher settlement, this can be considered a cofounding effect limiting the interpretation of our results. Nevertheless, while it is true that the surface topography created by the polymer and adhered to the solid surface could have been one of the proximate mechanisms deterring establishment of competent settling larvae, it is also possible that changes in the hydrophobic nature of the polymer also played a role (as shown in other species). In previous work [50], we have shown that larvae of most macro-foulers (barnacles, tunicates) settle abundantly over a very wide range of surface roughness, bracketing the roughness range presented in these experiments. Simple comparison among materials of similar surface roughness will therefore not provide much insight into the mechanisms by which PEDOT deters larval settlement. The primary experiments presented here were not designed to test those mechanisms of larval inhibition, but only to determine whether a polymer-covered solid surface elicits deterrent activity for macrofouling. We consider that the PEDOT effect on larvae is rather indirect, though affecting the composition and attributes of the biofilms, as shown by Aguirre and collaborators [27]. We encourage future experiments to focus on disentangling the mechanisms of action and considering a wide range of roughness and hydrophobicity and, ideally, independent control of the biofilm.

Notobalanus floscolus demonstrated to be a good candidate for settlement studies because of its high abundance in the field, low mortality of adults, continuous larval production and low mortality rate of nauplius (data not shown), which led to a sufficient number cyprids available for the settlement experiments. These biological characteristics, its frequent presence in all man-made materials deployed at sea [35,36] and its extensive geographic distribution make it an ideal model species to compare experimental results across biofouling studies. While *N. flosculus* showed relatively high abundance among the biofouling species of central Chile [36] another barnacle species, such as the giant barnacle *Austromegabalanus psittacus*, can attain much larger biomass and create more serious problems to the maritime industry. The size of *A. psittacus*, however, make this species more difficult to rear under laboratory conditions, larval stages are harder to cultivate, and its availability along the shore is sparser. Settlement behaviour of barnacles, consisting in an elaborate sequence of exploratory activity, is well known among different species of balanoid barnacles, particularly among the family Balanidae of which *Balanus amphitrite* and *Semibalanus balanoides* (not present in the Southeastern Pacific) have been amply used as model experimental species. The native *N. flosculus* could fulfil a similar role in science exploration along the Pacific shores.

In both settlement experiments, cyprids showed settlement preferences for plates with free-space availability such as the acrylic plates and plates with biofilm covers between 40 and 50% (between 2 weeks and 1 month of biofilm stage), which correspond to mid-successional microbial biofilm communities [2,15]. It has been reported that larvae of *B. amphitrite* preferred biofilm-free surfaces ([51]), but observations with other barnacle species suggest larvae settle on the biofilm [15,16,52]. In our study, higher settlement found on the 2-week- and 1-month-old biofilms than on the 1-week-old biofilm that had higher bare surface (see Figure 3), suggested that *N. flosculus* larvae preferred to settle on the established biofilms probably with more favourable chemical and mechanical conditions than bare space. It is important to highlight that in our experiments, there was a little mixing of cyprids age during the assay, because not all nauplius turned in cyprids at the same time. Mixing of cyprids prior the start the experiments strongly reduced this effect, yet a possible confounding effect of cyprids age on settlement remains. Specifically, caution should be taken when drawing conclusions about the difference in settlement between early and intermediate biofilm deposition ages. Nonetheless the results of low settlement on a more complete biofilm cover is clear, and suggest that later stages of microbial community succession are less suitable for larval settlement. While the association between biofilm age/density and larval settlement is generally positive [52–54], this is not always the case. For example, the settlement rate of the sea urchin *Tripneustes gratilla* larvae was not affected by the increasing age of a mixed consortium of bacteria [55].

Our experiments, as most others examining biofilm-larval interactions, could not separate between biofilm cover and biofilm age. As soon as the first microbial biofilm species get established on the surface, within hours of deployment in seawater [56], an ecological succession of highly diverse microbial species starts, changing diversity and composition over the course of weeks as the community advances towards late, more stable microbial composition [15]. Thus, differential *N. flosculus* settlement among different ages of biofilms are probably also or even mostly related to the changes in microbial community composition.

There is still much debate as to whether the microbial community can act as facilitator or inhibitor of macroscopic organisms (metazoans), such as barnacles [15], and clearly part of the answer is related to the state of the microbial succession with which larvae interact. This is today a very active and fascinating area of interdisciplinary research in which basic scientific breakthroughs could make major contributions to help solve the ages-old problem of marine biofouling.

5. Conclusions

In conclusion, the efficiency of PEDOT in reducing barnacle settlement was evident and promising and we showed that this organic coating is a good candidate as fouling-resistant strategy for specific application at sea. Further tests, however, are needed to confirm this potential deterrent efficacy against other fouling species, including other barnacles as the settlement behaviour of this taxonomic group highly depend on the substrate [15,16,36,52]. Future laboratory experiments should also be designed to examine in more detail the proximate mechanisms of deterrence of barnacle and other larvae by PEDOT coatings and its modulation by microbial biofilms, which will allow improvements in polymer designs. Beyond the laboratory testing, the final test of any antifouling strategy is the exposure of the material and technology to the diverse fouling community and environmental conditions encountered at sea where the application is to be deployed. This lies still ahead in the development of antifouling polymers.

Supplementary Materials: The following are available online at https://www.mdpi.com/2077-1312/9/1/82/s1,-Sea Voyage CDP 100 manufacturer instructions.

Author Contributions: Conceptualization, S.B., M.F. and S.A.N.; methodology, S.B. and F.A. and I.T.V.; formal analysis, S.B. and S.A.N.; resources, S.A.N.; data curation, S.B.; writing—original draft preparation, S.B.; writing—review and editing, all authors; funding acquisition, S.A.N. All authors have read and agreed to the published version of the manuscript.

Funding: This work was supported by the Marine Energy Research and Innovation Center, MERIC, an applied research centre funded by the Ministry of Energy and the Production Development Corporation, project CORFO 14CEI2-28228. Complementary funding was provided by ANID PIA/BASAL FB0002 to Sergio A. Navarrete.

Institutional Review Board Statement: Not applicable.

Informed Consent Statement: Not applicable.

Data Availability Statement: The data presented in this study are available on request from the corresponding author.

Acknowledgments: This work was possible thanks the valuable help of undergraduate and PhD students of the Pontificia Universidad Catolica de Chile and to the biofouling research group of the Marine Energy and Innovation Center (MERIC). We are particularly grateful to Patricio Manriquez from Centro de Estudios Avanzados en Zonas Aridas (CEAZA), Chile, for his contribution on the initial setting of the larval laboratory at the Estacion Costera de Investigaciones Marinas (ECIM) and to the student Clara Arboleda Baena for her wise suggestions on microbial biofilm formation.

Conflicts of Interest: The authors declare no conflict of interest.

References

1. Ayers, J.; Turner, H. The Principal Fouling Organisms. In *Marine Fouling and Its Prevention*; Redfield, A., Ketchum, B., Eds.; United States Naval Institute: Annapolis, MD, USA, 1952.
2. Callow, J.A.; Callow, M.E. Trends in the development of environmentally friendly fouling-resistant marine coatings. *Nat. Commun.* **2011**, 1–10. [CrossRef] [PubMed]
3. Want, A.; Crawford, R.; Kakkonen, J.; Kiddie, G.; Miller, S.; Harris, R.E.; Porter, J.S. Biodiversity characterisation and hydrodynamic consequences of marine fouling communities on marine renewable energy infrastructure in the Orkney Islands Archipelago, Scotland, UK. *Biofouling* **2017**, *33*, 567–579. [CrossRef] [PubMed]
4. Vinagre, P.A.; Simas, T.; Cruz, E.; Pinori, E.; Svenson, J. Marine biofouling: A European database for the marine renewable energy sector. *J. Mar. Sci. Eng.* **2020**, *8*, 495. [CrossRef]
5. Davidson, I.; Scianni, C.; Hewitt, C.; Everett, R.; Holm, E.; Tamburri, M.; Ruiz, G. Mini-review: Assessing the drivers of ship biofouling management—Aligning industry and biosecurity goals. *Biofouling* **2016**, *32*, 411–428. [CrossRef] [PubMed]
6. Loxton, J.; Macleod, A.K.; Nall, C.R.; McCollin, T.; Machado, I.; Simas, T.; Vance, T.; Kenny, C.; Want, A.; Miller, R.G. Setting an agenda for biofouling research for the marine renewable energy industry. *Int. J. Mar. Energy* **2017**, *19*, 292–303. [CrossRef]
7. Arinaga, R.A.; Cheung, K.F. Atlas of global wave energy from 10 years of reanalysis and hindcast data. *Renew. Energy* **2012**, *39*, 49–64. [CrossRef]
8. López, I.; Andreu, J.; Ceballos, S.; De Alegría, M.I.; Kortabarria, I. Review of wave energy technologies and the necessary power-equipment. *Renew. Sustain. Energy Rev.* **2013**, 413–434. [CrossRef]
9. Gentry, R.R.; Froehlich, H.E.; Grimm, D.; Kareiva, P.; Parke, M.; Rust, M.; Gaines, S.D.; Halpern, B.S. Mapping the global potential for marine aquaculture. *Nat. Ecol. Evol.* **2017**, *1*, 1317–1324. [CrossRef]
10. Lacoste, E.; Gaertner-Mazouni, N. Biofouling impact on production and ecosystem functioning: A Review for bivalve aquaculture. *Rev. Aquac.* **2015**, *7*, 187–196. [CrossRef]
11. Macleod, A.K.; Stanley, M.S.; Day, J.G.; Cook, E.J. Biofouling community composition across a range of environmental conditions and geographical locations suitable for floating marine renewable energy generation. *Biofouling* **2016**, *32*, 261–276. [CrossRef]
12. Clare, A.S.; Aldred, N. Surface colonisation by marine organisms and its impact on antifouling research. In *Advances in Marine Antifouling Coatings and Technologies*; Elsevier Ltd.: Amsterdam, The Netherlands, 2009; pp. 46–79. [CrossRef]
13. Qian, P.Y.; Dahms, H.-U. A triangle model: Environmental changes affect biofilms that affect larval settlement. In *Marine and Industrial Biofouling*; Springer: Berlin/Heidelberg, Germany, 2008; pp. 315–328. [CrossRef]
14. Hadfield, M.G. Biofilms and marine invertebrate larvae: What bacteria produce that larvae use to choose settlement sites. *Ann. Rev. Mar. Sci.* **2011**, *3*, 453–470. [CrossRef] [PubMed]
15. Dobretsov, S.; Rittschof, D. Love at first taste: Induction of larval settlement by marine microbes. *Int. J. Mol. Sci.* **2020**, *21*, 731. [CrossRef] [PubMed]
16. Aldred, N.; Nelson, A. Microbiome acquisition during larval settlement of the barnacle *Semibalanus Balanoides*. *Biol. Lett.* **2019**, *15*, 20180763. [CrossRef] [PubMed]
17. Maan, A.M.C.; Hofman, A.H.; Vos, W.M.; Kamperman, M. Recent developments and practical feasibility of polymer-based antifouling coatings. *Adv. Funct. Mater.* **2020**, *30*, 2000936. [CrossRef]
18. Salta, M.; Wharton, J.A.; Blache, Y.; Stokes, K.R.; Briand, J.-F. Marine biofilms on artificial surfaces: Structure and dynamics. *Environ. Microbiol.* **2013**. [CrossRef]
19. Salta, M.; Wharton, J.A.; Stoodley, P.; Dennington, S.P.; Goodes, L.R.; Werwinski, S.; Mart, U.; Wood, R.J.K.; Stokes, K.R. Designing biomimetic antifouling surfaces. *Philos. Trans. R. Soc. A Math. Phys. Eng. Sci.* **2010**, 4729–4754. [CrossRef]
20. Liu, L.; Li, W.; Liu, Q. Recent development of antifouling polymers: Structure, evaluation, and biomedical applications in nano/micro-structures. *Wiley Interdiscip. Rev. Nanomed. Nanobiotechnol.* **2014**, *6*, 599–614. [CrossRef]
21. Gomez-Carretero, S.; Nybom, R.; Richter-Dahlfors, A. Electroenhanced antimicrobial coating based on conjugated polymers with covalently coupled silver nanoparticles prevents *Staphylococcus Aureus* biofilm formation. *Adv. Healthc. Mater.* **2017**, *6*, 1700435. [CrossRef]
22. Mathiazhagan, A.; Joseph, R. Nanotechnology-A new prospective in organic coating—Review. *Int. J. Chem. Eng. Appl.* **2011**, 225–237. [CrossRef]
23. Yang, M.C.; Tsou, H.M.; Hsiao, Y.S.; Cheng, Y.W.; Liu, C.C.; Huang, L.Y.; Peng, X.Y.; Liu, T.Y.; Yung, M.C.; Hsu, C.C. Electrochemical polymerization of PEDOT-graphene oxide-heparin composite coating for anti-fouling and anti-clotting of cardiovascular stents. *Polymers* **2019**, *11*, 1520. [CrossRef]
24. Li, Y.; Ning, C. Latest research progress of marine microbiological corrosion and bio-fouling, and new approaches of marine anti-corrosion and anti-fouling. *Bioact. Mater.* **2019**, 189–195. [CrossRef] [PubMed]
25. Armelin, E.; Meneguzzi, Á.; Ferreira, C.A.; Alemán, C. Polyaniline, polypyrrole and poly (3,4-Ethylenedioxythiophene) as additives of organic coatings to prevent corrosion. *Surf. Coat. Technol.* **2009**, *203*, 3763–3769. [CrossRef]

26. Zhu, G.; Hou, J.; Zhu, H.; Qiu, R.; Xu, J. Electrochemical synthesis of poly(3,4-Ethylenedioxythiophene) on stainless steel and its corrosion inhibition performance. *J. Coat. Technol. Res.* **2013**, *10*, 659–668. [CrossRef]
27. Aguirre, J.; Daille, L.; Fischer, D.A.; Galarce, C.; Pizarro, G.; Vargas, I.; Walczak, M.; de la Iglesia, R.; Armijo, F. Study of poly (3,4-Ethylendioxythiphene) as a coating for mitigation of biocorrosion of AISI 304 stainless steel in natural seawater. *Prog. Org. Coat.* **2017**, *113*, 175–184. [CrossRef]
28. Smart Polymers and Their Applications | ScienceDirect. Available online: https://www.sciencedirect.com/book/9780857096951/smart-polymers-and-their-applications (accessed on 2 November 2020).
29. Del Valle, M.A.; Ramírez, A.M.; Hernández, L.A.; Armijo, F.; Díaz, F.R.; Arteaga, G.C. Influence of the supporting electrolyte on the electrochemical polymerization of 3,4-ethylenedioxythiophene. Effect on *p*- and n-doping/undoping, conductivity and morphology. *Int. J. Electrochem. Sci.* **2016**, *11*, 7048–7065. [CrossRef]
30. Romero, M.; del Valle, M.A.; del Río, R.; Díaz, F.R.; Armijo, F.; Dalchiele, E.A. Temperature effect on nucleation and growth mechanism of poly (o-anisidine) and poly (aniline) electro-synthesis. *J. Electrochem. Soc.* **2013**, *160*, G125–G134. [CrossRef]
31. Salgado, R.; del Rio, R.; del Valle, M.A.; Armijo, F. Selective electrochemical determination of dopamine, using a poly (3,4-Ethylenedioxythiophene)/polydopamine hybrid film modified electrode. *J. Electroanal. Chem.* **2013**, *704*, 130–136. [CrossRef]
32. Darmanin, T.; Guittard, F. Wettability of conducting polymers: From superhydrophilicity to superoleophobicity. *Prog. Polym. Sci.* **2014**, 656–682. [CrossRef]
33. Tapia, F.J.; Largier, J.L.; Castillo, M.; Wieters, E.A.; Navarrete, S.A. Latitudinal discontinuity in thermal conditions along the nearshore of central-northern Chile. *PLoS ONE* **2014**, *9*, e110841. [CrossRef]
34. Cifuentes, M.; Kamlah, C.; Thiel, M.; Lenz, M.; Wahl, M. Effects of temporal variability of disturbance on the succession in marine fouling communities in northern-central Chile. *J. Exp. Mar. Biol. Ecol.* **2007**, *352*, 280–294. [CrossRef]
35. Cifuentes, M.; Krueger, I.; Dumont, C.P.; Lenz, M.; Thiel, M. Does primary colonization or community structure determine the succession of fouling communities? *J. Exp. Mar. Biol. Ecol.* **2010**, *395*, 10–20. [CrossRef]
36. Navarrete, S.A.; Parragué, M.; Osiadacz, N.; Rojas, F.; Bonicelli, J.; Fernández, M.; Arboleda-Baena, C.; Perez-Matus, A.; Finke, R. Abundance, composition and succession of sessile subtidal assemblages in high wave-energy environments of central Chile: Temporal and Depth Variation. *J. Exp. Mar. Biol. Ecol.* **2019**, *512*, 51–62. [CrossRef]
37. Broitman, B.; Navarrete, S.; Smith, F.; Gaines, S. Geographic variation of southeastern Pacific intertidal communities. *Mar. Ecol. Prog. Ser.* **2001**, *224*, 21–34. [CrossRef]
38. Pacheco, A.S.; Laudien, J.; Thiel, M.; Oliva, M.; Heilmayer, O. Succession and seasonal onset of colonization in subtidal hard-bottom communities off northern Chile. *Mar. Ecol.* **2011**, *32*, 75–87. [CrossRef]
39. Darwin, C. A monograph of the subclass cirripedia, with figures of all the species. The lepadidæ or pedunculated cirripedes. *Ann. Mag. Nat. Hist.* **1853**, *12*, 444–448. [CrossRef]
40. Valdivia, N.; Heidemann, A.; Thiel, M.; Molis, M.; Wahl, M. Effects of disturbance on the diversity of hard-bottom macrobenthic communities on the coast of Chile. *Mar. Ecol. Prog. Ser.* **2005**, *299*, 45–54. [CrossRef]
41. Rueden, C.T.; Schindelin, J.; Hiner, M.C.; DeZonia, B.E.; Walter, A.E.; Arena, E.T.; Eliceiri, K.W. ImageJ2: ImageJ for the next generation of scientific image data. *BMC Bioinform.* **2017**, *18*, 529. [CrossRef]
42. Otsu, N. Threshold selection method from gray-level histograms. *IEEE Trans. Syst. Man Cybern.* **1979**, *9*, 62–66. [CrossRef]
43. Jonsson, P.R.; Wrange, A.L.; Lind, U.; Abramova, A.; Ogemark, M.; Blomberg, A. The barnacle balanus improvisus as a marine model—Culturing and gene expression. *J. Vis. Exp.* **2018**, *2018*, 57825. [CrossRef]
44. Pansch, C.; Jonsson, P.R.; Berglin, M.; Pinori, E.; Wrange, A.L. A new flow-through bioassay for testing low-emission antifouling coatings. *Biofouling* **2017**, *33*, 613–623. [CrossRef]
45. Quinn, G.P.; Keough, M.J. *Experimental Design and Data Analysis for Biologists*; Cambridge University Press: Cambridge, UK, 2002. [CrossRef]
46. Dahlström, M.; Jonsson, H.; Jonsson, P.R.; Elwing, H. Surface wettability as a determinant in the settlement of the barnacle balanus improvisus (DARWIN). *J. Exp. Mar. Biol. Ecol.* **2004**, *305*, 223–232. [CrossRef]
47. Petrone, L.; di Fino, A.; Aldred, N.; Sukkaew, P.; Ederth, T.; Clare, A.S.; Liedberg, B. Effects of surface charge and gibbs surface energy on the settlement behaviour of barnacle cyprids (*Balanus Amphitrite*). *Biofouling* **2011**, *27*, 1043–1055. [CrossRef] [PubMed]
48. Aldred, N.; Ista, L.K.; Callow, M.E.; Callow, J.A.; Lopez, G.P.; Clare, A.S. Mussel (*Mytilus Edulis*) byssus deposition in response to variations in surface wettability. *J. R. Soc. Interface* **2006**, *3*, 37–43. [CrossRef] [PubMed]
49. Callow, J.; Callow, M.; Ista, L.; Lopez, G.; Chaudhury, M. The influence of surface energy on the wetting behaviour of the spore adhesive of the marine alga *Ulva Linza* (Synonym *Enteromorpha Linza*). *J. R. Soc. Interface* **2005**, *2*, 319–325. [CrossRef] [PubMed]
50. Navarrete, S.; Parragué, M.; Osiadacz, N.; Rojas, F.; Bonicelli, J.; Fernández, M.; Arboleda-Baena, C.; Finke, R.; Baldanzi, S. Susceptibility of different materials and antifouling coating to macrofouling organisms in a high wave-energy environment. *J. Ocean Technol.* **2020**, *15*, 70–91.
51. Roberts, D.; Rittschof, D.; Holm, E.; Schmidt, A.R. Factors influencing initial larval settlement: Temporal, spatial and surface molecular components. *J. Exp. Mar. Biol. Ecol.* **1991**, *150*, 203–221. [CrossRef]
52. Wieczorek, S.; Clare, A.; Todd, C. Inhibitory and facilitatory effects of microbial films on settlement of balanus amphitrite amphitrite larvae. *Mar. Ecol. Prog. Ser.* **1995**, *119*, 221–228. [CrossRef]
53. Huggett, M.J.; Nedved, B.T.; Hadfield, M.G. Effects of initial surface wettability on biofilm formation and subsequent settlement of hydroides elegans. *Biofouling* **2009**, *25*, 387–399. [CrossRef]

54. Toupoint, N.; Mohit, V.; Linossier, I.; Bourgougnon, N.; Myrand, B.; Olivier, F.; Lovejoy, C.; Tremblay, R. Effect of biofilm age on settlement of mytilus edulis. *Biofouling* **2012**, *28*, 985–1001. [CrossRef]
55. Mos, B.; Cowden, K.L.; Nielsen, S.J.; Dworjanyn, S.A. Do cues matter? Highly inductive settlement cues don't ensure high post-settlement survival in sea urchin aquaculture. *PLoS ONE* **2011**, *6*, e28054. [CrossRef]
56. Daille, L.K.; Aguirre, J.; Fischer, D.; Galarce, C.; Armijo, F.; Pizarro, G.E.; Walczak, M.; de la Iglesia, R.; Vargas, I.T. Effect of tidal cycles on bacterial biofilm formation and biocorrosion of stainless steel AISI 316L. *J. Mar. Sci. Eng.* **2020**, *8*, 124. [CrossRef]

MDPI
St. Alban-Anlage 66
4052 Basel
Switzerland
Tel. +41 61 683 77 34
Fax +41 61 302 89 18
www.mdpi.com

Journal of Marine Science and Engineering Editorial Office
E-mail: jmse@mdpi.com
www.mdpi.com/journal/jmse

www.ingramcontent.com/pod-product-compliance
Lightning Source LLC
Chambersburg PA
CBHW061612210926
51850CB00009B/2299
* 9 7 8 3 0 3 6 5 7 5 5 3 7 *